Managing Forest Ecosystems

Volume 29

Aims & Scope

Well-managed forests and woodlands are a renewable resource, producing essential raw material with minimum waste and energy use. Rich in habitat and species diversity, forests may contribute to increased ecosystem stability. They can absorb the effects of unwanted deposition and other disturbances and protect neighbouring ecosystems by maintaining stable nutrient and energy cycles and by preventing soil degradation and erosion. They provide much-needed recreation and their continued existence contributes to stabilizing rural communities.

Forests are managed for timber production and species, habitat and process conservation. A subtle shift from multiple-use management to ecosystems management is being observed and the new ecological perspective of multi-functional forest management is based on the principles of ecosystem diversity, stability and elasticity, and the dynamic equilibrium of primary and secondary production.

Making full use of new technology is one of the challenges facing forest management today. Resource information must be obtained with a limited budget. This requires better timing of resource assessment activities and improved use of multiple data sources. Sound ecosystems management, like any other management activity, relies on effective forecasting and operational control.

The aim of the book series Managing Forest Ecosystems is to present state-of-the-art research results relating to the practice of forest management. Contributions are solicited from prominent authors. Each reference book, monograph or proceedings volume will be focused to deal with a specific context. Typical issues of the series are: resource assessment techniques, evaluating sustainability for evenaged and uneven-aged forests, multi-objective management, predicting forest development, optimizing forest management, biodiversity management and monitoring, risk assessment and economic analysis.

More information about this series at http://www.springer.com/series/6247

Susana Barreiro • Mart-Jan Schelhaas
Ronald E. McRoberts • Gerald Kändler
Editors

Forest Inventory-based Projection Systems for Wood and Biomass Availability

COST is supported by the EU Framework Programme Horizon 2020

EUROPEAN COOPERATION IN SCIENCE AND TECHNOLOGY

Springer

Editors
Susana Barreiro
Forest Research Centre (CEF), School of Agriculture
University of Lisbon
Lisbon, Portugal

Ronald E. McRoberts
Northern Research Station, U.S. Forest Service
Saint Paul, MN, USA

Mart-Jan Schelhaas
Wageningen Environmental Research (Alterra)
Wageningen, The Netherlands

Gerald Kändler
Forest Research Institute of Baden-Württemberg
Freiburg, Germany

ISSN 1568-1319 ISSN 2352-3956 (electronic)
Managing Forest Ecosystems
ISBN 978-3-319-56199-8 ISBN 978-3-319-56201-8 (eBook)
DOI 10.1007/978-3-319-56201-8

Library of Congress Control Number: 2017944189

Printed on acid-free paper

This Springer imprint is published by Springer Nature
The registered company is Springer International Publishing AG
The registered company address is: Gewerbestrasse 11, 6330 Cham, Switzerland

Preface

Forecasts of future wood availability have shown a deficit of wood supply compared to wood consumption in Europe. Recent policy targets for bioenergy and emphasis on the contribution of forests to the bio-economy have sparked interest in the availability and mobilization of wood from the forest. However, a wide variety of national-level systems for projecting future woody biomass supply exists and has led to an increasing need for harmonized international reporting.

The members of COST Action FP1001 (USEWOOD) have prepared and compiled a unique description of the projection systems commonly used by European countries. In addition to the member countries and institutions, the projection systems used in Canada and in the USA have also been described. Part I (Chaps. 1, 2, 3, 4 and 5) provides an introduction followed by three chapters on wood availability and related issues, an overview of the nationwide projection systems and a description of the country-level forest resource projection tools. Finally, the challenges for woody biomass projections, advantages and limitations in the use of different approaches for international reporting are discussed. Part II contains country descriptions of projection systems organized in the form of 22 separate chapters. The country chapters provide a contextualizing account of the national forests, forest inventory systems and the issues that triggered development of each tool. This is followed by structured descriptions of the projection systems used for official reporting, thereby allowing the reader to gain a precise understanding of the purposes, functions, limitations and potential of each tool. This book describes existing National Forest Inventory (NFI)-based tools and discloses the situation in countries where NFIs have recently been implemented but no NFI-based simulation tools have been developed.

The authors and editors hope that forest inventory experts, model developers, researchers and policy- and decision makers interested in the fields of assessment and modelling of wood supply worldwide find this information useful. Furthermore, one of the principal ideas of this book is to reduce the "noise" in the comparison of the different methodological approaches, thereby providing better insight on long-term trends expected for the supply and demand of forest products.

Lisbon, Portugal — Susana Barreiro
Wageningen, The Netherlands — Mart-Jan Schelhaas
Saint Paul, MN, USA — Ronald E. McRoberts
Freiburg, Germany — Gerald Kändler

Acknowledgements

This work is based on discussions and insights gained from COST Action FP1001 Improving Data and Information on the Potential Supply of Wood Resources: A European Approach from Multisource National Forest Inventories (USEWOOD) and Working Group 3: Predicting the Use of Wood Resources under Competitive Conditions. Special thanks are extended to the COST Office (Domain Forests, their Products and Services) which enabled transnational networking of researchers, in particular to Ms. Melae Langbein, Dr. Fatima Bouchama and Ms. Cassia Azevedo.

Recognition must also be given to all country chapter authors for their contributions and patience throughout the extensive revision process and for making this book possible. The editors of the book acknowledge the assistance of Antoine Colin and Clara Antón-Fernandez in the initial stage of revision of the country chapters. Finally, the authors further collectively acknowledge Barry Gardiner for his helpful comments that contributed to the consistency of the book as a whole.

We thank Florian Steierer from UNECE/FAO for his support with Part I, Chap. 1.

The chapter describing the European-level tools (Part I, Chap. 4) has been written with support from the Trees4Future project (Grant Agreement No. 284181) of the EU Seventh Framework Programme. All chapters in Part I have benefitted from the support of the SIMWOOD project (Grant Agreement No. 613762) of the EU H2020 Programme.

The continued development and maintenance of Canada's NFCMARS requires cooperation between provincial, territorial and federal government agencies. The authors of the Canadian chapter (Part II, Chap. 3) thank all members (past and present) of the National Forest Sinks Committee and their colleagues. They also thank other members of the Canadian Forest Service Carbon Accounting Team, in particular Eric Neilson for graphical production assistance. They thank Graham Stinson for helpful review comments. Funding for the analysis in their chapter was provided by the Government of Canada's Clean Air Agenda, Leadership for Environmental Advantage in Forestry, Panel on Energy Research and Development and in kind contributions from provincial and territorial governments.

We are very grateful for the support of Ineke Ravesloot from Springer for her support throughout the preparation process.

COST Information

Created in 1971, the European Cooperation in Science and Technology (COST) is the longest-running European framework supporting transnational cooperation among researchers, engineers and scholars across Europe with the aim of closing the gap between science, policymakers and society throughout Europe and beyond.

COST is a pan-European intergovernmental framework. Its mission is to enable breakthrough scientific and technological developments leading to new concepts and products and thereby contributing to strengthening Europe's research and innovation capacities. It allows researchers, engineers and scholars to jointly develop their own ideas and take new initiatives across all fields of science and technology, while promoting multi- and interdisciplinary approaches. COST aims at fostering a better integration of less research-intensive countries into the knowledge hubs of the European Research Area. The COST Association, an international not-for-profit association under Belgian law, integrates all management, governing and administrative functions necessary for the operation of the framework. The COST Association has currently 36 member countries, and Israel is a cooperating state (www.cost.eu).

Contents

Contributors

Book Editors

Susana Barreiro Forest Research Centre (CEF), School of Agriculture, University of Lisbon, Lisbon, Portugal

Gerald Kändler Forest Research Institute, Baden-Württemberg, Freiburg, Germany

Ronald E. McRoberts Northern Research Station, U.S. Forest Service, Saint Paul, MN, USA

Mart-Jan Schelhaas Wageningen Environmental Research (Alterra), Wageningen, The Netherlands

Editorial Board

Clara Antón-Fernández Norwegian Institute of Bioeconomy Research, Ås, Norway

Susana Barreiro Forest Research Centre (CEF), School of Agriculture, University of Lisbon, Lisbon, Portugal

Antoine Colin IGN, Direction Interrégionale Nord-Est, Champigneulles, France

Gerald Kändler Forest Research Institute of Baden-Württemberg, Freiburg, Germany

Mart-Jan Schelhaas Wageningen Environmental Research (Alterra), Wageningen, The Netherlands

Country Authors

Iciar Alberdi Department of Silviculture and Forest Management, INIA, Forest Research Centre, Madrid, Spain

Clara Antón-Fernández Norwegian Institute of Bioeconomy Research, Ås, Norway

Susana Barreiro Forest Research Centre (CEF), School of Agriculture, University of Lisbon, Lisbon, Portugal

Jean-Daniel Bontemps UMR LERFOB, AgroParisTech, INRA, Nancy, France

IGN, Laboratoire de l'Inventaire Forestier (LIF), Nancy, France

Vilis Brukas Southern Swedish Forest Research Centre, Swedish University of Agricultural Sciences, Alnarp, Sweden

Paolo Camin Federal Office for the Environment FOEN, Ittigen, Switzerland

Sandra A.P.P.M. Clerkx Wageningen Environmental Research (Alterra), Wageningen, The Netherlands

Antoine Colin IGN, Direction Interrégionale Nord-Est, Champigneulles, France

Sonia Condés Department of Natural Systems and Resources, School of Forestry, Technical University of Madrid, Madrid, Spain

Marius Dumitru Forest Research and Management Institute, Bucharest, Romania

Angel Ferezliev Forest Research Institute – BAS, Sofia, Bulgaria

Christoph Fischer Swiss Federal Institute for Forest, Snow and Landscape Research WSL, Birmensdorf, Switzerland

Stefan Frank International Institute for Applied Systems Analysis (IIASA), Ecosystems Services and Management Program (ESM), Laxemburg, Austria

Nicklas Forsell International Institute for Applied Systems Analysis (IIASA), Ecosystems Services and Management Program (ESM), Laxemburg, Austria

Fernando García-Robredo Department of Natural Systems and Resources, School of Forestry, Technical University of Madrid, Madrid, Spain

Patrizia Gasparini Council for Agricultural Research and Economics, Research Centre for Forestry and Wood (CREA-FL), Trento, Italy

Giacomo Grassi European Commission, Joint Research Centre, Directorate D – Sustainable Resources – Bio-Economy Unit, Ispra, Italy

Lars Graudal Department of Geosciences and Natural Resource Management, Section for Forest, Nature and Biomass, University of Copenhagen, Copenhagen, Denmark

Thomas Gschwantner Department of Forest Inventory, Federal Research and Training Centre for Forests, Natural Hazards and Landscape (BFW), Vienna, Austria

Mykola Gusti International Institute for Applied Systems Analysis (IIASA), Ecosystems Services and Management Program (ESM), Laxemburg, Austria

Petr Havlik International Institute for Applied Systems Analysis (IIASA), Ecosystems Services and Management Program (ESM), Laxemburg, Austria

Geerten Hengeveld Wageningen Environmental Research (Alterra), Wageningen, The Netherlands

Vivian Kvist Johannsen Department of Geosciences and Natural Resource Management, Section for Forest, Nature and Biomass, University of Copenhagen, Copenhagen, Denmark

Gerald Kändler Forest Research Institute Baden-Württemberg, Freiburg, Germany

Albertas Kasperavičius State Forest Service, Kaunas, Lithuania

Edgar Kaufmann Swiss Federal Institute for Forest, Snow and Landscape Research WSL, Birmensdorf, Switzerland

Georg Kindermann Department of Forest Growth and Silviculture, Federal Research and Training Centre for Forests, Natural Hazards and Landscape (BFW), Vienna, Austria

Bjarki Kjartansson Icelandic Forest Research, Mógilsá, Iceland

Kari T. Korhonen Natural Resources Institute Finland, Joensuu, Finland

Pál Kovácsevics National Food Chain Safety Office, Forestry Directorate, Budapest, Hungary

Miloš Kučera Forest Management Institute, Brandýs nad Labem-Stará Boleslav, Czech Republic

Gintaras Kulbokas State Forest Service, Kaunas, Lithuania

Aleksandras Stulginskis University, Kaunas, Lithuania

Andrius Kuliešis State Forest Service, Kaunas, Lithuania

Werner A. Kurz Natural Resources Canada, Canadian Forest Service, Pacific Forestry Centre, Victoria, BC, Canada

Thomas Ledermann Department of Forest Growth and Silviculture, Federal Research and Training Centre for Forests, Natural Hazards and Landscape (BFW), Vienna, Austria

Anders Lundström Department of Forest Resource Management, Division of Forest Resource Data, Swedish University of Agricultural Sciences, Umeå, Sweden

Gheorghe Marin Forest Research and Management Institute, Bucharest, Romania

Marian Lajos Mayr University of Hamburg, Centre of Wood Science, Hamburg, Germany

Ronald E. McRoberts Northern Research Station, U.S. Forest Service, Saint Paul, MN, USA

Juha M. Metsaranta Natural Resources Canada, Canadian Forest Service, Northern Forestry Centre, Edmonton, AB, Canada

Radoslav Milchev University of Forestry, Sofia, Bulgaria

Gintautas Mozgeris Aleksandras Stulginskis University, Kaunas, Lithuania

Gert-Jan Nabuurs Wageningen Environmental Research (Alterra), Wageningen, The Netherlands

Forest Ecology and Forest Management Group, Wageningen University and Research, Wageningen, The Netherlands

Thomas Nord-Larsen Department of Geosciences and Natural Resource Management, Section for Forest, Nature and Biomass, University of Copenhagen, Copenhagen, Denmark

Tuula Packalen Natural Resources Institute Finland, Joensuu, Finland

Alessandro Paletto Council for Agricultural Research and Economics, Research Centre for Forestry and Wood (CREA-FL), Trento, Italy

Edmundas Petrauskas Aleksandras Stulginskis University, Kaunas, Lithuania

Henry Phillips Forestry Consultant, Sligo, Ireland

Roberto Pilli European Commission, Joint Research Centre, Directorate D – Sustainable Resources – Bio-Economy Unit, Ispra, Italy

John Redmond Forest Service, DAFM, Wexford, Ireland

Uli Riemer Forest Research Institute, Baden-Württemberg, Freiburg, Germany

Torben Riis-Nielsen Department of Geosciences and Natural Resource Management, Section for Forest, Nature and Biomass, University of Copenhagen, Copenhagen, Denmark

Sandro Sacchelli Department of Agricultural, Food and Forest Systems Management (GESAAF), University of Florence, Florence, Italy

Ola Sallnäs Swedish University of Agricultural Sciences, Upsala, Sweden

Olli Salminen Natural Resources Institute Finland, Vantaa, Finland

Klemens Schadauer Department of Forest Inventory, Federal Research and Training Centre for Forests, Natural Hazards and Landscape (BFW), Vienna, Austria

Mart-Jan Schelhaas Wageningen Environmental Research (Alterra), Wageningen, The Netherlands

Erik Schou Department of Geosciences and Natural Resource Management, Section for Forest, Nature and Biomass, University of Copenhagen, Copenhagen, Denmark

Allan Sims Estonian University of Life Sciences, Tartu, Estonia

Estonian Environmental Agency, Tallinn, Estonia

Carolyn E. Smyth Natural Resources Canada, Canadian Forest Service, Pacific Forestry Centre, Victoria, BC, Canada

Arnór Snorrason Icelandic Forest Research, Mógilsá, Iceland

Nickola Stoyanov University of Forestry, Sofia, Bulgaria

Maria Stoyanova Forest Research Institute – BAS, Sofia, Bulgaria

Alain Thivolle-Cazat Institut Technologique FCBA, Grenoble, France

Esther Thürig Swiss Federal Institute for Forest, Snow and Landscape Research WSL, Birmensdorf, Switzerland

Margarida Tomé Forest Research Centre (CEF), School of Agriculture, University of Lisbon, Lisbon, Portugal

Stein Tomter Norwegian Institute of Bioeconomy Research, Ås, Norway

Mark Twomey Forest Service, DAFM, Cork, Ireland

Roberto Vallejo Ministry of Agriculture, Food and Environment, Madrid, Spain

Pieter Johannes Verkerk European Forest Institute, Joensuu, Finland

Holger Wernsdörfer UMR LERFOB, AgroParisTech, INRA, Nancy, France

Per-Erik Wikberg Department of Forest Resource Management, Division of Forest Resource Data, Swedish University of Agricultural Sciences, Umeå, Sweden

Abbreviations

BAU	Business as usual
CBM-CFS	Carbon Budget Model of the Canadian Forest Sector
COST	European Cooperation in Science and Technology
dbh	Diameter at breast height
EFDM	European Forest Dynamics Model
EFI	European Forest Institute
EFISCEN	European Forest Information Scenario Model
EFSOS	European Forest Sector Outlook Study
ENFIN	European National Forest Inventory Network
ETTS	European Timber Trend Studies
EU	European Union
EUROSTAT	Statistical Office of the European Commission
FAO	Food and Agriculture Organization of the United Nations
FAWS	Forest available for wood supply
FIA	US Forest Inventory and Analysis Program
FMP	Forest management plans
FRA	Forest Resources Assessment
G4M	Global Forest Model
GHG	Greenhouse gas
IPCC	Intergovernmental Panel on Climate Change
ITOC	Inventory to consumer
LUC	Land-use change
LULUCF	Land use, land-use change and forestry
MCPFE	Ministerial Conference on the Protection of Forests in Europe
NAI	Net annual increment
NFI	National Forest Inventory
SFI	Standwise forest inventory
UN-ECE	United Nations Economic Commission for Europe
UNFCCC	United Nations Framework Convention on Climate Change
YT	Yield tables

Part I
Current Approaches for Projecting Wood Availability

Chapter 1
Introduction

Susana Barreiro, Mart-Jan Schelhaas, Gerald Kändler, and Ronald E. McRoberts

1.1 Background

For millennia, forests have been a strategic resource for mankind, providing building material for houses, ships, and mining; household and industrial fuel; hunting grounds and grazing opportunities for cattle. Despite its strategic importance, forest management was often not sustainable. Many European countries have a long history of deforestation and overexploitation of forest resources. North America has also seen substantial deforestation since European colonization. The principles of sustainable forest management were formulated around 1700 by Von Carlowitz in Germany (1713). Centuries later the concept of sustainability became popular under the term "sustainable development" as coined by the Report of the World Commission on Environment and Development: Our Common Future, also known as the Brundtland report (http://www.un-documents.net/our-common-future.pdf). On larger scales, around 1990 with the meetings of the Ministerial Conference on the Protection of Forests in Europe (MCPFE, now known as Forest Europe) the

S. Barreiro (✉)
Forest Research Centre (CEF), School of Agriculture, University of Lisbon, Lisbon, Portugal
e-mail: smb@isa.ulisboa.pt

M.-J. Schelhaas
Wageningen Environmental Research (Alterra), Wageningen, The Netherlands
e-mail: martjan.schelhaas@wur.nl

G. Kändler
Forest Research Institute, Baden-Württemberg, Freiburg, Germany
e-mail: gerald.kaendler@forst.bwl.de

R.E. McRoberts
Northern Research Station, U.S. Forest Service, Saint Paul, MN, USA
e-mail: rmcroberts@fs.fed.us

S. Barreiro et al. (eds.), *Forest Inventory-based Projection Systems for Wood and Biomass Availability*, Managing Forest Ecosystems 29,
DOI 10.1007/978-3-319-56201-8_1

principles of sustainable management were established as guidelines for the stewardship of forests as natural resources. At national and regional scales, the idea of sustainable wood supply was the basic principle behind forest management and the restoration of forests in Europe since the middle of the nineteenth century. Many European countries have seen an expansion of their forest areas in the past 100–150 years due to active afforestation and natural expansion of forests on waste lands and marginal agricultural lands (Mather 2001). This phenomenon has also been observed in the southern United States of America (USA) during the period 1935–1975 by Rudel (2001) who describes how smaller yields on marginal agricultural lands led to conversion of these lands to forest.

Concerns about possible shortages of wood and overharvesting were major reasons that many countries began implementing forest resource monitoring systems. The first nation-wide assessments were initiated in the early twentieth century, mostly in Northern Europe. Additionally, many countries established research trials to investigate growth and yield in relation to forest management, resulting in the production of yield tables for the most commonly used tree species. These yield tables were an important tool for assessing forest productivity and harvesting possibilities, mostly at the stand scale. As a logical next step, governments and industries asked for projections of future wood availability and allowable harvest levels. In response, many countries developed projection systems that featured a wide array of mostly country-specific tools, models and simulators.

Over recent decades, societal appreciation has increased for the many functions and services that forests provide: nature conservation, recreation, protection of infrastructure and protection of soil and water resources. However, demands for these functions and services sometimes conflict with the wood production function and, therefore, must be included in studies on potential wood supply.

The situation with respect to forests and forestry within Europe varies enormously. Forest cover ranges from 1.9% in Iceland to 76% in Finland. Countries with large forest areas generally have well-developed forest industries, while the focus in countries with relatively little forest is more on conservation and recreational use. A large array of forest management systems is used, including the traditional clear-fell and replant system, shelterwood systems, group-and individual tree selection systems, coppicing and strict nature reserves. The number of tree species in the boreal zone is limited, but is much greater in mid-and southern Europe. Western European countries with little forest cover often have active forest area expansion policies, while area expansion in southern and eastern countries occurs spontaneously on abandoned agricultural lands.

Since the beginning of the twenty-first century, climate change has been considered the most important environmental issue in many European and North American countries. While forests are thought to be impacted by future climate change, at the same time they are seen as part of the solution. Forests play an important role in the global carbon cycle. Large quantities of carbon are exchanged every day between forests and the atmosphere, and large amounts of carbon are stored in biomass, soil and wood products. Among the six economic sectors identified by the United Nations Framework convention on Climate Change as sources of anthropogenic greenhouse gas (GHG) emissions, the Land Use, Land Use Change and Forestry (LULUCF) sector is the only terrestrial sector with the potential to remove GHG emissions from the atmosphere. Besides acting as a carbon sink, wood products can substitute for energy-intense materials such as aluminium and concrete and can easily be recycled. When used for generating energy, woody biomass can replace fossil fuels such as gas or coal. If sourced from sustainably managed forests, wood can be regarded as a renewable resource that contributes to reducing fossil carbon emissions.

European countries have adopted ambitious renewable energy standards with the aim of providing 20% of total energy consumption by 2020 from renewable resources of which 40–50% is expected to be delivered by bioenergy (Muys et al. 2013). The proposal for increasing the use of wood in strategies for mitigating the impacts of climate change raised the question of wood availability in Europe. Currently, the European annual harvest is about 66% of the annual increment (FOREST EUROPE 2015), theoretically indicating the possibility for increased harvesting. However, it is questionable how much of this unused potential is really available because of multiple challenges such as fragmented ownership, conflicts with other forest uses, accessibility, economic constraints, and discrepancy between required and available species and dimensions. Projections using European level forest scenarios with different sets of constraints point to considerable deficits in wood supply by 2020 when compared to expected wood demand (Mantau et al. 2010). In recent years, the role and importance of woody biomass has grown steadily, resulting in large imports of wood pellets from North America (Goh et al. 2013).

Overall, the pressure on European and North American forests is rising. Wood demand is increasing due to developments in the bioenergy sector and also in the bio-economy sector which is expected to consume more biomass in the future. At the same time, forests are expected to fulfil a multitude of other functions. The impacts of climate change on forests are expected to range from increased growth due to higher temperatures and longer growing seasons to increased mortality due to changes in precipitation patterns, shifts in tree species' ranges and/or changes in disturbance regimes. Forest resource projections can be used to investigate these impacts and to provide insight into the consequences of intended future policies. In this book we provide an overview of available projection tools and assess the degree to which they can respond to the challenges posed.

1.2 Forest Resources in Europe and North America

In Sects. 1.2.1, 1.2.2, 1.2.3, and 1.2.4 unless otherwise noted, all European estimates are from the 2015 FOREST EUROPE report (FOREST EUROPE 2015), all Canadian estimates are from State of Canada's Forests report (NRC-CFS 2015), and all estimates for the USA are from a report on the Forest Resources of the United States, 2012 (Oswalt et al. 2014).

1.2.1 Forest Area and Forest Available for Wood Supply

In 2015, Europe excluding the Russian Federation had an average forest cover of 32%, the same percentage as in the USA in 2012 but less than the nearly 35% of Canada that is covered by forests. For environmental, economic and social reasons, not all forests are available for wood supply. Central-West and Central-East Europe are the regions with the greatest volumes of growing stock (8.8 and 7.9 billion m^3, respectively) and the greatest (94%) and least (70%) shares of Forests Available for Wood Supply (FAWS), respectively. European forest area is 215 million ha, of which 150 million ha is available for wood supply. In the USA, of the 310 million ha of forests, 29.7 million ha (9.6%) of forests are classified as "Reserves" where timber harvesting is legally prohibited. Canada has 347 million ha of forest and reports 59% of their forest to be minimally affected by human activities (primary forests). These areas include protected forests and inaccessible forests.

1.2.2 Afforestation and Deforestation Trends

In Europe, total forest area expanded by nearly 700 thousand ha per year between 1990 and 2015, while the area available for wood supply expanded by about 50 thousand ha per year in the same period. Afforestation of agricultural land unsuitable for agriculture is one of the policy objectives most frequently reported in FOREST EUROPE 2015. The report on the State of Canada's Forest 2015 indicates a decline in forest area of about 0.33% from 1990 until 2010, mostly as a result of forest land converted to agricultural and urban uses. Most Canadian forests originate from natural regeneration, although planting initiatives are underway for a small proportion of the total forest area. In the USA, after the severe deforestation observed in the period 1630–1910, forest area has not only stabilized but since 2007 has been trending upward at a rate of about 1% per year.

1.2.3 Growing Stock and Fellings

The average growing stock in Europe increased from 126 m^3/ha in 1990 to 163 m^3/ha in 2015. In 2015, the total growing stock was 35 billion m^3 of which 84% is available for wood supply. In the USA, the growing stock distributions of softwood and hardwood vary by region. In the Northeast and Southeast regions hardwoods comprise most of the timber volume (6.9 and 6.3 billion m^3, respectively), whereas in the Rocky Mountains and Pacific Coast (including Alaska and Hawaii) regions, softwoods comprise most timber volume (4.02 and 7.03 billion m^3, respectively). Overall, since 2007 softwood growing stock has experienced a modest increase of approximately 3% (15.5 billion m^3). Canada's most productive species are found in the

Pacific Maritime ecozone with an average growing stock of 432 m^3/ha, whereas the forests in the Taiga Shield and Hudson Plains ecozones have the slowest growing species with 61 m^3/ha and 36 m^3/ha, respectively.

The balance between Net Annual Increment (NAI)[1] and annual fellings is traditionally among the most frequently used criteria for assessing the sustainability of forests in Europe. According to the FOREST EUROPE 2015 report approximately 66% of the NAI is utilized by fellings. The greatest felling rates were reported for Austria (94%) and Sweden (102%), while the smallest rates were reported for Ukraine (29%), Turkey (37%) and Italy (39%). In 2010, increments amounted to 839.7 million m^3 and fellings to 582.3 million m^3. Between 1986 and 2006, growing-stock removals in the USA remained fairly stable and totalled 364 million m^3 in 2011. Softwoods accounted for 235.5 million m^3 (65%) of growing stock removals in 2011, and hardwoods accounted for 128.4 million m^3 (35%). In Canada, harvests are regulated by Allowable Annual Cuts (AACs) calculated for the provinces and territories and have been relatively constant since 1990s. In 2013, the total volume of timber harvested was 148 million m^3, only two-thirds of the AAC.

1.2.4 Ownership and Landowner Characteristics

About half of European forests are publicly owned, but the share varies from 3% in Portugal to 90% in some Eastern European countries. A little more than half the forest land in the USA (58%) is privately owned, while the remaining 42% is controlled by Federal, State, and local governments. Ownership patterns also vary across the country with private ownerships dominating in the Northeast and Southeast and public ownerships dominating in the Rocky Mountains and along the Pacific Coast. Conversely, only 6% of Canada's forests are privately owned. Canadian provinces and territories own 90% of the forests and have responsibility for ensuring compliance with forest management plans. The remaining 4% represents forest lands owned by the federal government (Natural Resources Canada 2016).

In Europe the number of private holdings has increased over time to 18% in the period 1990–2015, probably as a result of active afforestation on private lands in Western Europe, restitution and privatisation processes in countries with formerly centrally planned economies and fragmenting of properties at inheritance. The average size of private holdings is much smaller than the public holdings, with the majority of the private holdings less than 10 ha.

The USA has an estimated 11 million private forest landowners, the majority of whom own properties of less than 4 ha. However, 67% of forest land is in holdings of at least 40 ha, and 22% is in holdings of at least 4000 ha and is owned by less than 1% of the owners, primarily corporations or investment organizations and primarily managed for commercial purposes.

---

[1] The net annual increment is the average annual volume over the given reference period of gross increment less that of natural losses on all trees to a minimum diameter of 0 cm (*dbh*). (According to UNECE: http://www.unece.org/forests/fra/definit.html)

1.3 Production and Consumption of Wood in Europe and North America

Europe and North America are major producers of roundwood. Together they produce about 1 billion m^3 annually (Fig. 1.1), 35–40% of the world total. The majority of the removals are industrial roundwood (82%), while fuelwood accounts for only 18%. Felling levels in the USA and Canada have been rather stable since 1990, but the 2008 recession caused a reduction of more than 25% and felling levels have yet to recover. In Europe, the felling level gradually increased over time until the recession. Thereafter it mostly stabilised, but fuelwood increased from 80 million m^3 in 1990 to 131 million m^3 in 2013. Due to the increased demand for bioenergy in Europe, the USA and Canada substantially increased their exports of wood pellets in recent years.

For the period 2010–2030, the European Forest Sector Outlook Study II (EFSOS II, UNECE/FAO 2011) projects a doubling of wood-based energy demand as a consequence of the policy targets for renewable energy, while demand for products is projected to increase only by 5–10% in the same period. To fulfil the extra bioenergy demand from domestic sources, in addition to harvesting more stemwood, mobilisation of harvesting residues (tops, branches and stumps) must be increased enormously. Other potential sources of biomass are short-rotation coppices and trees outside forests. The supply can be further increased if more post-consumer wood is mobilized. However, increased dependence on imports is likely. The North American Forest Sector Outlook Study (NAFSOS, UNECE/FAO 2012) projects industrial roundwood to return to pre-recession levels for Canada, and to increase above those levels by 10–20% by 2030 for the USA. Fuelwood production in the USA until 2030 is projected to triple or even quadruple.

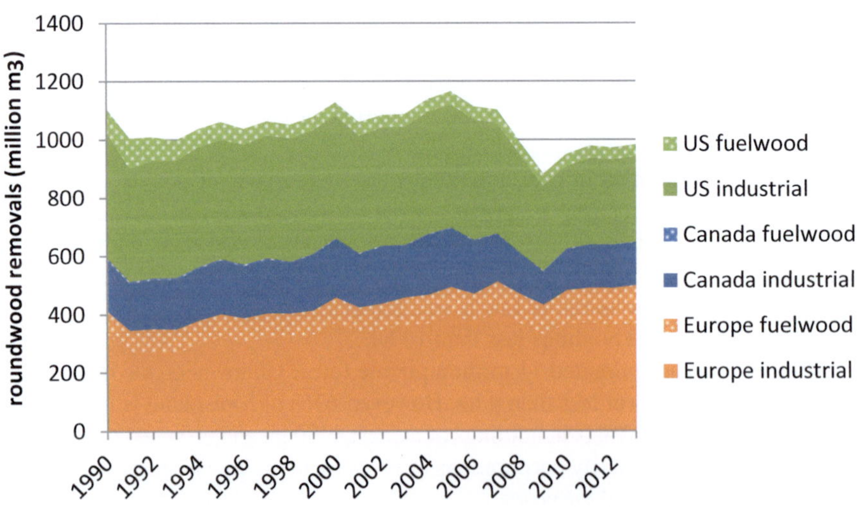

Fig. 1.1 Roundwood removals in Europe (excluding Russian Federation), USA and Canada for the period 1990–2013 (UNECE/FAO 2016)

Europe as a whole has been largely self-sufficient for the last 50 years, but there has been extensive trade in wood and wood products with other regions. The same is true for the USA, with domestic supply equal to 96% of industrial wood consumption (Oswalt et al. 2014). Canada is a net exporter of wood and wood products, with the USA as its largest trading partner followed by China (NRC-CFS 2015).

Harvest in Canada is mostly concentrated in the western parts of the country. In the USA, the harvest pattern has changed over time. While most wood was traditionally harvested in the Pacific Northwest region of the country, the Southeast is now the greatest supplier. Main harvesting regions in Europe are southern Sweden, southern Finland and the Baltic States, large plantations of *Pinus pinaster* in south-western France, plantations of Eucalypt in Portugal, and Central Europe where Norway spruce is an important commercial species.

1.4 Forest Inventory

Concerns for overharvesting produced the need for accurate information on the state of the forest and the rates at which the forest was growing and was being harvested. Forest inventory programs were established to satisfy this information demand. Early attempts at basic forest inventories had already been conducted by the end of the Middle Ages. The first forest inventories were local and aimed at assessing available timber resources for specific purposes. However, these inventories were not suitable for compiling information to assist forest policy and decision making at national levels. National-level inventories were started in the early twentieth century and have traditionally collected information about the status of the forest in terms of tree species composition, age class distribution and growing stock volume.

In the 1980s, the impact of air pollution on forests became evident. In Europe, forest health and crown condition monitoring networks were established on separate sampling grids, e g., Level I monitoring of the International Co-operative Programme on Assessment and Monitoring of Air Pollution Effects on Forests (known as ICP Forests, http://icp-forests.net). Moreover, the increasing significance of ecological functions of forests such as biodiversity required new monitoring concepts and studies. Thus, forest monitoring can be understood to cover all forest features beyond the target variables of traditional forest inventory with special emphasis on interactions between forest ecosystems and the environment. However, the separation between production-oriented forest inventories and environmentally-oriented monitoring programmes has become less distinct. The changing role of forests, combined with national and international reporting requirements, has altered the demand for forest information. In recent years ecological variables have been also included in traditional forest inventories, as well as variables aimed at quantification of carbon stocks and other forest functions.

Nowadays, nation-wide inventories are conducted in North America and in most European countries. In both Europe and North America, projection systems have been developed that rely on forest inventory data to characterize the initial state of the forest and often to develop and/or calibrate growth models around which the projection systems are built. Therefore, the detail and accuracy of the forest data generated by the inventory systems are crucial for the quality of the projections. Here we describe the two most commonly used methods for compiling nation-wide forest estimates: the Standwise Forest Inventory (SFI) approach and the sample-based National Forest Inventory (NFI) approach.

1.4.1 Standwise Forest Inventory

The idea of sustainability gradually emerged during the eighteenth century and finally led to the establishment of regular forest management. The theoretical and practical concepts of forest management planning were developed by the early German forest academics (e.g. Hartig 1795; Cotta 1804; Hundeshagen 1826). At that time, wood production was the main purpose of forestry. Management planning requires quantitative information on wood resources at the stand level in terms of growing stock volume and site productivity. Thus, the assessment of forest conditions was closely related to the elaboration of Forest Management Plans (FMP). Hence, management planning was established as an instrument to ensure sustainable utilization of forests, and Standwise Forest Inventory (SFI) was a basic principle for management planning.

Forest management implies the spatial division of forest areas into districts, compartments and stands. The area unit of an FMP inventory is normally the stand, which is considered to be reasonably homogeneous with respect to species composition, age, and site characteristics and is subject to specific management and silvicultural treatment. Standwise management is often closely linked to yield table-based assessments of growing stock, growth and yield. Hence, yield tables are also used for updating growing stocks without measurements. SFI includes a wide spectrum of methods ranging from visual assessment to sample-based surveys. The inventory cycle is normally 10 years; thus, every year about 10% of the forest area of an ownership is inventoried and the allowable cut for the subsequent 10-year period is calculated. Allowable cut is an implicit assessment of future wood supply at the stand level. Data obtained from standwise inventories, especially those based on ocular estimates, are regarded as less reliable than data from statistical inventories because they provide no estimates of uncertainty and may be subject to systematic deviations (e.g. Kangas et al. 2004; Haara and Leskinen 2009; Šmelko et al. 2008).

Small scale, stand-by-stand forest inventories were established in most European countries during the 19th and early 20th centuries. Depending on forest legislation, forest management planning was implemented in different manners according to ownership. FMPs were systematically implemented by the forest authorities, especially in state and communal forests.

Historically, forest management planning evolved differently in different regions of Europe. After World War II, most forests in the Eastern European communist countries were under state control. Forest management was subject to central planning which required a uniform forest management planning system to cover as much forest area as possible. In these countries, SFI became the basis for the preparation of FMPs for state forest enterprises (administrative units). Thus, forest authorities managed forests countrywide with uniform standards. Hence, it became possible to compile nationwide forest inventory data at the forest enterprise level for estimating total forest area. After the end of the communist era in the 1990s, centrally planned economies were abolished and economic and administrative structures changed substantially. Forests formerly under centralized state control were reprivatized, and organizational structures were adapted to include forest services. However, forest data continues to be stored in centralized databases under state supervision: Hungary, National Forestry Database NFD; Estonia, Forest Register FR; the Czech Republic, FMP database. Such databases are used to aggregate forest statistics at higher levels (regional units, national) and may be linked to other geo-referenced spatial data that provide additional information value, e.g., thematic maps of forest characteristics. With the political changes, responsibilities for implementing FMPs also changed to varying degrees among countries. In Hungary, state forest authorities are responsible for forest management planning as well as for supervision of forest management in private forests, whereas in Estonia, the state is only responsible for granting management planning licenses and only licensed companies are allowed to carry out forest management planning. In all cases, forest management planning instructions and rules are regulated by law and are under state inspection. Conversely, in other European countries, FMPs remain owner-specific instruments.

1.4.2 National Forest Inventory

National Forest Inventories (NFI) typically use systematic sampling designs based on grids placed over a forest map of the country and establish field sample plots at grid intersections with forest cover. The sample plots are usually in the form of concentric circles with fixed radii where smaller trees are measured only on the smaller radii circles. A few countries use variable radius sampling, also called angle count or Bitterlich sampling. Measurements on individual trees on sample plots always include tree species, diameter at breast height (*dbh*), and often additional variables such as height, log quality, and visual damage. Individual sample plots typically represent areas of 100–2000 ha, depending on the grid used. Methodologically, the advantages of NFIs are the uniform protocols applied nationwide which, in combination with proper quality assurance measures, provide consistent and comparable data. The statistical design allows use of unbiased estimators of totals and means for important forest attributes and uncertainty measures that facilitate calculation of confidence intervals. Another great advantage of NFIs is that they typically cover all lands that satisfy the country's forest definition, regardless of ownership or other cadastral categories. Data may be collected at defined intervals, often 5 or 10 years, although more commonly nowadays a proportion of plots are measured each year. Tomppo et al. (2010) provides a detailed description of NFI systems including ranges of grid dimensions, plot sizes and configurations, and minimum diameter thresholds characteristic of European and North American NFIs.

National Forest Inventories (NFI) were established in the early twentieth century in the Nordic countries: Norway in 1919, Finland in 1921, and Sweden in 1923. At that time forestry was a very important economic sector and there was a great need for information on the state and development of the forests as a natural resource at the national level. Other European countries followed after World War II. In the USA sample-based inventories date from 1928, although they were often implemented at state and regional levels which inhibited consistent and timely reporting across all states. However, in the 1990s the Forest Inventory and Analysis (FIA) program of the U.S. Forest Service standardized inventories with respect to plot configuration, sampling design, measurement protocols, and reporting requirements for the entire the country. Eastern European countries traditionally used SFI approaches, although recently NFIs have also been established in most of these countries; some countries maintain both inventory systems. However, for national and international wood availability studies the SFI data are still preferred as input data for projections.

1.5 International Reporting, Harmonisation Efforts and the Role of the USEWOOD Cost Action

NFIs have been the source of forest information for decades for purposes of assisting forestry and environmental planning, forest policy and industrial investment decisions. The desire to monitor the sustainable use of forests triggered the implementation of NFIs in central Europe (Tomppo et al. 2010). Deforestation, biodiversity losses, acid deposition, GHG emissions, and unsustainable forest management have all contributed to development of other monitoring programmes such as FOREST EUROPE (2015). Most European countries joined international treaties such as the United Nations Framework Convention on Climate Change (UNFCCC 2010) and the Kyoto Protocol and are obliged to monitor and report net carbon stock changes and GHG emissions by sources and sinks in the LULUCF sector. Over the years, good practice guidance documents have been prepared to assist countries in developing accurate and comparable inventories for the LULUCF sector. Notwithstanding, international reporting requirements revealed that the use of harmonized sets of definitions was required. For this reason considerable effort has been committed to developing methods for harmonized reporting in Europe. The European National Forest Inventory Network (ENFIN) has been the motivation underlying multiple European funded projects focused on harmonized reporting. These efforts have produced considerable progress that has motivated countries to voluntarily revise their definitions and their measurement protocols to minimize data differences resulting from different sampling designs, plot configurations, definitions and measurement protocols. Primary results have included development of common reference definitions and methods for producing harmonized estimates despite different national inventory features.

Recent renewable energy and climate change policies are expected to drive an increase in European consumption of wood for energy production. A reasonable question is whether European forests can sustainably produce sufficient woody biomass. Multiple studies on supply potentials have been carried out, but the approaches, terminology, constraints, assumptions and therefore the estimates and conclusions, vary considerably among studies. One difficulty in comparing results at the international level is that national definitions of wood categories vary widely because they depend on the species, the industry and the market. Simultaneously, a vast number of projection systems used for providing estimates of future wood availability have been developed in response to country-specific problems and national forest policy requirements. Some countries developed their projection systems; others adapted existing systems, while others rely on European simulators to carry out such analyses.

COST Action FP1001 (USEWOOD) facilitated research cooperation and coordination among European researchers focusing on NFIs, wood availability and projection of wood resources. USEWOOD included three Working Groups (WG): (1) WG1 focused on techniques for assessing and estimating the state of and changes in wood resources based on NFI data, definitions and harmonization; (2) WG2 focused on improving wood resources' estimates by integrating remotely sensed and NFI field data; and (3) WG3 focused on predicting the use of wood resources. The scope of WG3 included describing both the data used for projecting the potential supply of tree biomass under economic, social and ecological conditions and the different methods used. The descriptions will lead to clearer interpretations of projection results through a deeper understanding of the mechanisms behind the projections, their required inputs, underlying assumptions, scenario simulation capacity and limitations. Similar descriptions are also reported for data and methods currently used by Canada and the USA.

1.6 Organization of Book Contents

Part 1 of this book aims to synthesize current approaches for projecting wood availability with discussions ranging from the concept of wood availability to the challenges ahead. Part 2 includes a series of chapters describing the national approaches used in some European countries and North America (Fig. 1.2).

After the introduction (Chap. 1), Chap. 2 focusses on the concept of wood availability and the challenges in transferring forest inventory information to market availability. Chapter 3 provides descriptions of the types of growth models and additional simulation modules, how they are combined into forest simulators and their specific features. The chapter also presents an overview of the different projection systems described in more detail in the second part of the book. Chapter 4 describes the simulators that are currently used in Europe and discusses approaches for carrying out European-wide studies. Finally, Chap. 5 summarizes the driving forces affecting woody biomass availability and its projection, discusses the advantages and disadvantages of choosing each of the approaches described and the future challenges ahead.

Fig. 1.2 Map of Europe and North America showing country contributions in *dark grey*

References

Cotta H (1804) Systematische Anleitung zur Taxation der Waldungen. Johann Daniel Sander, Berlin

FOREST EUROPE (2015) State of Europe's Forests 2015

Goh CS, Junginger M, Cocchi M et al (2013) Wood pellet market and trade: a global perspective. Biofuels Bioprod Biorefin 7:24–42. doi:10.1002/bbb.1366

Haara A, Leskinen P (2009) The assessment of the uncertainty of updated stand-level inventory data. Silva Fennica 43(1):87–112. doi:10.142/4/Sf.219

Hartig GL (1795) Anweisung zur Taxation der Forste, oder zur Bestimmung des Holzertrags der Wälder. Heyer, Gießen

Hundeshagen JC (1826) Die Forstabschätzung auf neuen wissenschaftlichen Grundlagen. Heinrich Laupp, Tübingen

Kangas A, Kangas J, Heikkinen E, Maltamo M (2004) Accuracy of partially visually assessed stand characteristics: a case study of Finnish forest inventory by compartments. Can J For Res 34:916–930. doi:10.1139/x03-266

Mantau U, Saal U, Prins K, et al. (2010) EUwood - Real potential for changes in growth and use of EU forests. Final report. Hamburg, Germany: University of Hamburg, Centre of Wood Science. http://www.egger.com/downloads/bildarchiv/187000/1_187099_DV_Real-potential-changes-growth_EN.pdf. Accessed 5 July

Mather A (2001) The transition from deforestation to reforestation in Europe. In: Angelsen A, Kaimowitz D (eds) Agricultural technologies and tropical deforestation. CAB International, Wallingford

Muys B, Hetemäki L, Palahi M (2013) Sustainable wood mobilization for EU renewable energy targets. Biofuels Bioprod Biorefin 7:359–360. doi:10.1002/bbb.1421

Natural Resources Canada (2016) Forest land ownership. https://www.nrcan.gc.ca/forests/canada/ownership/17495. Accessed 15 July 2016

NRC-CFS, Natural Resources Canada, Canadian Forest Service (2015) The State of Canada's forests. Annual report 2015, p 76

Oswalt SN, Smith WB, Miles PD, Pugh SA (2014) Forest resources of the United States, 2012: a technical document supporting the forest service update of the 2010 RPA assessment, US Forest services, USDA

Rudel KT (2001) Did a green revolution restore the forests of the American South? In: Angelsen A, Kaimowitz D (eds) Agricultural technologies and tropical deforestation. CAB International, Wallingford

Šmelko Š, Šeben V, Bosela M, Merganič J, Jankovič J (2008) National forest inventory and monitoring of the Slovak Republik 2005–2006. National Forest Centre – Forest Research Institute, Zvolen

Tomppo E, Gschwantner T, Lawrence M, McRoberts RE (2010) National Forest Inventories Pathways for Common Reporting Springer Heidelberg ISBN 978-90-481-3232-4

UNECE/FAO (2011) The European Forest Sector Outlook Study II (EFSOS II). 2010–2030. UNECE/FAO, ECE/TIM/SP/28

UNECE/FAO (2012) The North American Forest Sector Outlook Study 2006–2030. UNECE/FAO, Geneva, ECE/TIM/SP/29

UNECE/FAO (2016) Forestry production and trade database. http://faostat3.fao.org/download/F/FO/E Accessed 15 July 2016

UNFCCC (2010) Report of the conference of the parties on its fifteenth session, held in Copenhagen from 7 to 19 December 2009. Part One: Proceedings http://unfccc.int/resource/docs/2009/cop15/eng/11.pdf Accessed 15 July 2016

von Carlowitz H (1713) Sylvicultura Oeconomica oder haußwirthliche Nachricht und Naturgemäßige Anweisung zur Wilden Baum-Zucht, Leipzig: Johann Friedrich Braun (2 Bände)

Chapter 2
Wood Availability

Mart-Jan Schelhaas and Marian Lajos Mayr

2.1 Harvest to Increment Ratio

Most national monitoring and projection studies were initiated in response to concerns regarding overexploitation of forests and the future availability of timber. In these studies, overexploitation was assessed using an indicator calculated as the balance between annual fellings and Net Annual Increment (NAI). Values of the indicator greater than 100% meant that more wood was being harvested than was growing in the period considered. Indeed, some inventories such as in Sweden in the 1850s showed overharvesting (see corresponding country chapter). The balance between fellings and NAI is simple to calculate and easy to understand, particularly for people outside the forest sector. However, it should only be applied over large spatial scales and long time periods. It does not make sense to apply it at stand level where long periods without intervention are interspersed with years with thinnings or regeneration fellings. Similarly there may be reasons why the harvest would exceed the increment for particular periods, even at the country scale. Examples include policies aimed at reducing the average growing stock (Swiss Agency for the Environment, Forests and Landscape 1999), sanitary fellings after natural disturbances and skewed age-class distributions requiring regeneration of large areas in a short period of time. Generally, the harvest to increment ratio was well below 100%

M.-J. Schelhaas (✉)
Wageningen Environmental Research (Alterra), Wageningen, The Netherlands
e-mail: martjan.schelhaas@wur.nl

M.L. Mayr
University of Hamburg, Centre of Wood Science, Hamburg, Germany
e-mail: marian.lajos.mayr@uni-hamburg.de

S. Barreiro et al. (eds.), *Forest Inventory-based Projection Systems for Wood and Biomass Availability*, Managing Forest Ecosystems 29,
DOI 10.1007/978-3-319-56201-8_2

for most European countries and the United States of America (USA) in the period 1950–2015 as is evident by the large accumulation of growing stock.

The harvest to increment ratio is also used as a simple way to assess the amount of wood that could potentially be available on the market (Hetsch 2009). Although this might give a rudimentary approximation if the age-class distribution is essentially balanced, it does not work if the age-class distribution is unbalanced as is the case in most European countries (Vilén et al. 2012). Countries with right-skewed age class distributions such as the Czech Republic and Germany have harvesting possibilities that could be much greater than the current increment, whereas harvest possibilities in countries with a left-skewed age-class distribution such as in Iceland may be much less than the increment. Estimates of wood availability should therefore take into account the age-class distribution and rotation lengths that are normally applied. Several countries including Canada and Lithuania have developed methods for calculating an Annual Allowable Cut (AAC) based on the actual situation in the forest and management prescriptions, sometimes in combination with other policy aims such as trying to influence the age-class distribution. Similarly, in the EFISCEN model (Chap. 4), the potential harvest is calculated from the actual age-class distribution and prescribed management regimes defined by the fraction per age-class that can be regenerated and the age range for which thinnings can be carried out. Applying this "instantaneous" potential would generally lead to a large harvest in the first time-step by logging all areas older than the specified rotation length. This would not be seen as a sustainable practice and is not very realistic. A long-term maximum potential sustainable supply is therefore estimated by re-running the model until a supply level is found that can be sustained for the next 50 years (Verkerk et al. 2011).

2.2 Constraints

Estimates of the potentially available wood supply all rely on assumptions regarding harvesting constraints. The concept Forest area Available for Wood Supply (FAWS) was introduced in the international reporting procedures to account for some of the possible constraints. The United Nations (2000) defines FAWS as follows:

> Forest where any legal, economic, or specific environmental restrictions do not have a significant impact on the supply of wood. Includes: areas where, although there are no such restrictions, harvesting is not taking place, for example areas included in long-term utilization plans or intentions.

Although the definition seems straightforward, an interpretation that is commonly accepted and implemented among countries is lacking (Alberdí et al. 2016). In particular, the economic restrictions may change over time, depending on wood prices and technological development. Still, the FAWS concept is widely accepted as a basis for estimation in studies on future wood availability (Hetsch 2009;

Verkerk et al. 2011). Most studies distinguish between two potentials: (1) the maximum physical potential an area could deliver, characterized as the bio-technical potential, and (2) the likely availability of wood and woody biomass given a range of additional constraints, characterized as the socio-economic potential (Hetsch 2009). For example, this approach is used in Switzerland in the form of the "onion model" (Chap. 26; Hofer et al. 2011) where each layer of the onion reduces the potentially available supply by subsequently adding constraints. Hetsch (2009) classified the constraints as technical, ecological and economical. Technical constraints entail lack of infrastructure, equipment, logistic factors and information deficiency. Ecological constraints include harvest restrictions in protected areas, regulations on biodiversity measures and harvesting techniques and limitations on residue removal. The ecological restrictions depend very much on social conventions and can be subject to changes associated with implementation of natural conservation programs. Economic constraints include demand and price constraints, high costs, mismatch between supply and demand (in terms of assortments) and underdeveloped markets. Additionally, the behaviour of forest owners and managers plays an important role. Although they are usually modelled as rational agents reacting to costs and prices, in practice their behaviour is difficult to predict, especially for non-industrial private owners. However, Rinaldi et al. (2015) recently attempted to integrate more diverse behavioural assumptions regarding forest owners when simulating forest resource development.

Although quantification of each of these constraints is far from easy, identification of the "real" future availability of woody biomass is important. Furthermore, the quantification process gives insight into why additional resources that are apparently available are difficult to mobilise in practice. For example, in Switzerland, these constraints are currently quantified by expert judgement, but in the future they should be replaced by estimates based on National Forest Inventory (NFI) data (see corresponding country chapter). Also, other countries such as France and Italy are attempting to estimate such constraints from NFI data (see corresponding country chapters). Hetsch (2009) did not quantify the constraints individually but assumed that 35% of the difference between current felling level and the maximum physical potential would be available. Verkerk et al. (2011) quantified potential availability in the European Union (EU) using three wood mobilisation scenarios (low, medium, high) with different assumptions regarding environmental, technical and social constraints. Matthews et al. (2015) estimated the maximum long-term increment European forests could sustain by assigning yield classes to all forests. The maximum potential wood supply from forests was taken as 75% of this long-term increment level, allowing for non-optimal rotation lengths and other constraints. Figure 2.1 illustrates how a realistic potential wood supply is obtained by firstly estimating a bio-technical potential (e.g. using the harvest/increment ratio) and then applying a set of technical, ecological and economic constraints (e.g. the onion model).

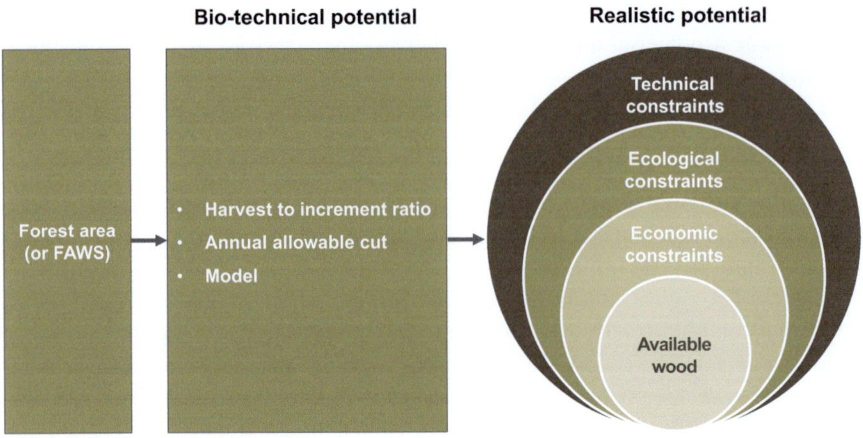

Fig. 2.1 Calculation steps from forest area to realistic potential supply. If based on FAWS, parts of the constraints are already applied before estimation of the bio-technical potential

2.3 From NFI Data to Market Availability

Potential supply levels can be estimated using the felling to increment ratio, allowable cut estimation methods or more elaborate methods based on constraints and NFI data. Mostly, these methods yield estimates of the potential felling level which are not the same as the potential volume of wood available for consumption on the wood market. For a range of reasons, (potential) felling statistics cannot be translated one-to-one to (potential) market availability. During felling, forwarding and transporting from the harvesting location to the mill site, losses occur that reduce the theoretical available potential. These losses can be further separated into harvesting losses, measurement losses, logistical losses and quality losses. Harvesting losses are difficult to determine but may significantly reduce the availability of the theoretical biomass potentials. Losses due to measurement methods and trade guidelines have an even greater impact on the available volume. Measurement losses result in underestimation of the volume that is actually harvested relative to the registered harvested volume. The difference in the form of losses between standing tree volumes and the available volumes of felled trees that reach the mill site is assumed to be around 10% (Mantau et al. 2016). Part of these losses may be recovered and still become available for use.

Furthermore, a direct comparison between actual and potential felling level for estimation of additionally available wood supply can be misleading. A part of the wood harvested and removed from the forest, mainly consisting of wood harvested for private purposes, is not covered by official statistics (e.g. Jochem et al. 2015). In countries with many small forest ownerships these removals may constitute a considerable amount. In general, they are legal, but in many cases are too small to be registered by official felling statistics. Quantification of this amount is therefore complicated. Estimation of the extent of these fellings can be via surveys of forest owners or by estimating it as the difference between other known wood flows. NFIs with permanent plots can estimate the total drain from trees that were reported to be harvested between two consecutive inventories. This estimate can be compared to the officially reported felling level after adjusting for differences in measurement methods. If the consumer side is completely covered by empirical studies (e.g. including fuelwood consumption of private households, biomass power plants, etc.), the total consumption of forest resources could be estimated using the wood resource balance (Mantau 2012).

COST Action FP1001 (USEWOOD) assisted in the development of an estimation framework to facilitate the exchange of data between NFI specialists and wood market analysts. A tool called ITOC (from Inventory to Consumer biomass avail-

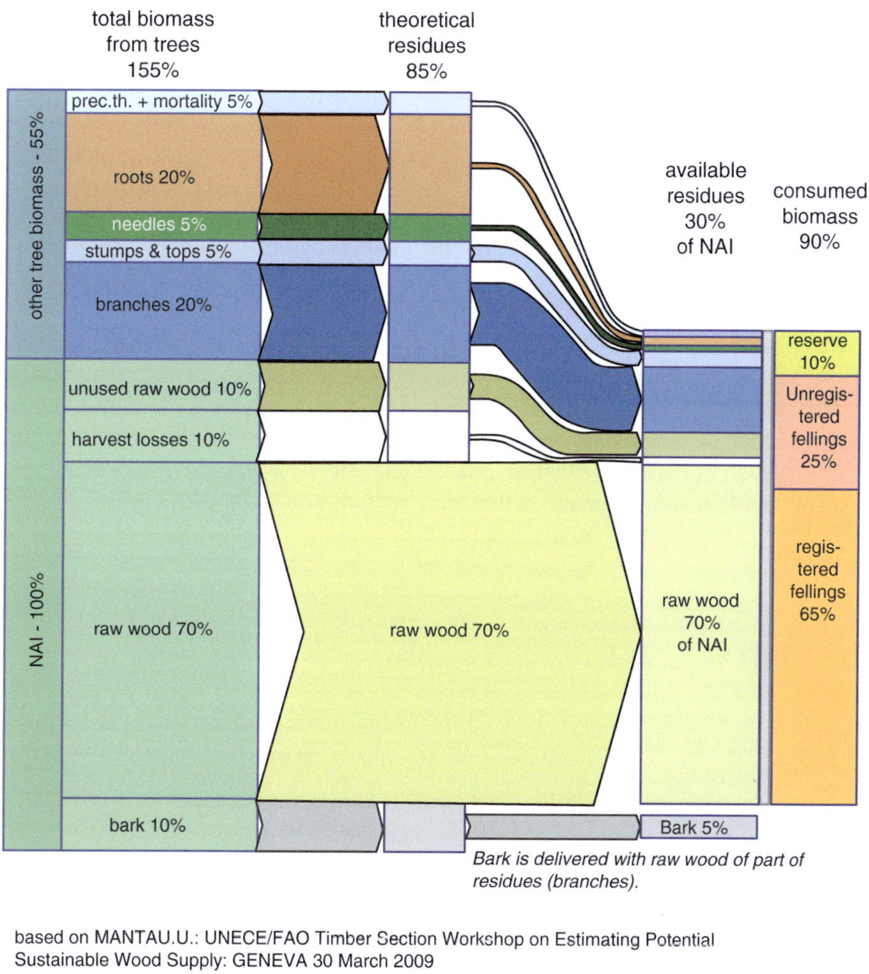

Fig. 2.2 Outline of the ITOC model calculation steps (Based on Mantau U: UNECE/FA)

ability) (Mantau et al. 2016) primarily focuses on bridging the gap between NFIs as data providers and assessments of use categories according to consumer needs. It aims at developing a transparent system for data exchange that is acceptable for forest inventory experts as well as for wood market analysts. The tool is flexible in that it allows application in a wide range of countries with different levels of detail. It can deal with NFI data as well as potential supply estimates obtained from resource projections. Default values are provided, but can be replaced by country-specific information when available. The tool follows a stepwise procedure to convert potential harvestable volume into potentially available woody biomass, quantifying at each step the conversions or expansion factors needed and the amount of the losses that can be recovered (Fig. 2.2). After having defined the consumer biomass availability, the final step is to assess the biomass reserve available in a country, i.e. the amount of biomass that is available in addition to the current removals. The ITOC tool therefore balances the resulting biomass potentials with data from corresponding felling statistics in the countries.

2.4 Other Considerations

The majority of fresh wood coming onto the market is supplied from FAWS. Besides the stem wood of the felled trees, branches, foliage and stumps can also be extracted. Residue extraction is subject to technical, economic and ecological constraints. Furthermore, additional woody biomass is available from other sources, such as trees along roads, maintenance of rural and urban landscape elements, maintenance of tree orchards and gardens, and even from forest not available for wood supply as a by-product of maintenance measures. Other sources are by-products from the industry, such as shavings, sawdust, black liquor, etc. Although these may be important sources of woody biomass, in this book we focus on methods for estimating the potential woody biomass supply from FAWS.

References

Alberdi I, Michalak R, Fischer C et al (2016) Towards harmonized assessment of European forest availability for wood supply in Europe. For Policy Econ 70:20–29. doi:10.1016/j.forpol.2016.05.014

Hetsch S (2009) Potential sustainable wood supply in Europe, Geneva timber and forest discussion paper, vol 52, United Nations Economic Commission for Europe, Food and Agriculture Organization of the United Nations, Geneva

Hofer P, Altwegg J, Schoop A et al (2011) Holznutzungspotenziale im Schweizer Wald. Auswertung von Nutzungsszenarien und Waldwachstumsentwicklung. Bundesamt für Umwelt, Bern. Umwelt-Wissen Nr. 1116

Jochem D, Weimar H, Bösch M et al (2015) Estimation of wood removals and fellings in Germany: a calculation approach based on the amount of used roundwood. Eur J For Res 134(5):869–888

Mantau U (2012) Holzrohstoffbilanz Deutschland: Entwicklungen und Szenarien des Holzaufkommens und der Holzverwendung von 1987 bis 2015. Zentrum Holzwirtschaft, Hamburg

Mantau U, Gschwanter T, Paletto A et al (2016) From inventory to consumer biomass availability – the ITOC-model. Ann For Sci. doi:10.1007/s13595-016-0582-1

Matthews R, Mortimer N, Lesschen JP et al (2015) Carbon impacts of biomass consumed in the EU: quantitative assessment. Final project report, project: DG ENER/C1/427. Forest research: Farnham. https://ec.europa.eu/energy/sites/ener/files/documents/EU%20Carbon%20Impacts%20of%20Biomass%20Consumed%20in%20the%20EU%20final.pdf Accessed 14 Nov 2016

Rinaldi F, Jonsson R, Sallnäs O, Trubins R (2015) Behavioral modelling in a decision support system. Forests 6(2):311–327. doi:10.3390/f6020311

Swiss Agency for the Environment, Forests and Landscape (1999) The Swiss forest – taking stock, interpretation of the second national forest inventory in terms of forest policy, Swiss agency for the environment, Forests and Landscape, Bern

United Nations (2000) Forest resources of Europe, CIS, North America, Australia, Japan and New Zealand (Industrialized temperate/boreal countries). UNECE/FAO contribution to the global forest resource assessment 2000, Main report (Geneva timber and forest study papers, No. 17). United Nations, New York/Geneva

Verkerk PJ, Anttila P, Eggers J et al (2011) The realisable potential supply of woody biomass from forests in the European Union. For Ecol Manag 261:2007–2015

Vilén T, Gunia K, Verkerk PJ et al (2012) Reconstructed forest age structure in Europe 1950–2010. For Ecol Manag 286:203–218. doi:10.1016/j.foreco.2012.08.048

Chapter 3
Projection Systems in Europe and North America: Concepts and Approaches

Susana Barreiro and Margarida Tomé

3.1 Background

Forests are long-lived biological systems that fulfil many roles including providing renewable raw materials and energy, maintaining biodiversity, and protecting land and water resources. Originally, forestry and forest management focused mainly on timber production and harvestable standing biomass, and the first prediction tools were simple yield tables. As forest management became more complex and with the need to account for social, economic and environmental values, yield tables started being replaced by growth and yield models. With the introduction of the 'sustainable multifunctional management' concept, improved growth and yield models were developed to additionally include non-wood forest products and services (Rennolls et al. 2007) which led to further improvements in modelling capability. Because forests have always been subject to natural and human-induced disturbances that lead to continuous changes, growth projections that incorporate disturbance factors are required to support decision making (Peng 2000). As unanticipated events accumulate and societal and environmental conditions change, goals and objectives also change. Consequently growth models must evolve to better reflect current knowledge of ecosystem functions and to better exploit current technology (Rennolls et al. 2007).

In this book a *forest model* is a dynamic quantitative representation of the forest, at any level of complexity, based on a set of (sub-)models or modules that together predict the dynamics of the forest as defined by the values of a set of variables that characterize the forest at a given moment (Tomé and Faias 2011). For practical applications, forest models should be implemented in computer programs with user-friendly interfaces – usually designated as forest simulators. A *forest simulator* is a computer tool that is based on a set of forest models and makes long-term pre-

S. Barreiro (✉) • M. Tomé
Forest Research Centre (CEF), School of Agriculture, University of Lisbon, Lisbon, Portugal
e-mail: smb@isa.ulisboa.pt; magatome@isa.ulisboa.pt

S. Barreiro et al. (eds.), *Forest Inventory-based Projection Systems for Wood and Biomass Availability*, Managing Forest Ecosystems 29,
DOI 10.1007/978-3-319-56201-8_3

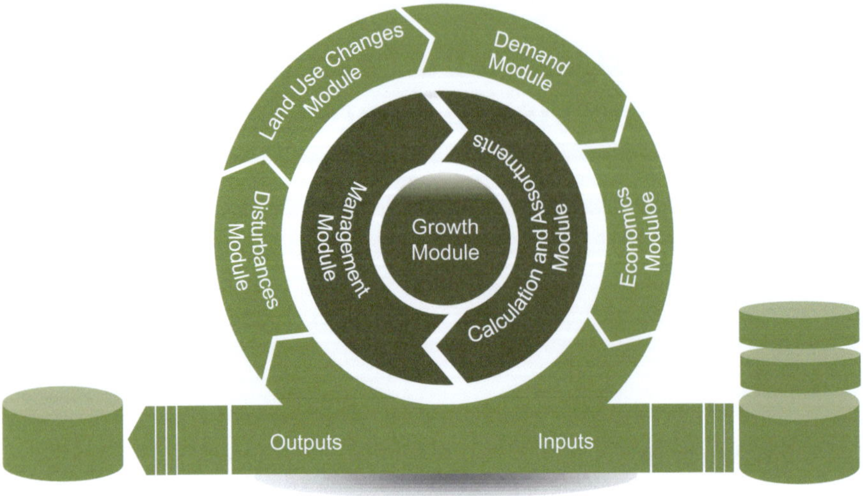

Fig. 3.1 Simplified overview of a projection system

dictions of forest status within a well-defined region under specific climate scenarios, forest policies and/or management alternatives (Tomé and Faias 2011). Forest simulators predict timber outputs, the state of the forest, and may additionally predict non-wood outputs for each point in time. The core of any forest simulator is the growth module responsible for updating the values of state variables. This module comprises the suite of fitted growth functions/sub-models, in the case of empirical models, or algorithms representing processes, in the case of process-based models. The value of each state variable in the following instant in time is dynamically predicted based on the present characteristics of the stand and environment. The calculation module contains fitted functions/sub-models and other components that permit the estimation of other tree and stand variables. This module is static which means that all variables refer to the same instant in time. For simplicity reasons, sub-models will be referred to as models. Silvicultural practices influence stand development and long-term site properties. Besides management, simulations are usually driven by additional external drivers such as disturbances, demand, or land use changes, all of which are usually implemented in individual modules (Fig. 3.1). The term *projection system* designates regional/country tools that range from complex forest simulators that combine national forest inventory (NFI) data with growth functions/models and management modules to more general approaches that combine standwise forest inventory (SFI) data and yield tables to estimate forest growth.

Projection systems can be linked to other models running in a chain or they can run in parallel with other models.

Over recent decades, scientists have faced challenges in building projection systems that accommodate the multitude of factors that affect forest growth and that are necessary to assist policy decision-making. Research teams from many countries have

Table 3.1 Overview of the projection systems from Europe and North America included in this book

Country	Projection system acronym	References
Northern Europe		
Finland	MELASIM	Siitonen et al. (1996) and Redsven et al. (2013)
Sweden	HUGIN	Lundström and Söderberg (1996)
Norway	AVVIRK2000	Eid and Hobbelstad (2000)
Central Europe		
Austria	CALDIS	Kindermann (2010a, b)
	PROGNAUS	Sterba et al. (1995) and Ledermann (2006)
Germany	WEHAM	Rock et al. (2013)
Switzerland	*Massimo 3*	Kaufmann (2001, 2011)
France	MARGOT	Wernsdörfer et al. (2012)
	Matrix – Age	Alvarez-Marty (1989)
West-Central Europe		
Denmark	DK Simulator	Nord-Larsen and Suadicani (2010)
Netherlands	EFISCEN	Sallnäs (1990) and Schelhaas et al. (2007)
	ForGEM	Kramer et al. (2010) and Kramer and Van der Werf (2010)
Iceland	Icelandic-Simulator	Snorrason (2006)
Ireland	FORECAST approach	Phillips (2011)
Eastern Europe		
Lithuania	*Kupolis*	Petrauskas and Kuliešis (2004)
Bulgaria	FRAM	Kostov (1993)
Czech Republic	THP	–
Estonia	*Various tools*	*Not applicable*
Hungary	Hungarian approach	–
Romania	–	–
Southern Europe		
Portugal	StandsSIM.dd	Barreiro and Tomé (2011, 2012)
Italy	*r.green.biomassfor*	Garegnani et al. (2015)
Spain	*Various tools*	*Not applicable*
North America		
Canada	CBM-CFS3	Kurz et al. (2009) and Kull et al. (2011)
USA	FVS	Stage (1973)

developed new tools to address their needs, or in some cases have chosen to adapt tools developed for other countries; the result is a wide variety of projection systems.

This chapter aims to summarize the differences and similarities among the projection systems from Europe and North America as described in the second part of the book (Table 3.1). Some of the projection systems do not operate at the national scale, but are developed for a specific region and/or certain tree species (e.g. Spain, Estonia), or a specific stand structure (e.g. Czech Republic). Some countries are still in the process of developing growth models and projections systems (e.g. Iceland, Romania). Further details on each projection system can be found in the second part of this book.

3.2 History of Projections

In Sweden, the first projections were made more than a century ago and showed that the increasing demand for sawn wood could only be met if harvest considerably exceeded increment. Currently, projections aim to provide public authorities, organizations and industry a broad base for strategic decision-making. In the USA, projection-based forest resource assessments were initiated in the 1970s and have been updated at 10-year intervals. In Canada, energy concerns led to the first forecasts of biomass as an energy source nearly a century ago. In the meantime, the forest carbon accounting needs under the United Nations Framework Convention on Climate Change and the Kyoto Protocol have been the motivation for simulating the dynamics of forest carbon stocks. There is no tradition in making nation-wide wood availability projections regularly in most European countries. The large diversity of geomorphological features and climatic conditions found in some of these countries and the heterogeneous forests with large numbers of tree species that they produce have hindered the development of country-wide projection systems. Austrian projection systems are based on the Forest Vegetation Simulator (FVS, Stage 1973) developed in the USA, whereas the Canadian Carbon Budget Model (CBM-CFS3, Kurz et al. 2009; Kull et al. 2011) was adapted for Italian conditions (Pilli et al. 2013, see Chap. 4). In France, large-scale forest resource simulators have only started being developed in the last two decades. In Portugal, where plantation forestry has strongly been favoured, the first projection studies were carried out in 1990s for maritime pine stands using a system including optimization based on linear programming (Bento 1994). In West-Central European countries where forest areas comprise less than 15% of the total area, projections of future wood availability have been triggered more recently by the need to report carbon estimates for the Kyoto Protocol, to increase forest biomass production and to ensure the sustainable use of wood.

3.3 Analysing the Structure of the Projection Systems Used in Europe and North America

3.3.1 Tools and Their Applicability Ranges

Projection systems can operate at multiple spatial scales (Tomé and Faias 2011). *Stand*-based systems focus on simulating single stands. *Landscape*-based systems simulate all the stands in a region on a stand-by-stand basis with stands spatially described in a geographic information system (GIS) and outputs provided for the entire landscape. Management Unit-based systems are a variant of landscape-based systems developed to assist in preparing a common management plan for a well-defined region. *Regional/National*-based systems are usually based on forest inventory data, which can be aggregated according to a spatial grid or a set of polygons connected to a GIS, and produce outputs by forest type for the entire region. The connection to a GIS facilitates computation of indicators that include spatial relationships and facilitates testing of the effects of spatial restrictions such as maximum or minimum harvested areas or maximization of edges. Spatial conditions can be incorporated into the drivers, for instance to include distance to mills in harvesting algorithms; ecological conditions in the areas that are abandoned or planted; or the spatial propagation of disturbance events. Notwithstanding, it is not uncommon to find that a tool does not fit well into any of the above categories, and many of the projection systems described in the second part of the book can be classified as "non-spatial regional/national simulators" depending on their spatial scales and application ranges. Examples include the age- or diameter-class based matrix models that are applied to all forests in a region without considering their spatial configuration. Several projection systems are "stand simulators" with the ability to process multiple plots/stands.

Additionally, projection systems may only forecast the evolution of a stand until harvest or may produce forecasts for longer time intervals considering changes in forest management approaches. When long-term projections are produced, multiple harvesting and stand regeneration methods can be considered. The lengths of projections range from 10 to 120 years, and the simulation time-steps range from 1 to 20 years (Table 3.2).

3.3.2 Growth Modules

3.3.2.1 Model Philosophy

The first prediction tools were simple *Yield Tables* developed to predict yields for even-aged stands. These tables assume density to be constant at full stocking, normal stocking or average levels, and can only include two variables. Conversely, growth and yield models consider density as a dynamic component of the stand projection. Despite their simplicity, increment estimates obtained from yield tables are still included in some projection systems commonly used in Eastern European systems.

Forest growth models can be based on different philosophies. *Empirical Growth and Yield Models* describe behaviors without trying to identify the underlying causes. They provide biologically realistic predictions that are accurate within the limits of sampling and measurement accuracy (Vanclay 1994). However, these models lack flexibility and the capacity to simulate environmental stresses (Landsberg 2003) and are not valid for new silvicultural management treatments (Reed 1999; Reed et al. 2003) that differ from those observed during the period when the measurements on which they are based were made (Landsberg 2003). Conversely, *Process-based models* describe the physiological processes that lead to forest development (Landsberg 2003). However, these models contain many, often poorly known, parameters for their projections to be considered as reliable as empirical projections. Additionally, these models require values for input variables such as detailed climate and soil conditions that are not readily available from forest

Table 3.2 Application range and spatial and temporal scales of the projection systems

Projection system acronym	Projection length[a]	Time-step[a]	Spatial scale/application range
Northern Europe			
MELASIM (FI)	30	5	Plot-based management driven simulator applicable to the whole country
HUGIN (SE)	100	5	Plot-based management driven simulator applicable to the whole country
AVVIRK2000 (NO)	100	10	Stand-based simulator applicable to region-country level by post-aggregation of results at stand, landscape and region levels
Central Europe			
CALDIS (AT)	10–200	1	Plot-based simulator applicable to region-country level with post-aggregation of results at management unit, landscape or region
PROGNAUS (AT)	10–200	5	Plot-based simulator applicable to region-country level with post-aggregation of results at management unit, landscape or region
WEHAM (DE)	40	5	Plot-based management driven simulator applicable to the whole country by post-aggregation of results
Massimo 3 (CH)	100	10	Plot-based management driven simulator applicable to the whole country by post-aggregation of results
MARGOT (FR)	20–30	1–5	Matrix model at the national scale
Matrix – Age (FR)	20–30	1	Matrix model at the national scale
West-Central Europe			
DK Simulator (DK)	100–500	1	Matrix model at the national scale
EFISCEN (NL)	50–60	5	Matrix model at the national scale
ForGEM (NL)	20–30	mth	Plot/stand management driven simulator applicable to region-country level by post-aggregation of results
Icelandic-Simulator (IS)	120	1	Plot/stand management driven simulator applicable to region-country level by post-aggregation of results
FORECAST approach (IE)	80	1	Stand-based simulator
Eastern Europe			
Kupolis (LT)	100	1–20	Management unit stand-based simulator applicable to region-country level by post-aggregation of results
FRAM (BG)	–	–	Management unit simulation applicable to region-country level by post-aggregation of results
THP (CZ)	30	10	
Various tools (EE)	Felling age	1	
(HU)	30–100	10	
(RO)	10	10	

(continued)

Table 3.2 (continued)

Projection system acronym	Projection length[a]	Time-step[a]	Spatial scale/application range
Southern Europe			
StandsSIM.dd (PT)	100	1	Gridded demand-driven regional simulator running for the whole country
r.green. biomassfor (IT)	–	–	Spatialized Projection System for bioenergy and timber availability assessment given ecological, technical, economic, and sustainable production constraints. Developed for the Province of Trento. Requires volume as input
Various tools (ES)	–	–	Stand-based simulators not applicable to regional levels
North America			
CBM-CFS3 (CA)	100	1	Requires volume as input
FVS (US)	100	1–5	Plot-based management driven simulator applicable to the entire country

[a]Projection length and time-step unit is years, except where marked with mth (month)

inventories, whereas empirical models use field measurements often available from forest inventories as input. For this reason, process-based models have only rarely been used by forest managers as management tools at stand-level and have not been incorporated into large scale forest projection systems. *Hybrid models* combine process-based elements to predict productivity by addressing the effects of a changing environment (Monserud 2003) with statistical descriptions of stand structure (for further details on hybrid models see Mäkelä 2009; Kimmins et al. 2010). In several Eastern European countries, projection methods estimate standing volume at the end of each time-step by starting with standing volume at the beginning of the time step, adding the increment which is usually estimated from yield tables depending on stand density and site index, and subtracting the drain. For the remaining countries, except for the Netherlands where the ForGEM process-based model is used, all projection systems are based on empirical growth models (Table 3.3).

Burkhart and Tomé (2012), following Munro (1974), classify empirical growth models with respect to the projection unit. *Whole-stand models* project number of trees, basal area and/or volume per stand, generally based on stand age, stand density and site index in the case of even-aged stands. *Size-class models* project the number of trees by diameter-classes, which are subsequently converted to volume and/or biomass by diameter-class using tree taper, volume or biomass prediction models. The future distributions by diameter-class can be based on a probability density function (usually designated as diameter distribution models, which are usually complemented by site index curves and models for basal area growth and for the prediction of the number of trees), or on transitions between classes (matrix model). *Individual-tree models* project tree growth (e.g. diameter and height increments) for individual trees (eventually tree size classes) as a function of the present

Table 3.3 Characteristics of the growth modules in the projection systems

Projection system acronym	Empirical Yield table	 Whole stand	 Size class (matrix)	 Individual-tree distance independent	Process-based Individual-tree distance dependent	Climate sensitive
Northern Europe						
MELASIM (FI)	–	–	–	Yes	–	Yes
HUGIN (SE)	–	–	–	Yes	–	Yes
AVVIRK2000 (NO)	–	Yes	–	–	–	No
Central Europe						
CALDIS (AT)	–	–	–	Yes	–	Yes
PROGNAUS (AT)	–	–	–	Yes	–	No
WEHAM (DE)	–	–	–	Yes	–	No
Massimo 3 (CH)	–	–	–	Yes	–	No
MARGOT (FR)	–	–	Yes (d)	–	–	No
Matrix – Age (FR)	–	–	Yes (a)	–	–	No
West-Central Europe						
DK Simulator (DK)	–	–	Yes (a)	–	–	No
EFISCEN (NL)	–	–	Yes (a)	–	–	No
ForGEM (NL)	–	–	–	–	Yes	Yes
Icelandic-Simulator (IS)	Yes	–	–	–	–	No
FORECAST approach (IE)	Yes	Yes	–	–	–	No
Eastern Europe						
Kupolis (LT)	–	Yes	–	–	–	No
FRAM (BG)	Yes	–	–	–	–	No
THP (CZ)	Yes	–	–	–	–	No
Various tools (EE)	–	–	–	Yes	–	No
(HU)	Yes	–	–	–	–	No
(RO)	Yes	–	–	–	–	No
Southern Europe						
StandsSIM.dd (PT)	–	Yes	–	Yes	–	Yes
r.green.biomassfor (IT)	–	–	–	–	–	–
Various tools (ES)	Depends on the tool					
North America						
CBM-CFS3 (CA)	Yes	–	–	–	–	No
FVS (US)	–	–	–	Yes	–	Yes

Yes (d) represents diameter-class matrix model
Yes (a) represents age-class matrix model

tree size and stand-level variables such as age, stand density and site index. These models can be further subdivided into individual-tree distance independent or distance dependent models, depending on whether individual tree locations are or are not used. Distance independent individual-tree models are commonly used, especially in northern and central Europe (Table 3.3). Both AVVIRK2000 and *Kupolis* use whole-stand models that are implemented by predicting the average tree of a stand and extrapolating for the stand using the number of trees. StandsSIM.dd uses whole-stand or individual-tree models, depending on the tree species. Matrix models are the preferred type of size-class models in projection systems and are applied at the regional or national scale. In France, two matrix models are used: the first is a diameter-class model governed by a Markov transition matrix with constant recruitment (MARGOT), whereas the second is based on within-strata age distributions of the forested area. The EFISCEN system in the Netherlands uses growth models with increment predicted as a function of age, while species-specific yield tables are used for estimating Danish growth and thinning volumes. The growth modules can target individual tree basal area increment (HUGIN, *Massimo 3*, PROGNAUS, CALDIS) or diameter increment (MELASIM, WEHAM, StandsSIM.dd). Height increment is predicted in MELASIM, PROGNAUS, CALDIS while in WEHAM and in StandsSIM.dd height is predicted as a function of diameter and stand variables and in HUGIN as a function of top height.

3.3.3 Calculation Modules and the Assortments

All systems produce outputs in terms of wood volume, either directly predicted in the growth module or from other variables, with different methodologies depending on the type of model, in the calculation module. This conversion may include height and the use of taper models, such as in PROGNAUS and CALDIS and WEHAM. Projection systems usually partition the harvested volume by thinning, regeneration cuttings and clear-cut, and may separate volume further by assortments. Some systems estimate woody assortments within the calculation module while others have a separate module, sometimes called a drain or a grading module. PROGNAUS and CALDIS also include models that describe quality parameters, branch diameter, and the presence of wood decay to partition trees into assortments. StandsSIM.dd uses species-specific models (systems of compatible volume, volume ratio and taper model), while Ireland relies on assortment tables. With the MELASIM system, thinning and final-felling can be constrained to remove different combinations of roundwood, energy wood and waste wood.

Additionally, some projection systems explicitly consider losses, e.g. harvest losses, low wood quality, logs left in the forest or by the road side. For example, the Icelandic system estimates the potential wood removals overbark using a 10% discount to account for harvesting residues and wood not meeting quality requirements. StandsSIM.dd, MELASIM and FVS differentiate between forest residues left on site and those removed. In FVS, these amounts not only reduce harvest volumes but also contribute to increasing the fuel load which is taken into account when simulating fire (Dixon 2002). Also, harvesting constraints may be included in the projection system. In Ireland, volumes are adjusted to exclude for example thinning small forest areas with a potentially uneconomic forest road requirement or final harvest for accessibility reasons. In Switzerland, the "onion model" by Hofer et al. (2011) is combined with *Massimo 3* to account for ecological and socioeconomic constraints (see also Chap. 2).

Biomass and carbon quantities can be estimated at individual-tree or whole-stand level using models, biomass expansion factors (BEF) and/or wood and carbon densities. The number of carbon pools distinguished differs among the projection systems but usually includes several above-ground living carbon pools such as stemwood, branches and foliage and at least one belowground carbon pool. In addition, some systems also model carbon in litter and/or soil such as FVS and CBM-CFS3, or harvested wood products such as *Massimo 3*.

3.3.4 Management Modules

3.3.4.1 Thinning

Most projection systems described in this book are management driven with thinning and final-felling operations considered by nearly all of them (Table 3.4). Thinning and final felling procedures greatly differ, mostly depending on the type of model used: individual-tree, whole-stand or matrix. The simpler models have thinning schedules that are fixed with respect to frequency and intensity, such as the Icelandic system and in matrix models such as EFISCEN. Also, final felling in these approaches is usually determined by age, but it may be determined by demand as in StandsSIM.dd.

In other models, the decision to thin a stand may depend on stand-level variables such as basal area increment since last thinning (*Massimo 3*); stand density (*Kupolis*, PROGNAUS and CALDIS), tree size (PROGNAUS and CALDIS) or a combination of variables such as dominant height and basal area (MELASIM); tree species, age and dimension of the trees (WEHAM); or on a priority function (HUGIN).

The thinning intensity may depend on stand structure (*Massimo 3*) or stand density (*Kupolis*). Thinning intensity can be expressed as basal area to be removed (HUGIN, StandsSIM.dd), thinned or post-thinned stand density, volume or basal area (Irish FORECAST approach, StandsSIM.dd) or may be based on thinning schedules (HUGIN) or yield-tables (WEHAM). Similarly, StandsSIM.dd allows the use of the Wilson factor (Wilson 1946), a relative spacing measure function of dominant height and stand density.

Selection of individual trees to be removed can be based on tree size (WEHAM, *Massimo 3*) and may mimic thinning methods such as thinning from above (FVS), from below (FVS), or systematic thinning. PROGNAUS and CALDIS include algorithms responsible for mimicking thinning patterns observed in NFI data. The FVS

Table 3.4 Components of the assortments and management modules

Projection system acronym	Woody assortments/ losses	Carbon/soil carbon	Management Woody removal operations	Other silvicultural treatments with impact on growth	Stand regeneration
Northern Europe					
MELASIM (FI)	Yes/yes	Yes/yes**	Yes	No	Yes
HUGIN (SE)	Yes/yes	Yes/yes**	Yes	Yes	Yes
AVVIRK2000 (NO)	Yes*/–	–	Yes	No	–
Central Europe					
CALDIS (AT)	Yes*/–	Yes/yes**	Yes	–	Yes
PROGNAUS (AT)	–	–	Yes	–	–
WEHAM (DE)	Yes/yes	Yes/–	Yes	–	–
Massimo 3 (CH)	Yes/yes	Yes/yes**	Yes	–	Yes
MARGOT (FR)	Yes/yes	Yes/–	Yes	–	Yes
Matrix – Age (FR)	Yes/yes	Yes/–	Yes	–	Yes
West Central Europe					
DK Simulator (DK)	Yes/yes	Yes/–	Yes	No	Yes
EFISCEN (NL)	No/yes	Yes/yes**	Yes	No	Yes
ForGEM (NL)	Yes/yes	Yes/–	Yes	No	Yes
Icelandic-Simulator (IS)	–/yes	Yes/–	Yes	No	–
FORECAST approach (IE)	Yes/yes	–	Yes	No	–
Eastern Europe					
Kupolis (LT)	Yes/yes	Yes/–	Yes	No	Yes
FRAM (BG)		–	–	–	–
THP (CZ)	–	–	Yes	–	–
Various tools (EE)	Yes/–	–	Yes	–	*In some tools*
(HU)	–	–	Yes	–	–
(RO)	–	–	–	–	–
Southern Europe					
StandsSIM.dd (PT)	Yes/yes	Yes/–	Yes	No	Yes
r.green.biomassfor (IT)	Yes (–)	Yes/yes	Yes	No	No
Various tools (ES)	*In some tools*	*In some tools*	Yes	–	*In some tools*
North America					
CBM-CFS3 (CA)	Yes	Yes/yes	Yes	No	Yes
FVS (US)	Yes/yes	Yes/yes	Yes	Yes	Yes

Yes* represents assortments estimated as a post-processing operation;
Yes** represents soil carbon estimated using an independent soil carbon model

model considers numerous thinning options: (1) selection of specific trees for removal; (2) selection of classes of trees for removal; and (3) selection of trees for density control. When density control thinning is applied, a species-specific removal priority is assigned to each tree based on tree size (*dbh*) and special tree status, after which trees with the greatest removal priorities are harvested until a target density expressed in terms of basal area or stand density is met.

In most systems the effect of thinning on the remaining trees is implicitly included as increased growth due to reduced competition. However, some systems include explicit responses, such as HUGIN for which the effect of thinning is estimated using thinning response functions developed with experimental data (Jonsson 1974).

Multiple approaches are used to simulate thinning in age-class matrix models. For example, a thinning rate can be applied as a percentage of the growing stock per unit area and a clear-cut rate can be applied as a percentage of the forest area in the age-class; a thinning frequency corresponding to about a tenth of stand age can be applied; or a probability can be applied whereby a thinning is carried out as a function of stand age. In EFISCEN, thinning is simulated by moving a specified number of hectares in a specified volume class to a smaller volume class, with the difference between volume classes being the volume thinned.

3.3.4.2 Final Felling

Final felling tends to be implemented based on rotation age for even-aged stands and can be combined with a target diameter for both even- and uneven-aged stands. Priority functions can be combined with final felling criteria for both individual-tree models and whole-stand models. In HUGIN only stands that have reached 90% of the rotation age are ranked for final felling using priority rules such as functions of relative age priority, net value growth, or volume increment percentage. A similar method is used with StandsSIM.dd which allows users to define minimum harvest ages and assign different final-felling priorities according to stand structure and stand age in the case of even-aged stands. AVVIRK2000 allows application of different final felling criteria such as a user-defined harvest age or the relative annual value increment less than a user defined percentage. In *Massimo 3* the probability of a tree being cut is predicted with logistic regression models using stand and site characteristics and harvest conditions as predictor variables (Thürig et al. 2005). In StandsSIM.dd which is a demand driven simulator, a harvesting probability is assigned to each NFI plot/stand in the input according to rules defined by the user. Probabilities are based on stand structure and a minimum age to harvest that must also be defined. A stand is harvested when a random number is less than or equal to the assigned probability, and stands are harvested until the demand defined in the scenario is met for each simulation step.

In age-class matrix models such as EFISCEN, final felling probability is expressed as a function of stand age. The area to be clear-felled is moved from its respective age class to the regeneration class, while the average volume of the original age class is considered as the final-felled volume.

3.3.4.3 Stand Regeneration

Most systems include a module for stand regeneration after clear-cut or regeneration fellings (Table 3.4). Systems may target artificial regeneration (CALDIS, DK simulator, StandsSIM.dd), natural regeneration (Massimo 3), or include both options (CBM-CFS3, *Kupolis*, MELASIM and HUGIN). Regeneration by planting usually allows the choice of tree species and planting density. Sometimes site preparation operations prior to planting can be included as well. CBM-CFS3 represents planting through reductions in the regeneration delay in growth and-yield curves or through switching to a different growth-and-yield curve. Usually observed data from permanent plots, NFI plots or regeneration trials are used to mimic natural regeneration patterns, or, in some cases, data for an existing plot can be imputed to initialize the previously harvested stand (FVS, *Massimo 3*, HUGIN). The process-based ForGEM model is able to model regeneration explicitly by seed dispersal and seed inflow from outside.

3.3.4.4 Other Silvicultural Treatments

A wide range of additional silvicultural treatments and operations can be considered. However, only a few systems simulate the impact of silvicultural treatments such as fertilization or draining. Several approaches can be used to include the response to fertilization. The HUGIN system simulates the effect of fertilization, and soil drainage can be simulated whenever stands meet conditions specified by the user and both are expressed by an increase in site index. The FVS system considers different types of multipliers affecting both diameter and height growth (Dixon 2002) to simulate the effects of declining growth due to senescence or silvicultural treatments such as fertilization.

3.3.5 Additional External Drivers Modules

3.3.5.1 Natural Disturbance Simulation Modules

Disturbance can be defined as an event in time that disrupts an ecosystem, community, or population structure by changing available resources. Multiple aspects of disturbance events including intensity, extent, spatial and temporal probability of occurrence, are usually collectively characterized as the disturbance regime (Perera et al. 2015). Natural disturbances rarely eliminate all individuals from the stand and, depending on the type of disturbance, reductions in growth and sometimes stand structure changes can be expected. Projection systems tend to include the main natural disturbances affecting forests in the country for which they have been developed (Table 3.5).

Table 3.5 Additional external drivers' modules in the different projection systems

Projection system acronym	Other	Disturbances	Land use changes	Economics
Northern Europe				
MELASIM (FI)	Climate	–	–	Yes
HUGIN (SE)	Climate	Browsing	Afforestation	No
AVVIRK2000 (NO)	–	–	–	Yes
Central Europe				
CALDIS (AT)	Wood/biomass demand	–	–	Yes
PROGNAUS (AT)	–	–	–	Yes
WEHAM (DE)	–	–	–	No
Massimo 3 (CH)	–	Storms	–	–
MARGOT (FR)	No	No	–	–
Matrix – Age (FR)	No	No	–	–
West-Central Europe				
DK Simulator (DK)	FnAWS Set aside areas	Wind throw	Afforestation	Yes
EFISCEN (NL)	–	No	Afforestation/ deforestation	No
ForGEM (NL)	–	No	No	No
Icelandic-Simulator (IS)	–	–	Afforestation	No
FORECAST approach (IE)	–	Wind throw Diseases	–	No
Eastern Europe				
Kupolis (LT)	–	–	–	Yes
The descriptions of the projection systems of the other countries in this region make no mention to additional external drivers				
Southern Europe				
StandsSIM.dd (PT)	Wood/biomass demand	Fire	Afforestation Deforestation	Yes
r.green.biomassfor (IT)	–	–	–	yes
Various tools (ES)	–	–	–	*In some tools*
North America				
CBM-CFS3 (CA)	–	Fire Pests	Afforestation Deforestation	–
FVS (US)	Climate	Fire Pests	Afforestation	yes

Risk agents such as wildfire and wind are often assigned annual probabilities of occurrence that depend on probability distributions modelled by a Poisson distribution, or by applying historic time series. Wildfire converts living trees to standing and lying dead wood. After catastrophic windthrow events, large trees may survive either intact or damaged, whereas smaller trees can be converted to logs and debris covering the forest floor (Franklin et al. 2002). The current capability to simulate the occurrences and severity of disturbances is very limited. Therefore, the simulation of natural disturbances can be based on probabilities estimated from long series of observations, tree ring analysis and/or expert opinion. Details on how such events are triggered and how they affect stand structure and composition can also be simulated.

Risk agents such as insects are more complex to simulate because periods with minimal impacts alternate with outbreak events. Impacts can be restricted to the year of the outbreak or last for multiple years. Modelling insect impacts may require not only modelling population build-up over time, but also interactions with other disturbance agents. If a stochastic approach is used in disturbance modelling, Monte Carlo simulations can provide an average prediction as well as an uncertainty estimate. Spatial-based simulators require even more complex approaches that consider not only the total area affected annually, but also the number, intensity and distribution of events as well as the geographic coordinates for the disturbed areas.

Disturbances can be simulated using specific models that run independently or are integrated into the projection system. Alternatively, simpler approaches implemented in specific modules can be used such as a volume reduction to discount losses resulting from windthrow or diseases. This reduction can be based on historical data, or on a simple assumption as is done in the Irish approach.

In Canada, probability distribution functions estimated from historic records of fire and insect (seven major pests) outbreaks are used to generate time-series of areas of future disturbance. A Monte Carlo approach is used to generate a range of disturbance time-series for fire and insects which are used as input to CBM-CFS3. The disturbance impacts caused by fire and insects are simulated using matrices representing proportional carbon transfers from living to dead pools within the ecosystem or out of the ecosystem, either by combustion or harvest. Mortality and growth reduction can also be considered. Information on the timing of future outbreaks, outbreak intervals, lengths and distribution of the type of impact are also required.

In the USA, the FVS system also takes fire into account. Users schedule a fire, and the fire module computes its intensity as well as its effects on stand development and management actions. The module uses elements from fire behaviour and fire effects models, although it does not simulate fire spread or the probability of fire occurrence (Dixon 2002). FVS also considers the interaction between specific insects or pathogens and stand and tree development. Depending on the agent, different approaches can be used with the most common being an independent model which is automatically invoked when damage codes are present in the FVS tree data input resulting in a mortality estimate. This estimate is then compared to the estimate of mortality from the FVS base model, and the greater of the two values is used as the estimate of mortality. Alternatively, when no model is available, an outbreak is scheduled when a random number is less than or equal to an estimated probability in which case tree mortality is increased based on the stand's hazard rating (Dixon 2002).

In the European systems described, disturbances are simulated using simpler approaches that vary depending on the type of growth model integrated in the projection system. The Danish simulator includes the effects of windthrow directly in the transition probabilities from one age class to the next. These probabilities are defined based on the observations between consecutive forest inventories. Shorter rotation ages have been considered as a preventive measure against the frequent wind-throws, bark beetle attacks, and root rot debilitation.

Massimo 3 allows simulation of storm scenarios using models fitted from NFI data that provide probabilities of fellings resulting from natural disturbances. The probabilities can be selectively altered to simulate the effects of various levels of disturbance (Thürig et al. 2005). HUGIN includes models to predict damage from moose browsing in young forests. The probability of damage can be changed to simulate changes in moose density.

Fire is the only disturbance considered in the StandsSIM.dd system where the user is allowed to define the series of annual burnt areas for the simulation period. Stands are burnt when a random number is less than or equal to a probability threshold. Burnt areas are followed by harvest, and a user-defined percentage of salvage wood with industrial use is considered.

3.3.5.2 Climate Change Modules

Process-based models are the most suitable tools for assessing impacts of changes in climate because climate variables used as input in these models drive the physiological processes that express tree or stand growth. Because these models are quite demanding in terms of inputs, most projections use empirically based projection systems for which the environment is usually expressed by site index, either alone or in combination with models integrating climatic variables. Empirical models that include climate are sometimes assumed to be able to handle climate change effects. However, these empirical models become less reliable if they are applied outside the climate range for which they were developed. Climate variables are used as predictors to facilitate responses to short-term climate changes in the CALDIS and StandSIM.dd systems.

Within the HUGIN system, the positive effects of climate change on height and diameter growth are simulated differently for young trees and adult trees (taller than 7 m). For young trees, site index is linearly increased annually to a maximum of 4 m after 100 years, whereas for adult trees, a species-specific multiplier is applied to tree diameter growth. The multiplier assumes percent increases that vary with tree species and region (Berg et al. 2007). Only positive temperature-related effects are considered but the effects of precipitation, wind and damage agents may compensate for the positive increase in temperature.

In the MELASIM system, species-specific models express the effects of climate change (e.g. increase in temperature and atmospheric carbon dioxide) on tree volume growth for each time-step. The FinnFor physiological model which is based on photosynthesis, respiration and transpiration, was used to simulate different climate scenarios, and the simulation results were used to develop the transfer functions.

FVS permits site index to be changed for individual species or species groups to account for climate change. Stand-BGC (Milner and Coble 1995) is a climate driven model that simulates photosynthesis, respiration, and evapotranspiration on a daily time-step and that allocates carbon to plant tissue pools (leaf, stem, and roots) on a yearly time-step (Dixon 2002). A FVS module integrates FVS and Stand-BGC to produce a hybrid simulator, FVS-BGC (McMahan et al. 2002). At the end of each growth cycle, simulated tree dimensions in FVS are updated using either the FVS or BGC growth estimates, as directed by the user.

3.3.5.3 Economic Modules

Economic modules may consist of independent economic models that run separately from the projection system without direct interaction with the growth simulation. In North America, economic analyses in the FVS system are carried out using the economic model CHEAPO II (Medema and Hatch 1982). A special FVS output file is used as input, and CHEAPO II runs independently without any dynamic interaction with FVS (Dixon 2002).

It is also possible to find projection systems that have an economic module for which wood (and eventually other wood and non-wood) revenues and costs for silvicultural operations and treatments are considered at each time-step. In Europe, a wide range of approaches is used. With PROGNAUS and CALDIS, the costs associated with various harvesting techniques are estimated using harvesting models and compared to the revenue. Revenues are computed by applying wood price scenarios to the assortments resulting from each harvested tree. Harvesting interventions that do not reach the profit margin can be ignored. A different approach is used in the stand-level based systems, StandsSIM.dd and *Kupolis*, for which the costs for all scheduled operations are taken into account. In the Portuguese system, default costs are available and can be changed by users. The economic module computes the costs for all operations and revenues based on the assortments resulting from thinning and final-felling operations, thus allowing calculation of net present value (NPV). Similarly *Kupolis* calculates the costs and revenues for operations beginning with planting and ending with movement of timber logs to roadsides for transport.

3.3.5.4 Land Use Changes Modules

Land use change (LUC) modules can consider both afforestation and deforestation or just one of them. Multiple approaches can be used to simulate LUC. With the Canadian system, CBM-CFS3, afforestation events can be considered by the user by selecting the tree species and selecting forest as the land cover. To consider afforestation in the age-class French matrix simulator, a net influx of new areas is assigned to the first age-class and a similar approach is used in the Danish simulator by defining the afforested area and the tree species. With StandsSIM.dd the area of new plantations defined in the scenario for each simulation step is translated into a number of grid cells (stands) to be planted and a species-specific site index is assigned to each stand. Deforestation is simulated by preventing some harvested stands (grid cells) from being replanted until the deforestation criterion defined in the scenario is met. A harvested stand is abandoned if a random number is less than or equal to the deforestation probability threshold which is greater for smaller site index values. The HUGIN system also considers afforestation of agricultural land not in use.

In spatial-based systems all units – forest and non-forest – must be classified in terms of current land use and assigned probabilities of conversion to other land uses. Users select the conditions for simulation units to be converted from forest to other land uses and for those to be afforested. In the latter case, users must define the new forest type and species composition. Alternatively, some restrictions may be implemented in the module. For example, a set of species-specific site productivity indices can be assigned to each simulation unit as well as the probability of conversion of a simulation unit to each forest type and the probability of conversion to another land use. The user can define the probability by selecting the criteria on which probabilities should be based such as proximity to industrial forest plants or to urban areas, accessibility, land use classification of the adjacent grid-cells, and past productivity indices. In non-spatial systems, afforestation and deforestation modules are less complex and simply define the additional forest area and/or the proportion of area to be converted from forest to other land use. However, restrictions can also be added to make the simulation more realistic such as specifying that less productive stands have greater probabilities of being abandoned.

3.3.6 Inputs and Outputs

Projection systems use forest inventory information to characterize forest conditions at the beginning of a simulation. Most of the systems described in this book use data collected by NFI programmes; however, SFI remains the main source of input information for Eastern European countries (Table 3.6).

Data requirements are closely related to the type of growth models used. Projection systems that include stand-level growth models require stand-level variables such as stand density, basal area, dominant height, age and site index for even-aged stands to characterize forests. If the growth module in the projection system includes individual-tree growth models, tree-level variables such as species, diameter, and height are required and can be combined with dominant height curves and in some cases a projection model for the number of trees. Additional information from various sources including NFI surveys, legal sources, and expert opinion may also be needed to run the management modules. When size-class models are used, regardless of whether they are of the matrix type, specific levels of data aggregation are required; examples include the Danish simulator for which stands are grouped into age-classes and MARGOT for which trees are grouped into diameter-classes.

Additionally, some management sub-modules such as for stand regeneration might require the use of permanent plot data to populate the recently harvested stands. Management prescriptions can be provided in the form of files and/or simple instructions.

The more complex the projection system, the more input information is required. To construct the scenarios under analysis, all the other drivers' modules require input information in the form of area for a given disturbance, area of land use change, and volume to be harvested; probabilities of occurrence for specific events; or tree and stand codes that trigger specific activities such as pest or disease outbreaks. Economic modules also require input in the form of costs for silvicultural treatments and other operations, climatic data series for climatic modules, or average values for specified variables. Additionally, spatial information from GISs can also be used by some tools, either to define environmental zones or to establish accessibility classes.

Table 3.6 Types of different inputs required by projection systems

	Type of inputs				
Projection system acronym	National forest inventory	Standwise forest inventory	Permanent plots	Global information systems	Other
Northern Europe					
MELASIM (FI)	×	–	–	×	×
HUGIN (SE)	×	–	×	–	×
AVVIRK2000 (NO)	×	–	–	–	×
Central Europe					
CALDIS (AT)	×	–	–	–	×
PROGNAUS (AT)	×	–	–	–	×
WEHAM (DE)	×	–	–	–	×
Massimo 3 (CH)	×	–	×	–	×
MARGOT (FR)	×	–	–	×	×
Matrix – Age (FR)	×	–	–	–	×
West-Central Europe					
DK Simulator (DK)	×	–	–	–	×
EFISCEN (NL)	×	–	–	–	×
ForGEM (NL)	×	–	–	–	×
Icelandic-Simulator (IS)	×	–	–	–	×
FORECAST approach (IE)	–	×	–	–	×
Eastern Europe					
Kupolis (LT) and all other tools	–	×	–	–	×
Southern Europe					
StandsSIM.dd (PT)	×	–	–	–	×
r.green.biomassfor (IT)	×	–	–	×	×
Various tools (ES)	*Depend on the tool*				
North America					
CBM-CFS3 (CA)	×	–	–	×	×
FVS (US)	×	–	–	–	×

Projection systems can produce a large number of outputs for different aggregation levels, and all can usually be directly exported to databases or text files. Outputs can be single tree lists (e.g. PROGNAUS, WEHAM) that can be post-processed or a range of ready-made tables. Growing stock, thinning and felling amounts, increment, biomass and carbon stocks are the most common outputs. Additionally, the distribution of these amounts by tree species or species groups, by assortment classes, age- and/or diameter-classes is also possible. Some systems can further produce outputs by region, and other tools can produce specific outputs that can be used as input to other tools. For example, *Kupolis* produces outputs that can be further explored using GISs, and multiple projection systems including CALDIS, *Massimo 3*, and MELASIM produce outputs that can be used as input to the *YASSO* carbon soil model (Liski et al. 2005).

3.4 Conclusions

The need for tools capable of accommodating the multitude of factors that affect forest growth led scientists to develop new or adapt existing projection systems to assist decision making. The development of such systems depends on the input data and growth models available but is mainly driven by the needs and political targets in each country. The structural diversity and underlying assumptions behind the existing projection systems hinder the comparability of wood availability projection results among different countries. Projections are done with more or less complex tools using a multitude of approaches.

The projection of forest growth in Europe and North America is achieved using yield tables, which is common in Eastern European countries, whole-stand models that are used by a few projection systems, tree-level empirical growth models that are the most common choice; or process-based growth models which are the least preferred, probably because of greater data requirements. All projection systems consider the effect of silvicultural treatments on growth. Thinning and final-cuts are the most widely implemented treatments, whereas the impacts of treatments such as fertilization are rarely taken into account. At the same time, the importance of wood availability leads several systems to estimate assortments and/or discount losses (e.g. harvest losses, quality losses) in various ways. In addition, other drivers affecting wood availability such as disturbances (e.g. storms, pests, fire) are considered in some systems and are simulated in different ways. A quantitative comparison of the different approaches would be very valuable to gain insight as to how different implementations of growth, management and the effect of external drivers affect the projections. Uncertainty analyses are also indispensable in this respect, but have hardly been conducted so far. Furthermore, many of the projection systems are poorly documented, which is a prerequisite for a deeper understanding of the functioning of these systems.

References

Alvarez-Marty S (1989) La méthode des générations dans l'étude de la ressource d'une forêt équienne. Afocel-Armef, Informations Forêt 3:135–146

Barreiro S, Tomé M (2011) SIMPLOT: simulating the impacts of fire severity on sustainability of eucalyptus forests in Portugal. Ecol Indic 11:36–45

Barreiro S, Tomé M (2012) Analysis of the impact of the use of eucalyptus biomass for energy on wood availability for eucalyptus forest in Portugal: a simulation study. Ecol Soc 17(2):14

Bento JMRS (1994) Oferta sustentada de material lenhoso de pinheiro bravo – uma aplicação a nível nacional. PhD dissertation Universidade de Trás-os-Montes e Alto Douro, 274 p

Burkhart HE, Tomé M (2012) Modeling forest trees and stands. Springer, Dordrecht/New York, 446 p

Dixon GE (2002) Essential FVS: a user's guide to the Forest Vegetation Simulator. Internal Report. US Department of Agriculture, Forest Service, Forest Management Service Center, Fort Collins, 226 p (Revised: November 2, 2015)

Eid T, Hobbelstad K (2000) AVVIRK-2000: a large-scale forestry scenario model for long-term investment, income and harvest analyses. Scand J For Res 15:472. doi:10.1080/028275800750172736
Franklin JF, Spiesb TA, van Pelta R, Careyc AB, Thornburghd DA, Berge DR, Lindenmayerf DB, Harmong ME, Keetona WS, Shawh DC, Biblea K, Cheni J (2002) Disturbances and structural development of natural forest ecosystems with silvicultural implications, using Douglas-fir forests as an example. For Ecol Manag 155:399–423
Garegnani G, Geri F, Zambelli P, et al. (2015) A new open source DSS for assessment and planning of renewable energy: r.green. In: Proceedings of FOSS4G Europe, Como 2015, 14–17 June 2015. Geomatics Workbooks, pp 39–50. ISSN: 1591-092X
Hofer P, Altwegg J, Schoop A et al (2011) Holznutzungspotenziale im Schweizer Wald. Auswertung von Nutzungsszenarien und Waldwachstumsentwicklung. Umwelt-Wissen Nr. 1116:80 S. Bundesamt für Umwelt, Bern.
Jonsson B (1974) The thinning response of Scots pine (Pinus sylvestris) in northern Sweden. Skogshögskolan, Inst för skogsproduktion, Rapp o upps no 28, 41pp
Kaufmann E (2001) Prognosis and management scenarios. In: Brassel P, Lischke H (eds) Swiss national forest inventory: methods and models of the second assessment. Swiss Federal Research Institute (WSL), Birmensdorf, pp 197–206
Kaufmann E (2011) Nachhaltiges Holzproduktionspotenzial im Schweizer Wald. Schweiz Z Forstwes 162(9):300–311. doi:10.3188/szf.2011.0300
Kimmins JP, Blanco JA, Seely B, Welham C, Scoullar K (2010) Forecasting forest futures: a hybrid modelling approach to the assessment of sustainability of forest ecosystems and their values. Earthscan, London. doi:10.1080/00207233.2011.552232
Kindermann G (2010a) Weiterentwicklung eines Kreisflächenzuwachsmodells – Refining a basal area increment model. In: Deutscher Verband Forstlicher Forschungsanstalten Sektion Ertragskunde. Jahrestagung vom 17–19 Mai 2010, Möhnesee, pp 82–95
Kindermann G (2010b) Eine klimasensitive Weiterentwicklung des Kreisflächenzuwachsmodells aus PrognAus – A climate sensitive refining of the basal area increment model in PrognAus. Centralblatt für das gesamte Forstwesen. Austrian J For Sci 127(3–4):147–178
Kostov G (1993) The Bulgarian Forest Resources Assessment Model – some base data and results. SUAS. Department of Operational Efficiency, College of Forestry, Garpenberg
Kramer K, van der Werf DC (2010) Equilibrium and non-equilibrium concepts in forest genetic modelling: population- and individually-based approaches. For Syst 19:100–112. doi:10.5424/fs/201019S-9312
Kramer K, Degen B, Buschbom J, Hickler J, Thuiller W, Sykes MT, Winter WD (2010) Modelling exploration of the future of European beech (*Fagus sylvatica* L.) under climate change--range, abundance, genetic diversity and adaptive response. For Ecol Manag 259:2213–2222. doi:10.1016/j.foreco.2009.12.023
Kull SJ, Rampley GJ, Morken S et al (2011) Operational-scale carbon budget model of the Canadian forest sector (CBM-CFS3) version 1.2: user's guide. Canadian Forest Service, Edmonton
Kurz WA, Dymond CC, White TM et al (2009) CBM-CFS3: a model of carbon-dynamics in forestry and land-use change implementing IPCC standards. Ecol Model 220:480–504
Landsberg J (2003) Modelling forest ecosystems: state of the art, challenges, and the future directions. Can J For Res 26:1174–1186
Ledermann T (2006) Description of PrognAus for Windows 2.2. In: Hasenauer H (ed) Sustainable forest management – growth models for Europe. Springer, Berlin/Heidelberg, pp 71–78
Liski J, Palosuo T, Peltoniemi M, Sievänen R (2005) Carbon and decomposition model Yasso for forest soils. Ecol Model 189:168–182. doi:10.1016/j.ecolmodel.2005.03.005
Lundström A, Söderberg U (1996) Outline of the Hugin system for long-term forecasts of timber yields and possible cut. In: Päivinen R, Roihuvuo L, Siitonen M (eds) Proceedings no. 5: in large-scale forestry scenario models: experiences and requirements. European Forest Institute, Joensuu, pp 63–77
Mäkelä A (2009) Hybrid models of forest stand growth and production. In: Dykstra DP, Monserud RA (eds) Forest growth and timber quality: crown models and simulation methods for sus-

tainable forest management. Proceedings of an international conference. U.S. Department of Agriculture, Forest Service, Pacific Northwest Research Station, Portland, pp 43–47
McMahan AJ, Milner KS, Smith EL (2002) FVS-BGC: a process-model extension to the Forest Vegetation Simulator. U.S. Department of Agriculture, Forest Service, Forest Health Technology Enterprise Team, Fort Collins, 51p
Medema EL, Hatch CR (1982) Computerized help for economic analysis of Prognosis Model outputs: a user's manual. Contribution no. 227. University of Idaho, Forest, Wildlife, and Range Experiment Station, Moscow, 72 p
Milner KS, Coble DW (1995) The ground surface vegetation model: a process-based approach to modeling vegetative interactions. Unpublished manuscript
Monserud RA (2003) Evaluating forest models in a sustainable forest management context. FBMIS 1:35–47
Munro D (1974) Forest growth models – a prognosis. In: Fries J (ed) Growth models for tree and stand simulation. Royal College of Forestry Res Notes 30, Stockholm, pp 7–21
Nord-Larsen T, Suadicani MK (2010) Træbrændselsressourcer fra danske skove over ½ ha: opgørelse og prognose 2010. Skov & Landskab, Københavns Universitet. Arbejdsrapport / Skov & Landskab, no. 113
Peng C (2000) Growth and yield models for uneven-aged stands: past, present and future. For Ecol Manag 132:259–279
Perera AH, Sturtevant BR, Buse LJ (2015) Simulation modeling of forest landscape disturbances: an overview. In: Perera AH, Sturtevant BR, Buse LJ (eds) Simulation modeling of forest landscape disturbances. Springer, Geneva, pp 1–15
Petrauskas E, Kuliešis A (2004) Scenario-based analysis of possible management alternatives for Lithuanian forests in the 21st century. Balt For 10:72–82
Phillips H (2011) All Ireland roundwood production forecast 2011–2028. COFORD, Dublin
Pilli R, Grassi G, Kurz WA et al (2013) Application of the CBM-CFS3 model to estimate Italy's forest carbon budget, 1995–2020. Ecol Model 266:144–171
Redsven V, Hirvelä H, Härkönen K, Salminen O, Siitonen M (2013) MELA2012 reference manual, 2nd edn. The Finnish Forest Research Institute. ISBN: 978-951-40-2451-1
Reed D (1999) Ecophysiological models of forest growth: uses and limitations. In: Amaro A, Tomé M (eds) Empirical and process based models for forest tree and stand growth simulation. Edições Salamandra, Novas Tecnologias, Lisboa, pp 305–311
Reed DD, Amaro A, Amateis R, Huang S, Tomé M (2003) Emerging trends and future directions: a workshop synthesis. In: Amaro A, Reed D, Soares P (eds) Modelling forest systems. CAB International, Oxford, pp 389–394
Rennolls K, Tomé M, McRoberts RE, Vanclay JK, LeMay V, Guan BT, Gertner GZ (2007) Potencial contributions of statistics and modelling to sustainable forest management: review and synthesis. In: Reynolds KM, Thomson AJ, Kohl M, Shannon MA, Ray D, Rennolls K (eds) Sustainable forestry: from monitoring and modeling to knowledge management & policy science. CABI, Wallingford/Cambridge, pp 314–341
Rock J, Bösch B, Kändler G (2013). WEHAM 2012 – Waldentwicklungs- und Holzaufkommensmodellierung für die dritte Bundeswaldinventur. In: Klädtke J, Kohnle U (eds) Deutscher Verband Forstlicher Forschungsanstalten. Sektion Ertragskunde. Jahrestagung, Rychnov nad Kneznou/Tschechien, pp 127–133. ISSN: 1432-2609. http://sektionertragskunde.fvabw.de/2013/Beitrag_13_17.pdf. Accessed 14 May 2016
Sallnäs O (1990) A matrix growth model of the Swedish forest. Studia Forestalia Suecica. Sveriges lantbruksuniversitet, Uppsala. ISBN: 91-576-4174-9
Schelhaas MJ, Eggers J, Lindner M, Nabuurs GJ, Pussinen A, Päivinen R, Schuck A, Verkerk PJ, van der Werf DC, Zudin S, (2007) Model documentation for the European Forest Information Scenario model (EFISCEN 3.1.1). Wageningen, Alterra, Alterra report 1559, EFI technical report 26, Joensuu, 118p. http://www.efi.int/files/attachments/publications/alterrarapport1559.pdf. Accessed 5 July 2015

Siitonen M, Härkönen K, Hirvelä H, Jämsä J, Kilpeläinen H, Salminen O, Teuri M (1996) MELA Handbook—1996 edn. Metsäntutkimuslaitoksen tiedonantoja 622, p 452

Snorrason A (2006) Langtímaspá um kolefnisbindingu nýskógræktar. Skógræktarritið 58–64 (in Icelandic)

Stage AR (1973) Prognosis Model for stand development. Res. Paper INT-137. U. S. Department of Agriculture, Forest Service, Intermountain Forest and Range Experiment Station, Ogden, 32p

Sterba H, Moser M, Monserud RA (1995) PROGNAUS – Ein Waldwachstumssimulator für Rein- und Mischbestände. Österreichische Forstzeitung 106(5):19–20

Thürig E, Palosuo T, Bucher J, Kaufmann E (2005) The impact of windthrow on carbon sequestration in Switzerland: a model-based assessment. For Ecol Manag 210:337–350. doi:10.1016/j.foreco.2005.02.030

Tomé M, Faias S (2011) Report describing the regional simulators and the European simulator. EFI Technical Report 69. European Forest Institute, Finland, 65pp

Vanclay JK (1994) Modelling forest growth and yield – applications to mixed tropical forests. CAB International, Oxon

Wernsdörfer H, Colin A, Bontemps J-D, Chevalier H, Pignard G, Caurla S, Leban J-M, Herve J-C, Fournier M (2012) Large scale dynamics of a heterogeneous forest resource are driven jointly by geographically varying growth conditions, tree species composition and stand structure. Ann For Sci 69:829–844

Wilson FG (1946) Numerical expression of stocking in term of height. J For 44(10):758–761

Chapter 4
Forest Resource Projection Tools at the European Level

Mart-Jan Schelhaas, Gert-Jan Nabuurs, Pieter Johannes Verkerk, Geerten Hengeveld, Tuula Packalen, Ola Sallnäs, Roberto Pilli, Giacomo Grassi, Nicklas Forsell, Stefan Frank, Mykola Gusti, and Petr Havlik

4.1 Introduction

Europe has a long history of assessing the future development of its forest sector. Outlook studies for the European forest sector have been regularly produced via joint efforts of the Food and Agriculture Organization (FAO) of the United Nations and the United Nations Economic Commission for Europe (UNECE). The first study was published in 1953 (UN-ECE/FAO 1953). After two World Wars, European forests were severely over-exploited, while the demand for roundwood was expected to increase due to European post-war reconstruction activities. Since

M.-J. Schelhaas (✉) • G. Hengeveld
Wageningen Environmental Research (Alterra), Wageningen, The Netherlands
e-mail: martjan.schelhaas@wur.nl; geerten.hengeveld@wur.nl

G.-J. Nabuurs
Wageningen Environmental Research (Alterra), Wageningen, The Netherlands

Forest Ecology and Forest Management Group, Wageningen University and Research, Wageningen, The Netherlands
e-mail: gert-jan.nabuurs@wur.nl

P.J. Verkerk
European Forest Institute, Joensuu, Finland
e-mail: hans.verkerk@efi.int

T. Packalen
Natural Resources Institute Finland, Joensuu, Finland
e-mail: tuula.packalen@luke.fi

O. Sallnäs
Swedish University of Agricultural Sciences, Upsala, Sweden
e-mail: ola.sallnäs@slu.se

S. Barreiro et al. (eds.), *Forest Inventory-based Projection Systems for Wood and Biomass Availability*, Managing Forest Ecosystems 29,
DOI 10.1007/978-3-319-56201-8_4

then, outlooks have been produced at approximately 10-year intervals. These outlooks were referred to as European Timber Trend Studies and abbreviated as ETTS I to V. After 2000, the series was renamed the European Forest Sector Outlook Study and abbreviated as EFSOS. All studies were aimed at projecting wood demand from industry and consumers and whether and how this demand could be satisfied with the available forest resources. Over time, the focus widened from a pure wood-balance perspective towards a perspective that recognizes the importance of maintaining the full suite of forest goods and services (Nabuurs et al. 2014). The studies were aimed at a broad audience including forest industry and policy makers at multiple levels.

All studies were initiated by the FAO/UN-ECE Secretariat but were carried out in close cooperation with national correspondents. Resource projections in the ETTS studies were delivered by the national correspondents using their own projection tools or expert estimates. To increase the consistency of the outlook studies, assessments in EFSOS I and II used the same set of models for all countries and, as much as possible, internationally published data from sources such as the FAOSTAT databases and the Forest Europe assessments (FOREST EUROPE 2011, 2015). Both EFSOS I (UNECE/FAO 2005) and II (UNECE/FAO 2011) applied the forest resource model EFISCEN (Sallnäs 1990; Nabuurs et al. 2007; Schelhaas et al. 2007; Verkerk et al. 2014).

Recently, several other forest models and forest sector models have been developed for large-scale European forest resource assessments: the European Forest Dynamics Model (EFDM, Packalen et al. 2014), the GLOBIOM model (Havlik et al. 2014), Global Forest Model (G4M, Kindermann et al. 2008b; Gusti 2010) and the Carbon Budget Model of the Canadian Forest Sector (CBM-CFS3) that was adapted for European conditions (Pilli et al. 2013). Development of these models follows a growing interest in forest resources with respect to climate change and their potential role in supplying biomass for the bio-economy. This chapter describes the large-scale forest resource simulators EFISCEN, EFDM, CBM-CFS3, and GLOBIOM/G4M.

R. Pilli • G. Grassi
European Commission, Joint Research Centre, Directorate D – Sustainable Resources – Bio-Economy Unit, Ispra, Italy
e-mail: roberto.pilli@jrc.ec.europa.eu; giacomo.grassi@jrc.ec.europa.eu

N. Forsell • S. Frank • M. Gusti • P. Havlik
International Institute for Applied Systems Analysis (IIASA), Ecosystems Services and Management Program (ESM), Laxemburg, Austria
e-mail: forsell@iiasa.ac.at; frank@iiasa.ac.at; gusti@iiasa.ac.at; havlik@iiasa.ac.at

4.2 European Forest Information Scenario Model

4.2.1 Background

The European Forest Information Scenario model (EFISCEN) is a large-scale forest resource projection system using an area-based matrix model. The state of the forest is represented by the distribution of forest area over a matrix of age and volume classes. Processes such as growth, aging, mortality, thinning and final felling are implemented as transitions between cells in the matrix. The forests to be simulated can be divided into multiple strata distinguished by attributes such as tree species, region, ownership or site class. These strata are commonly called ‘forest types’, each of which is represented as a separate matrix with its own dynamics.

The core of the model was developed in Sweden (Sallnäs 1990) with the first European scale application by Nilsson et al. (1992) who studied the consequences of large-scale European forest decline. The model has been re-programmed and developed further at the European Forest Institute (EFI) and Alterra, part of Wageningen University and Research. Over time, the model has been used to study the effects of more nature-oriented management (Nabuurs et al. 2002), the impacts of climate change on increment rates (Pussinen et al. 2009), adaptation to climate change (Schelhaas et al. 2015), mitigation of climate change effects (Böttcher et al. 2012), future wood availability (Verkerk et al. 2011), and biodiversity and ecosystem services (Verkerk et al. 2014).

4.2.2 Data Sources

For each forest type, required EFISCEN inputs are the area, average growing stock volume and current annual increment per age-class. This information can be obtained either from National Forest Inventories (NFI) based on statistical sampling for the entire country or from Standwise Forest Inventories (SFI) for local forest management planning. The data may be delivered by the respective national institutes in aggregated form, so there is no need to deliver plot information or disclose NFI plot locations. Also, individual countries can select the forest types and age-class sizes to be assessed.

The input data are used to construct the initial distribution of area over the matrix for each forest type. From the increment data, growth models are constructed to predict 5-year relative increments as a function of age. Alternatively, growth models can be estimated from yield tables or other sources. Each forest type is assigned a management regime defined in terms of the probability that a thinning or final harvest can be carried out as a function of stand age. These regimes can be derived from national management guidelines, can be determined in consultation with national experts, and sometimes can be inferred from the current age-class distribution.

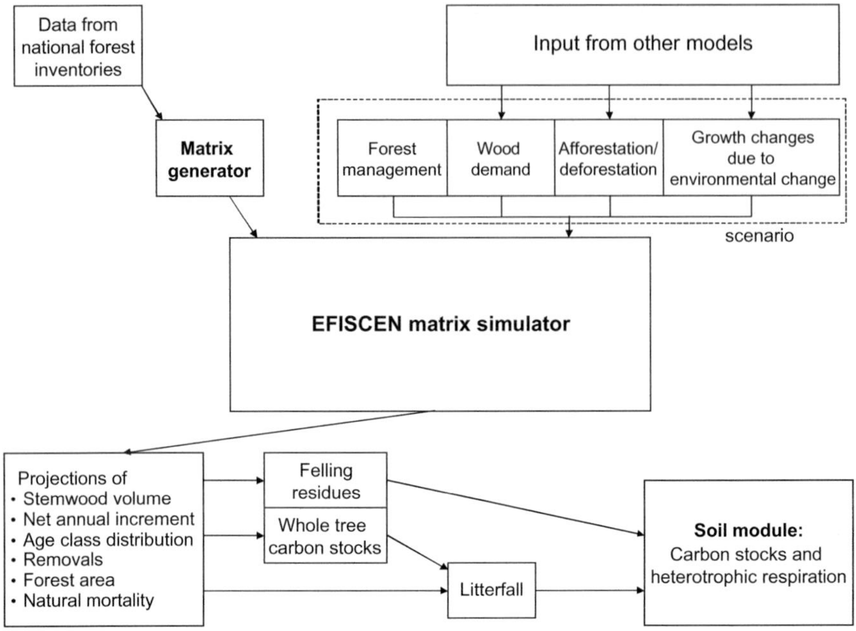

Fig. 4.1 Outline of the EFISCEN model and its components (Schelhaas et al. 2007)

For each 5-year time-step, the timber demand for the simulated area must be defined separately for thinnings and final fellings. For all forest types, the potential harvest is calculated from allowed thinning and felling possibilities per age class, combined with simulated age-class distributions. The defined timber demand is then obtained from the different forest types according to their relative shares in the potential harvest. Initially, demand was estimated from historic felling levels, assuming a stable level or a fixed future increase per 5-year time-step. For more recent applications (Böttcher et al. 2012), demand for domestic timber has been estimated from trade models. Similarly, future development of forest area can either be predicted from simple trend projections or from land-cover projection models. Optionally, EFISCEN can change the growth models over time to reflect changes in increment as a result of factors such as changing environmental conditions. These changes are usually derived from more detailed process-based models.

4.2.3 Methods

A schematic overview of EFISCEN is presented in Fig. 4.1.

A scenario in EFISCEN can be set by changes in one or more of the following components:

- Future wood demand
- Changes in forest area

- Changes in growth rates (for example due to changes in environmental conditions)
- Forest management regimes (thinning ages, rotation length, removal of harvest residues and stumps)
- Tree species conversion after final felling

Each component can be defined using simple assumptions with the additional feature that coupling to external models is possible for the first three components. As an alternative to specifying the future wood demand, EFISCEN can be run by specifying the share of the maximum available harvest volumes as defined by the management regimes that should be extracted in each time-step. In this case, for example, 50% of all forest areas older than 100 years will be harvested each time-step, rather than the area of forest in the same age classes needed to fulfil the required demand.

Projections use a 5-year time-step, usually for periods as long as 50–60 years. Longer projections are possible but are generally only used to present qualitative outcomes such as projecting conversions in tree species distribution. Results are usually presented at the aggregated national level for attributes such as tree species and age-class distributions, development of average growing stock, and average increment. Projection results are usually interpreted as differences in development between the scenarios, rather than precise quantifications of the future state of the forest. An uncertainty assessment for four European countries is presented by Meyer et al. (2005). They found a coefficient of variation (CV) of 2.1% for the initial tree carbon stock estimate at the national level for Sweden and Finland, and 4.8% for Spain. The greater value for Spain was explained by the large number of tree species in this country, so species had to be grouped. CV of stock changes amounted to 11–27%, with greater uncertainties in countries where stock changes were relatively small in comparison to the stocks. CVs in soil carbon stocks were in the range of 45–48% and for soil changes in carbon stock 21–34%.

Basic output of the EFISCEN core model consists of growing stock, increment, actual harvest, natural mortality, tree species distribution and age-class distribution. By applying wood density, carbon content and biomass distribution models, whole-tree carbon stocks are estimated in an extension of the model. Standard parameterisations are available for different regions in Europe but can also be easily adapted for particular countries. The same distribution models are used to quantify the amount of harvest residues that are generated in the felling operations, and these can be partly removed if desired. Residues left in the forest, together with litterfall calculated from litterfall rates, form the input to the built-in soil carbon model YASSO (Liski et al. 2005). YASSO calculates soil carbon stocks as a function of litter input quantity and quality and temperature and precipitation indices averaged over the region. Outputs can be further processed in separate modules such as a harvested wood products module (Eggers 2002), risk indices (Schelhaas et al. 2010), recreation attractiveness scores (Edwards et al. 2012) and wood assortments (Hanewinkel et al. 2013).

EFISCEN operates at the level of administrative regions. Spatially explicit information is not needed to run the model. However, spatial NFI information can be used in the data preparation phase to stratify by factors such as slope or altitude.

Alternatively, spatial information layers can be incorporated into the EFISCEN model by downscaling the input data using a tree species map, and then aggregating it again using spatial information for stratification. An example is the approach by Verkerk et al. (2014) that uses a harvest intensity map to distinguish among less intensively and more intensively managed forests in Europe. Furthermore, EFISCEN outputs can be linked to spatial information. For example, EFISCEN biomass potentials have been combined with spatially explicit information on harvesting constraints (Verkerk et al. 2011), and EFISCEN results have been linked with tree species maps to disaggregate EFISCEN results to the grid level (Elbersen et al. 2012; Crouzat et al. 2015).

4.2.4 Concluding Remarks for EFISCEN

EFISCEN has been successfully used for a wide range of projects and applications for more than two decades. For a complete overview of projects and publications, see the EFISCEN webpage (http://efiscen.efi.int/). All results for EFSOS II simulations are available at the national level for individual countries (http://www.unece.org/efsos2). To further improve the transparency of the tool and to increase its user group, EFISCEN 4.1 is available as an open source tool at https://github.com/EuropeanForestInstitute/efiscen.

Part of the success of EFISCEN is due to its flexibility with respect to input data. In particular, it is not very data-intensive, can accommodate different aggregation levels, and can accommodate different types of forests and different information sources. Often the information published by standard governmental statistical series is sufficient to initialise the model.

While the core of the model has remained unchanged, important improvements have been made in data handling, standardising parameterisations and automating simulation set-ups. Extensions to the original model allow simulation of a suite of indicators such as dead wood, carbon, fire and wind risk indices and attractiveness for recreation. The tool can be used to gain insight into forest resource development, woody biomass potentials, climate change mitigation and ecosystem service provisioning.

The model is designed for even-aged, managed forests, whereas Europe's forests have tended to become more mixed, more richly structured and less intensively managed. Furthermore, due to developments in remote sensing, computing power and data policies, easily accessible information is becoming available with much greater spatial detail and accuracy. Because EFISCEN was not designed as a spatially explicit model, such developments cannot be easily utilised. Therefore, a new, spatially explicit model is being designed (EFISCEN Space) which will be applicable to all types of forest and can provide projections at the European scale.

4.3 European Forestry Dynamics Model

4.3.1 Background

The European Forestry Dynamics Model (EFDM) originated from collaboration between the European Joint Research Centre (JRC) and the European National Forest Inventory Network (ENFIN). EFDM (Packalen et al. 2014) was jointly developed for the Forest Resources and Climate Unit of the JRC by the Finnish Forest Research Institute (Metla, now LUKE) and the Swedish University of Agricultural Sciences (SLU).

Many European countries have their own forest resource projection tools that are used for national and international outlook studies. ENFIN has been working towards harmonised forest information reporting by countries in COST Actions E43 and FP1001 (USEWOOD) and continues the work within the Horizon 2020 project DIABOLO. EFDM was developed as an option for harmonized pan-European forest scenario modelling, especially for countries that do not yet have a model of their own.

EFDM was designed to use detailed national-level input data, thus facilitating use of the model by institutes having access to NFI data. For the model implementation, a matrix modelling approach (Sallnäs 1990; Nilsson et al. 1992) was selected because it provides a simple but flexible solution to assess the changes in the state of forest resources under diverse forest growth and management conditions. For even-aged forests, the matrix is typically defined by age and volume classes, whereas for uneven-aged forests it is defined by tree number and volume classes (Sallnäs et al. 2015).

4.3.2 Data Sources

EFDM is a Markov model defined by combinations of factors and activities and their corresponding transition probabilities. EFDM is applied with NFI data using the following steps: (1) classification of the state space (age and volume, or stem number and volume) and factor combinations (cf., "forest types" in EFISCEN); (2) forest initial state matrices by factor combinations derived from the NFI data; (3) basic ("no management") transition probabilities; (4) transition probabilities for each activity (e.g., thinning or final felling) to be predicted; (5) activity probabilities based on the NFI data, national statistics or as specified by the analyst and (6) output requests.

The forest initial state matrices for different factor combinations can be derived through a simple classification and aggregation routine using NFI plot data. Estimation of the basic ("no management") transition probabilities matrix can be based on (1) pairwise observations of state of the forest (e.g., as derived from permanent NFI sample plots), (2) growth observations, (3) growth and yield models or (4) expert knowledge. Ideally, transition probabilities are estimated from a large number of observed transitions. In the absence of a sufficient number of observations, a recursive Bayesian filter can be applied (Särkka 2013). Activity probabilities can differ between factor combinations ("forest types").

4.3.3 Methods

EFDM is an area-based matrix model for which forest areas (not trees or stands) move between elements of a set of fixed states. In the simulation, a set of activities such as no management, thinning, and final felling must be defined. Each of these activities has a probability of occurrence, and each activity is linked to a separate transition probability matrix. Consequently, there is a transition matrix for each factor combination and each activity.

EFDM can be used to generate output matrices for the future state of forests in case of "no management", "text-book management" or "business-as-usual" scenarios. EFDM also estimates the volume of wood harvested by factors and activities.

EFDM is implemented as a generic, platform-independent, free and open source software system. To fulfil the European Commission's software requirements, EFDM software complies with the European Union Public Licence (EUPL) which requires acknowledgement of authors and allows re-use and re-distribution of software. The EFDM software for even-aged forests (v.2) has been released by JRC at https://forestwiki.jrc.ec.europa.eu/efdm/index.php/Main_Page.

4.3.4 Concluding Remarks for EFDM

EFDM is a simple, generic, flexible and expandable framework that will help in harmonizing national results for purposes such as greenhouse gas inventories or the Global Forest Resources Assessment. Further, EFDM is open source and thus improves the credibility of scenario modelling at national and European levels by facilitating transparency in documentation, evaluation of modelling results as well as collaborative development of new features. Finally, EFDM is free and supports capacity building, especially in countries that do not have their own modelling tools.

In the first phase of model development, the concept was tested in five countries covering different ecological and socio-economic conditions in Europe (Packalen et al. 2014). All countries considered EFDM a feasible modelling approach at national level, especially for tackling issues where traditional models have difficulties such as uneven-aged forestry or management under risk. More thorough testing of the basic concept for uneven-aged forests continues (Sallnäs et al. 2015).

4.4 Carbon Budget Model (CBM-CFS3)

4.4.1 Background

The Carbon Budget Model (CBM-CFS3) was developed by the Canadian Forest Service (Kurz et al. 2009). CBM-CFS3 is an inventory-based, yield-data driven model that simulates stand- and landscape-level carbon (C) dynamics for above- and belowground biomass, dead wood, litter and soil. The model was validated at regional and national scales in Canada (Kurz and Apps 1999; Stinson et al. 2011) and Russia (Zamolodchikov et al. 2008), and since 2009, the Joint Research Centre (JRC) of the European Commission has been testing the CBM-CFS3 for application in EU countries.

4.4.2 *Overall Model Description*

The CBM-CFS3 spatial framework conceptually follows IPCC Reporting Method 1 in which the Spatial Units (SPUs) are defined by their geographic boundaries. SPUs are further characterized by age, area, and as many as 10 classifiers including administrative and ecological information, links to the appropriate yield curves, and other parameters defining forest composition, management strategy and other management information. Each forest stand is assigned to an individual SPU.

During a model run, a library of yield tables (YT) defines gross merchantable volume production by age-class for each species. These yields represent the volume in the absence of natural disturbances and management practices. Disturbance and management factors that reduce the gross merchantable volume are applied during the model run. Species-specific, stand-level models (Boudewyn et al. 2007) convert the merchantable volume production into aboveground biomass components. Belowground biomass, its increment and annual turnover are estimated using the models provided by Li et al. (2003). Annual dead wood and foliage input are estimated as percentages (i.e., turnover rate) of standing biomass stock.

To estimate the decomposition rate for each dead organic matter (DOM) pool, CBM-CFS3 adjusts the base decomposition rates, defined for 10 °C, to correspond to the mean annual temperature defined for each SPU. The model uses an initialization process to estimate the size of all DOM pools at the start of the simulation, assuming DOM pools to be in an equilibrium. During this initialization phase, CBM-CFS3 can apply the same set of YTs selected for the main simulation or a different set specifically selected from an historical library of YTs.

During the actual run, the model applies a set of natural and anthropogenic disturbances such as fire, insects or storms and partial or clear-cut harvesting. For each disturbance, users define: (1) the amount, type and intensity by year and SPU; and (2) criteria such as forest type, age, or other classifier values that define the eligibility of stands for each disturbance (Kull et al. 2011). The impact of each disturbance on each C pool is defined by the user based on a disturbance matrix that quantifies proportions of C transferred among different forest pools in the model framework, transferred to the forest products sector (i.e., removed from the forest, as harvest wood products) or released to the atmosphere (e.g., via fires). Afforestation and deforestation can also be represented with specific disturbance types with their own disturbance matrices and transitions to and from the forest land.

The model predicts annual C stocks and fluxes such as the annual C transfers among pools, from pools to the atmosphere and to the forest product sector, as well as ecological indicators such as Net Primary Production (NPP) and Net Ecosystem Production (NEP).

4.4.3 Specific Implementation of the CBM-CFS3 for the European Countries

CBM-CFS3 was adapted for specific forest management conditions in Europe by the JRC. Where possible, default models to convert merchantable volume into aboveground biomass were replaced by country-specific models provided by NFIs or obtained from the literature. If no data were available, models were used from other countries and/or similar forest types, defined according to the main species.

The CBM-CFS3 model assumes that the values reported in the YTs represent gross merchantable wood volume (Kurz et al. 2009). For application of CBM-CFS3 to European countries, two sets of YTs were generally applied (Pilli et al. 2013). The first set, characterized as the historical library, was based on standing volume which reflects the net standing volume including the impacts of past silvicultural activities. This library was applied by the initialization procedure to provide the starting aboveground volume and biomass for each stand resulting from past management practices and disturbance events. The second set of tables, characterized as the current library, was based on the current annual increment (CAI) which represents the gross volume yield of each stand. This library was applied during the model run to estimate the current gross volume and increment potential for each stand. During the model run, this volume is directly reduced by management and natural disturbance events. This reduction is a relevant issue for the application of CBM-CFS3 and potentially for other yield-data driven models to European countries.

Yield data-driven models such as CBM-CFS3 cannot be directly applied to uneven-aged forest stands where no YTs are available. Because uneven-aged forests cover about 30% of the forest area in Europe, addressing this issue is relevant for the application of CBM-CFS3 to European countries. To overcome this limitation, the default model design was adapted to the tree selection system using the volume and increment data provided by NFIs for uneven-aged forests. For this purpose, all the uneven-aged forest area was allocated to a reference age-class with the average volume equal to the volume reported by the NFI for these stands. Starting from this age-class, a decreasing percentage increment was applied to the subsequent older age-classes. After time corresponding to species-specific cutting cycles defined at country level had elapsed, each uneven-aged stand was assumed to be thinned and reverted back to the initial reference age-class (Pilli et al. 2013). This approach was tested through multiple simulations with varying parameters including (1) a faster (but decreasing) re-growth phase during the first period following the partial cut and (2) a decreasing growth phase during the following years.

4.4.4 Concluding Remarks for CBM-CFS3

CBM-CFS3 was successfully adapted to specific forest management conditions in Europe (e.g., uneven-aged forests), validated at regional level (Pilli et al. 2014a) and applied in one country case to estimate the C balance for forest management (Pilli et al. 2013) and afforestation/reforestation activity reporting under the Kyoto Protocol (Pilli et al. 2014b). CBM-CFS3 is currently applied by the JRC for 26 EU countries, mainly using information provided by NFIs and forest management plans. The main limitation of the current version of the CBM-CFS3 model is the difficulty in simulating the impacts of environmental changes such as climate on forest growth because the model does not explicitly simulate the impacts of environmental variations on yields.

4.5 The GLOBIOM and G4M Models

4.5.1 Background

The Global Forest Model (G4M)[1] was developed by the International Institute for Applied Systems Analysis (IIASA) in 2008 to assess the mitigation potential from halting deforestation activities. By comparing the economic value of alternative land uses, the model simulates the impact of wood demand projections and carbon prices on forestry activities (afforestation, deforestation and forest management). Because the G4M model lacked the capability to project market developments (external projections of wood demand are required) the Global Biosphere Management Model (GLOBIOM)[2] was developed in 2011 at IIASA, providing market developments and a fully integrated forest and agricultural sector modelling framework. At its core, the GLOBIOM model is a spatially explicit land use model that projects the developments of the forestry and agricultural markets, international trade, impacts on land use, and CO_2 emissions for the LULUCF sector. Nowadays, the GLOBIOM and G4M models are used in conjunction to benefit from their respective strengths.

The G4M and GLOBIOM models are currently used to study the effects of climate change and adaptation of management on forests (Kindermann et al. 2013), forest resource developments over time (Böttcher et al. 2012), EU-wide LULUCF developments (European Commission 2013), soil organic carbon mitigation potentials (Frank et al. 2015), and woody biomass energy potentials (Lauri et al. 2014). The models have a long history of use in a policy context and providing EU member state projections of forest harvest levels and LULUCF projections. As an example, the models have been used to develop the EU28-wide LULUCF reference scenario, in consultation with national experts and cross-checked per country through a consultation process (European Commission 2013). The two models have also been used for Reducing Emissions from Deforestation and forest Degradation (REDD) assessments in tropical forest regions (Herrero et al. 2013), bioenergy sustainability assessments (Forsell et al. 2016), and even represent the land use part of the IIASA integrated assessment modelling framework MESSAGE-GLOBIOM (McCollum et al. 2014).

[1] See also: www.iiasa.ac.at/G4M

[2] See also: www.iiasa.ac.at./GLOBIOM

4.5.2 Overall G4M Model Description

G4M was developed by IIASA (Kindermann et al. 2008a; Gusti 2010) to estimate the availability and cost of woody biomass resources (Gusti and Kindermann 2011), and is used in conjunction with GLOBIOM to estimate the impact of forestry activities on biomass and carbon stocks. The G4M model estimates forest area change, carbon sequestration and emissions from forests, impacts of carbon incentives and supply of biomass for energy and non-energy uses. By comparing the income of managed forest (difference of wood price and harvesting costs plus income by storing carbon in forests) with income by alternative land use at the same place, an afforestation or deforestation decision is made. Because G4M is spatially explicit at $0.5 \times 0.5^{\circ}$ resolution, different levels of deforestation pressure can be handled at a spatially explicit level.

The available woody biomass resources are estimated by G4M for each forest area unit determined by mean annual increments, which are based on net primary productivity (NPP) maps from Cramer et al. (1999) and from downscaling techniques described in Kindermann et al. (2008a). This information is then combined with data from national sources such as NFIs to provide further and more detailed information concerning biomass stocks and forest age structure. When available, age structure and stocking degree are used for adjusting increment estimates.

The main forest management options considered by G4M are variations of thinning levels and rotation lengths. The rotation length can be user-defined or the model can estimate optimal rotation lengths to maximize increment, stocking biomass or harvestable biomass. Thinning is applied to all managed forests, and the stands are thinned to maintain a specified stocking degree. To fulfil harvest estimated by GLOBIOM, the harvest amount per country is estimated in G4M by choosing a set of rotation lengths starting from the length that maintains current biomass stocks. If total harvests are less than the wood demand, the model changes management per grid cell starting from the most productive forest to a rotation length that optimizes forest increment and thus allows for more harvest. The rotation length is updated for each 5-year time-step. If harvest is still too small and there is unmanaged forest available, the unmanaged forest will be taken under management. If total harvests are greater than the demand, the model will change management to maximize biomass rotation length, i.e. to manage forests for carbon sequestration.

The model can also consider the use of a carbon price to limit emissions and increase the forest carbon sink, using a cost curve algorithm. Introducing a carbon price incentive means that the forest owner is paid for the carbon stored in forest living biomass if its amount is greater than a baseline without a carbon price incentive, or pays a tax if the amount of carbon in forest living biomass is less than the baseline.

G4M calculates the optimal combination of the following mitigation measures:

- Reduction of deforestation area;
- Increase of afforestation area;
- Change of rotation length of existing used forests in different locations;
- Change of harvest intensity (amount of biomass extracted in thinning and final felling activity).

4.5.3 Overall GLOBIOM Model Description

The GLOBIOM model (Havlík et al. 2014) is a global recursive dynamic partial equilibrium model of the forest and agricultural sectors, for which economic optimization is based on spatial equilibrium modelling (Takayama and Judge 1971). A systematic overview of the GLOBIOM model framework is provided in Fig. 4.2. The model projects developments on the forestry and agricultural markets, international trade, impacts on land use, and emissions and removals for the LULUCF sector. In terms of forestry, the model not only covers the development of the forest resources, but also projects the development of forest based industries, and the use of woody biomass commodities for energy and material use. The model is based on a bottom-up approach for which the supply side of the model is built-up from the land cover, land use, and management systems at the bottom to production and markets at the top. Market equilibrium for forest and agriculture products is computed by allocating land use among production activities to maximize the sum of

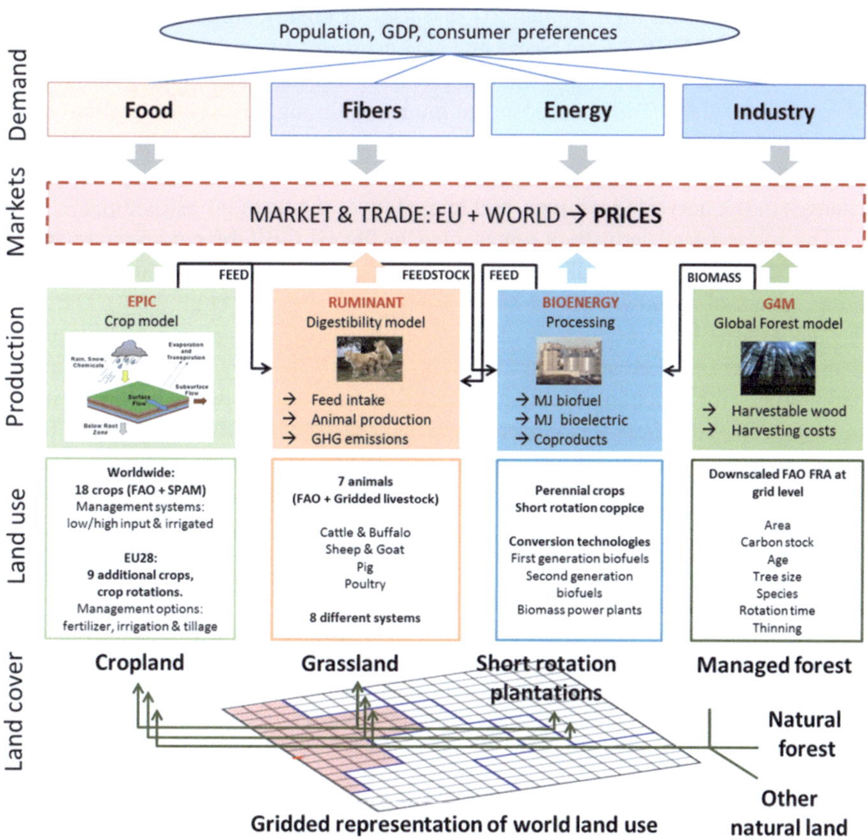

Fig. 4.2 GLOBIOM modelling framework

producer and consumer surplus, subject to resource, technological and policy constraints. The level of production in a given area is determined by the forest or agriculture productivity in that area which, in turn, depends on management and suitability, by market prices that reflect the level of demand, and by the conditions and costs associated with conversion of the land, to expansion of the production and, when relevant, to international market access. Trade is modelled following the spatial equilibrium approach, which means that the trade flows are balanced out between different specific geographical regions. This allows tracking of bilateral trade flows between individual regions.

In terms of forest, total forest area in GLOBIOM is calibrated according to FAO Global Forest Resources Assessments (FRA) and divided into managed and unmanaged forest using a downscaling routine based on human activity impact on forest areas (Kindermann et al. 2008a).[3] G4M provides the available woody biomass resources in the form of mean annual increments for each forest area unit which are allocated to commercial roundwood, non-commercial roundwood and harvest losses in GLOBIOM.[4] In addition to stem wood, available woody biomass resources also include branches and stumps; however, environmental and sustainability considerations constrain their availability and use for energy purposes. Harvest costs for forests in GLOBIOM are based on G4M estimates through the use of spatially explicit constant unit costs that include planting, logging, and chipping in the case of logging residues. Transport costs are modelled through the use of regional level constant elasticity transport cost functions, which approximate the short run availability of woody biomass in each region but are shifted over time in response to the changes in the harvested volumes and related investments in infrastructures.

The forest-based industries represented in the GLOBIOM model cover seven final products: chemical pulp, mechanical pulp, sawnwood, plywood, fiberboard, other industrial roundwood, and household fuelwood (Lauri et al. 2014). Demand for the various final products is modelled using regional level constant elasticity demand functions. Forest industrial products are produced by Leontief production technologies for which input-output coefficients are based on the engineering literature (e.g. Global Forest Resources Assessment 2010). By-products of these technologies (bark, black liquor, sawdust, and sawchips) can be used for energy production or as raw material for pulp and fibreboard (Lauri et al. 2014).

The GLOBIOM model also covers the wood demand for energy. Biomass use for energy production (heat, electricity, and biofuels) is commonly specified as a sce-

[3] The term "unmanaged forests" refers to all forest areas that do currently not contribute to wood supply, based on economic decision rules in the model. However, they may still be a source for collection and production of non-wood goods (e.g. food, wild game, ornamental plants). Forests that are used in a certain period to meet the wood demand, so-called managed forests, are modelled to be managed for woody biomass production. This implies a certain rotation time, thinning events and final harvest.

[4] Commercial roundwood is stemwood that is suitable for industrial roundwood (sawlogs, pulplogs and other industrial roundwood). Harvest losses and non-commercial roundwood are stemwood that is unsuitable for industrial roundwood. The difference between harvest losses and non-commercial roundwood is that the former has unwanted stemwood sizes, while the latter has unwanted wood characteristics.

nario specific input, based on the POLES, MESSAGE or PRIMES energy sector models (Havlík et al. 2011; Reisinger et al. 2013), but other estimates can also be used. The GLOBIOM model thereby projects which feedstocks will be used to fulfil the energy demand categories specified by the energy sector model. To fulfil the energy demand, GLOBIOM represents multiple conventional and advanced biofuels feedstocks, as well as production pathways for electricity and heat production (Havlík et al. 2011). This allows competition for biomass resources to be considered, also taking into account competition between the various sectors in terms of the demand for food, feed, timber, and energy. Plantations are also covered in GLOBIOM in the form of energy crop plantations dedicated to the production of wood for energy purposes. Plantation yields are based on NPP maps and the model's own calculations as described in Havlík et al. (2011).

By including not only the forestry sector but also the agriculture, livestock and bioenergy sectors, the GLOBIOM model allows for a full cross-sectorial assessment of market developments; see Havlík et al. (2014) for further details concerning the modelling of these sectors. The full coverage of the land use sectors, allows for a full account of all agriculture and forestry GHG sources. In terms of emissions, GLOBIOM accounts for 10 sources of GHG emissions based on IPCC accounting guidelines (Havlík et al. 2011).

4.5.4 Concluding Remarks on G4M and GLOBIOM

G4M and GLOBIOM have been peer-reviewed in various European and international projects and scientific publications (a.o. Forsell et al. 2016; Kindermann et al. 2008a; Kindermann et al. 2013; Lauri et al. 2014). The two models have a long history of use in a policy context and in consultation with EU member state countries to provide projections that are validated by country experts. The fact that the models are also being used and applied in multiple key countries improves the representation of international linkages.

Part of the strength of the GLOBIOM model is that it is global and explicitly links countries and regions through trade of commodities. Thus, the model can consider demand and supply developments on a global level and can assess the impact of national policies in terms of trade, leakage and rebound effects. More importantly, it allows analysis of the EU forest and forest industries from a global perspective and projection of the competitiveness of EU feedstocks production and commodities on the international markets. The model thereby enables assessment of the future share of EU domestic production where future import would originate, and the effect of such trade on importing as well as exporting countries. In addition, the GLOBIOM model is designed to integrate the forest and the agriculture sectors by representing both with a reasonable level of detail without compromising either, thus being a key model to assess the increasing interlinkages and feedbacks between these two sectors. While the forest-based industries were initially represented in a fairly simplified manner, key industrial sectors have been decomposed with great detail and the flow of material and by-products between individual industries has been improved. This has improved the GLOBIOM model's capability to account for additional aspects such as the circular economy and capture the particularities of carbon implications of resource demand. While linking integrated models is always challenging, work is continuously progressing to further improve the linkages between GLOBIOM and G4M with the ultimate objective to fully merge the two models into a single modelling framework.

4.6 The Role of European Forest Resource Tools

Many countries have developed their own systems for projecting forest resources and wood availability, and they generally issue regular reports of studies that use these tools. Although these studies are helpful for developing national policies, they cannot provide a consistent assessment for larger regions such as the entire EU or Europe as a whole. Individual national-scale studies differ considerably in timing, underlying methodology and scenarios, and reports are not issued for all countries in the region. However, a clear demand for consistent projections at European scale still remains. Projections are needed for multiple purposes including assessing impacts of climate change and global economic developments and for developing policies at the EU-level. Users of these studies include industries, non-governmental organisations, countries, the EU and UNECE/FAO.

Development of a scenario study generally consists of the following steps: (1) definition of the scope of the study, (2) choice of the simulator to be used, (3) data acquisition, (4) definition of the scenarios, (5) set-up of the simulator and implementation of the scenarios, (6) production of the simulations and (7) analysis of and reporting on the outcomes. Some of these steps are interlinked, and they do not necessarily need to be implemented in this order. Ideally, simulators are chosen or developed depending on the question at hand, but the available simulators usually limit the scope of the study, whereas the availability of data limits the choice of simulator.

Developing projections at the European scale involves dealing with a multitude of countries, each with its own particular NFI design, definitions, projection methods, national policies and forest circumstances. When working at the European level, options for most of the steps mentioned above range from optimal national level involvement to a strict centralised approach. The scope of the study is usually decided externally. Each country can use its own simulator (such as the ETTS studies) or a common simulator (such as EFSOS). The required input data can be prepared at national level, can be delivered centrally as aggregated data or as detailed plot measurements, or can be obtained from published data sources. Optionally, data can be harmonised among countries by enforcing common definitions. Usually a common set of scenarios is used, but the scenarios can be applied more flexibly or more strictly at the country level. All the earlier ETTS studies had a business-as-usual scenario, but the scenarios were not harmonised over the countries. Later studies featured greater harmonisation of scenario elements such as assumptions on economic growth, how to assess forest area expansion, and implementation of policies. Setting up the model, implementation of the scenarios and running the actual simulations can be done both nationally and centrally. When countries use their own simulator, simulations will be done at national level. If a common simulator is used, countries can be instructed how to use it, set up the simulator and do the simulations themselves, or everything can be done centrally. The final analysis is usually done centrally, but results may be submitted for commentary to the countries.

Involving local or national experts in model development or model application enhances the optimal use of specific knowledge, tools, and expertise. However, it is a very time-consuming process, and the results are usually not well harmonised as a result of factors such as different simulators, definitions, and scenarios. The other extreme is that everything is done centrally. Such studies can be carried out relatively quickly but may ignore the local or national expertise and may lead to results that are misinterpreted or even not accepted by the local or national experts. In European studies, a balance should therefore be sought between harmonisation with respect to simulators, definitions and scenarios and accommodating national peculiarities, national involvement and time required to produce results. This balance will depend very much on the type of study at hand and resources available.

European forest resource simulators cannot replace national simulators which may be better adapted to local circumstances. However, European forest resource simulators can provide consistent and comparable projections at European level. Furthermore, they can be applied by individual countries when no nation-wide tools are available (see country chapters for Bulgaria and Netherlands; Pilli et al. 2013). For example, 14 European countries relied on combined projections of EFISCEN and G4M to estimate their forest management reference levels in the context of carbon accounting rules for the LULUCF sector for the UNFCCC (http://unfccc.int/bodies/awg-kp/items/5896.php). Only a few studies have been published that compare results for the European resource simulators among themselves and with results from national simulators (Nabuurs et al. 2000; Nuutinen and Kellomäki 2001; Oosterbaan et al. 2007; Groen et al. 2013; Pilli et al. 2016a,b). Such comparisons are useful for all simulators involved and will lead to better mutual understanding and in the longer term to improved simulations.

References

Böttcher H, Verkerk PJ, Gusti M et al (2012) Projection of the future EU forest CO2 sink as affected by recent bioenergy policies using two advanced forest management models. GCB Bioenergy 4:773–783

Boudewyn P, Song X, Magnussen S, Gillis MD (2007) Model-based, volume-to-biomass conversion for forested and vegetated land in Canada. Canadian Forest Service, Victoria, Canada (Inf. Rep. BC-X-411) http://cfs.nrcan.gc.ca/pubwarehouse/pdfs/27434.pdf Accessed 20 Sept 2016

Cramer W, Kicklighter DW, Bondeau A et al (1999) Comparing global models of terrestrial net primary productivity (NPP): overview and key results. Glob Chang Biol 5:1–15

Crouzat E, Mouchet M, Turkelboom F et al (2015) Assessing bundles of ecosystem services from regional to landscape scale: insights from the French Alps. J Appl Ecol 52:1145–1155

Edwards DM, Jay M, Jensen FS et al (2012) Public preferences across Europe for different forest stand types as sites for recreation. Ecol Soc 17(1):27. doi:10.5751/ES-04520-170127

Eggers T (2002) The impacts of manufacturing and utilisation of wood products on the European carbon budget. European Forest Institute, Joensuu

Elbersen B, Staritsky I, Hengeveld GM et al (2012) Atlas of EU biomass potentials. Deliverable 3.3: Spatially detailed and quantified overview of EU biomass potential taking into account the main criteria determining biomass availability from different sources. Biomass Futures http://www.biomassfutures.eu/work_packages/WP3%20Supply/D_3_3__Atlas_of_technical_and_economic_biomass_potential_FINAL_Feb_2012.pdf. Accessed 20 Sept 2016

European Commission (2013) EU energy, transport and GHG emissions: trends to 2050. Scenario, Reference

FOREST EUROPE (2011) State of Europe's Forests 2011

FOREST EUROPE (2015) State of Europe's Forests 2015

Forsell N, Korosuo A, Havlík P et al (2016) Study on impacts on resource efficiency of future EU demand for bioenergy. Task 3: modelling of impacts of an increased EU bioenergy demand on biomass production, use and prices, 109p

Frank S, Schmid E, Havlík P et al (2015) The dynamic soil organic carbon mitigation potential of European cropland. Glob Environ Chang 35:269–278

Global Forest Resources Assessment (2010) Global Forest Resources Assessment, main report. Food and Agricultural Organization of the United Nations, Rome

Groen TA, Verkerk PJ, Böttcher H et al (2013) What causes differences between national estimates of forest management carbon emissions and removals compared to estimates of large-scale models? Environ Sci Pol 33:222–232

Gusti M (2010) An algorithm for simulation of forest management decisions in the global forest model. Artif Intell N4:45–49

Gusti M, Kindermann G (2011) An approach to modeling landuse change and forest management on a global scale. In: Kacprzyk J, Pina N, Filipe J (eds) SIMULTECH-2011. Proceedings of 1st international conference on simulation and modeling methodologies, technologies and applications, Noordwijkerhout, 29–31 July 2011: SciTePress – Science and Technology Publications, Setúbal, pp 180–185

Hanewinkel M, Cullmann DA, Schelhaas MJ et al (2013) Climate change may cause severe loss in the economic value of European forest land. Nat Clim Chang 3:203–207

Havlík P, Schneider UA, Schmid E et al (2011) Global land-use implications of first and second generation biofuel targets. Energ Policy 39:5690–5702

Havlík P, Valin H, Herrero M et al (2014) Climate change mitigation through livestock system transitions. Proc Natl Acad Sci 111:3709–3714

Herrero M, Havlík P, Valin H et al (2013) Biomass use, production, feed efficiencies, and greenhouse gas emissions from global livestock systems. Proc Natl Acad Sci 110(52):20888–20893

Kindermann G, Obersteiner M, Sohngen B et al (2008a) Global cost estimates of reducing carbon emissions through avoided deforestation. PNAS 105:10302–10307

Kindermann GE, McCallum I, Fritz S, Obersteiner M (2008b) A global forest growing stock, biomass and carbon map based on FAO statistics. Silva Fenn 42(3):387

Kindermann G, Schorghuber S, Linkosalo T et al (2013) Potential stocks and increments of woody biomass in the European Union under different management and climate scenarios. Carbon Balance Manag 8:2

Kull S, Kurz WA, Rampley G et al. (2011) Operational-scale carbon budget model of the Canadian Forest Sector (CBM-CFS3) Version 1.2: User's guide. Canadian Forest Service, Northern Forestry Centre

Kurz WA, Apps MJ (1999) A 70-year retrospective analysis of carbon fluxes in the Canadian forest sector. Ecol Appl 9:526–547

Kurz WA, Dymond CC, White TM et al (2009) CBM-CFS3: a model of carbon-dynamics in forestry and land-use change implementing IPCC standards. Ecol Model 220:480–504

Lauri P, Havlík P, Kindermann G et al (2014) Woody biomass energy potential in 2050. Energ Policy 66:19–31

Li Z, Kurz WA, Apps MJ, Beukema SJ (2003) Belowground biomass dynamics in the Carbon Budget Model of the Canadian Forest Sector: recent improvements and implications for the estimation of NPP and NEP. Can J For Res 33:126–136
Liski J, Palosuo T, Peltoniemi M, Sievänen R (2005) Carbon and decomposition model Yasso for forest soils. Ecol Model 189:168–182
McCollum D, Krey V, Kolp P et al (2014) Transport electrification: a key element for energy system transformation and climate stabilization. Clim Chang 123(3):651–664
Meyer J, Vilén T, Peltoniemi M et al. (2005) Uncertainty estimate of the national level biomass and soil carbon stock and stock change. CarboInvent Project Deliverable 6.3
Nabuurs GJ, Schelhaas MJ, Pussinen A (2000) Validation of the European Forest Information Scenario Model (EFISCEN) and a projection of Finnish forests. Silva Fenn 34(2):167–179. http://dx.doi.org/10.14214/sf.638
Nabuurs GJ, Goede DM, Michie B et al (2002) Long term international impacts of nature oriented forest management on European forests – an assessment with the EFISCEN model. J World Forest Resource Manag 9:101–129
Nabuurs GJ, Pussinen A, van Brusselen J, Schelhaas MJ (2007) Future harvesting pressure on European forests. Eur J For Res 126:391–400
Nabuurs GJ, Schelhaas MJ, Hendriks CMA, Hengeveld GM (2014) Can European forests meet the demands of the bioeconomy in the future? Wood supply alongside environmental services. In: Innes J, Nikolakis W (eds) Forests and globalization: challenges and opportunities for sustainable development. The Earthscan Forest Library, Routledge, Oxon/New-York
Nilsson S, Sallnäs O, Duinker P (1992) A report on the IIASA forest study: future forest resources of Western and Eastern Europe. The Parthenon Publishing Group, Carnforth
Nuutinen T, Kellomäki S (2001) A comparison of three modelling approaches for largescale forest scenario analysis in Finland. Silva Fenn 35(3):299–308
Oosterbaan A, van den Berg CA, Schelhaas MJ (2007) Ontwikkelingen in vraag en aanbod van rondhout in Nederland en aangrenzend gebied en mogelijke knelpunten en kansen voor de bos- en houtsector in de periode 2005–2025. Alterra rapport 1510, Wageningen
Packalen T, Sallnäs O, Sirkiä S et al (2014) The European Forestry Dynamics Model: concept, design and results of first case studies. Publications Office of the European Union, EUR 27004. doi:10.2788/153990
Pilli R, Grassi G, Kurz WA et al (2013) Application of the CBM-CFS3 model to estimate Italy's forest carbon budget, 1995–2020. Ecol Model 266:144–171
Pilli R, Grassi G, Cescatti A (2014a) Historical analysis and modeling of the forest carbon dynamics using the Carbon Budget Model: an example for the Trento Province (NE, Italy) – in Italian, with summary in English. Forest@ 11:20–35
Pilli R, Grassi G, Moris JV, Kurz WA (2014b) Assessing the carbon sink of afforestation with the Carbon Budget Model at the country level: an example for Italy. iForest 8:410–421
Pilli R, Grassi G, Kurz WA et al (2016a) Modelling forest carbon stock changes as affected by harvest and natural disturbances. I. Comparison with countries' estimates for forest management. Carbon Balance Manag 11:5
Pilli R, Grassi G, Kurz WA et al (2016b) Modelling forest carbon stock changes as affected by harvest and natural disturbances II. EU-level analysis. Carbon Balance Manag 11:20
Pussinen A, Nabuurs GJ, Wieggers HJJ et al (2009) Modelling long-term impacts of environmental change on mid- and high-latitude European forests and options for adaptive forest management. Forest Ecol Manag 258:1806–1813
Reisinger A, Havlik P, Riahi K et al (2013) Implications of alternative metrics for global mitigation costs and greenhouse gas emissions from agriculture. Clim Chang 117(4):677–690
Sallnäs O (1990) A matrix growth model of the Swedish forest. Stud For Suec 183:1–23
Sallnäs O, Berger A, Räty M, Trubins M (2015) An area-based matrix model for uneven-aged forests. Forests 6:1500 1515. doi:10.3390/f6051500

Särkkä S (2013) Bayesian filtering and smoothing (PDF). Cambridge University Press. http://becs.aalto.fi/~ssarkka/pub/cup_book_online_20131111.pdf

Schelhaas MJ, Eggers J, Lindner M et al. (2007) Model documentation for the European Forest Information Scenario model (EFISCEN 3.1). Alterra report 1559, Wageningen, EFI technical report 26, Joensuu

Schelhaas MJ, Hengeveld G, Moriondo M et al (2010) Assessing risk and adaptation options to fires and windstorms in European forestry. Mitig Adapt Strateg 15:681–701. doi:10.1007/s11027-010-9243-0

Schelhaas MJ, Nabuurs GJ, Hengeveld GM et al (2015) Alternative forest management strategies to account for climate change-induced productivity and species suitability changes in Europe. Reg Environ Chang 15:1581–1594

Stinson G, Kurz WA, Smyth CE et al (2011) An inventory-based analysis of Canada's managed forest carbon dynamics, 1990 to 2008. Glob Chang Biol 17:2227–2244

Takayama T, Judge GG (1971) Spatial and temporal price and allocation models. North-Holland Publishing Company, Amsterdam/London

UNECE/FAO (1953) European timber trends and prospects. FAO, Geneva

UNECE/FAO (2005) European Forest Sector Outlook Study: main report. United Nations, Geneva, ECE/TIM/SP/20

UNECE/FAO (2011). The European Forest Sector Outlook Study II (EFSOS II). 2010–2030. UNECE/FAO

Verkerk PJ, Antilla P, Eggers J et al (2011) The realisable potential supply of woody biomass from forests in the European Union. Forest Ecol Manag 261:2007–2015

Verkerk H, Lindner M, Helming J et al (2014) Identification of pathways to consolidated visions of future land use in Europe. VOLANTE Deliverable 11:3

Zamolodchikov DG, Grabovsky VI, Korovin GN, Kurz WA (2008) Assessment and projection of carbon budget in forests of Vologda Region using the Canadian model CBM-CFS Lesovedenie 6:3–14 (in Russian, with summary in English)

Chapter 5
Future Challenges for Woody Biomass Projections

Klemens Schadauer, Susana Barreiro, Mart-Jan Schelhaas, and Ronald E. McRoberts

5.1 Understanding the Driving Forces Affecting Woody Biomass Availability and Its Projection

The importance of forests for satisfying social, economic, and ecological demands continues to increase. National Forest Inventories (NFIs) are the main tool for responding to the information requests related to these demands. National and international reporting obligations mirror these needs for information, but due to a lack of political coordination, NFIs use a huge variety of data collection protocols and reporting formats including non-matching terms and definitions. These circumstances characterize the challenges linked to international information requests for estimates of future wood and biomass supply.

Woody biomass supply information is crucial for the wood processing industry and for environmental policy makers. Decisions are made on several levels, from the local scale where the decisions on biomass supply are strongly determined by forest management, to the regional, national, and European scales where support is

K. Schadauer (✉)
Department of Forest Inventory, Federal Research and Training Centre for Forests, Natural Hazards and Landscape (BFW), Vienna, Austria
e-mail: klemens.schadauer@bfw.gv.at

S. Barreiro
Forest Research Centre (CEF), School of Agriculture, University of Lisbon, Lisbon, Portugal
e-mail: smb@isa.ulisboa.pt

M.-J. Schelhaas
Wageningen Environmental Research (Alterra), Wageningen, The Netherlands
e-mail: martjan.schelhaas@wur.nl

R.E. McRoberts
Northern Research Station, U.S. Forest Service, Saint Paul, MN, USA
e-mail: rmcroberts@fs.fed.us

S. Barreiro et al. (eds.), *Forest Inventory-based Projection Systems for Wood and Biomass Availability*, Managing Forest Ecosystems 29,
DOI 10.1007/978-3-319-56201-8_5

needed for forest and forest-related policies on issues such as bioenergy targets, incentives, and subsidy policies. The simplest and often requested approach is based on the assumption that recent Net Annual Increment (NAI) is the best parameter for estimating future wood and biomass supply; however, this approach can be misleading (Chap. 2). Although it has been common knowledge in forest management planning for more than 100 years that NAI is just one among several variables for estimating future wood supply, conveying this knowledge to the decision makers outside the expert community has been challenging. Obviously, the driving forces that cause major deviations between NAI and the future supply are very complex and interlinked, and an exhaustive study is necessary to understand them fully.

First, a sound definition and a procedure for estimating the area of Forests Available for Wood Supply (FAWS) is important. The latest State of European Forests Report clearly showed that a definition for FAWS with a commonly accepted interpretation is still lacking (FOREST EUROPE 2015; Vidal et al. 2016; Alberdi et al. 2016). Some European countries have a small area of FAWS when compared to the total forest area. For example, in Bulgaria the share of FAWS is only 58% of total forest area. Other countries such as Switzerland consider nearly all forests as FAWS and report a share of 97%. It is unlikely that these deviations originate from actual differences in the structure of the forest. At least a portion of the deviations can be attributed to differences in the national definitions. The land use class FAWS is not used in the United States of America (USA) or Canada, although the concept is expressed in other land use classes. In the USA, one criterion for the land use class characterized as timberland is that these lands must not be withdrawn from timber utilization. More than 210 million ha (60%) of the 330 million ha of forest land in the USA is classified as timberland of which 70% is in private ownerships (Oswalt et al. 2014). In Canada, more than 234 million ha (56%) of the 417 million ha of forested territory is classified as commercial forest, nearly all of which is in public ownerships and of which 0.4% is annually harvested (Natural Resources Canada 2015). For future projections we must acknowledge that any FAWS estimate might change over time due to nature protection measures on one hand, and to forest road building in remote and previously inaccessible areas on the other hand.

Multiple factors influence the potential future woody biomass supply. Classical influences are the distributions of age and yield classes which are included in all national projection systems (Chap. 3). In addition, differing ownership structures and differing management regimes among different categories of forest owners affect the future supply. Although most projection systems recognise different ownership classes, it remains a challenge to model the behaviour of owners, especially small private owners. Also, market behaviour on both the demand and supply sides resulting from external influences (e.g., small forest owners or economic crisis) affect wood prices and harvest costs which, in turn, might be influenced by changes in harvesting technologies. However, there are only a few national projection systems that are able to estimate harvesting costs, and even fewer that are actually coupled to a market model that includes future demand and trade (Chap. 3). Because forest growth and development are strongly linked to climate conditions, climate change could be considered a source of variability over time. The effects of future temperature and precipitation changes are relatively easy to handle in modelling, whereas the potential for increasing storm and fire events are more difficult to accommodate in a scientifically robust manner. However, at the moment most national projection systems do not include direct effects of climate change on tree growth (Chap. 3). Empirical models are not very reliable for predicting future interactions between soil and climate variables for which there are no historical observations. Process-based models would be more suitable for such a task, but so far only Netherlands reported the use of a process-based model (see Chap. 20). Nevertheless, application of country-wide projection systems based on eco-physiological models is particularly demanding because these models often require a wide range of soil and climate data that are not available for the locations of inventory plots that serve as the source of most other input data. Finally, tree species composition changes in response to changing climate should be included in long-term wood supply projections. Although many projection systems allow replanting with other species after final felling, almost none of them model natural regeneration under a changing climate (Chap. 3).

Although the foregoing list of influences on future wood supply is not complete, the challenging character of the mission to include all these effects and their interrelations in the political discussions is bold and demanding. Even for experts, inclusion of all these driving forces into tools for estimating future woody biomass supply is challenging. Therefore, the preferable approach is to focus on scenario analyses under defined conditions rather than simple growth and mortality predictions.

5.2 Woody Biomass Scenario Analyses

Approaches to scenario analyses are quite diverse in both Europe and North America. One reason is that countries individually choose which of the previously mentioned factors to include in or exclude from their projection systems (see also Chap.3). Often the main forces driving the choices can be traced to the motivation for the tool development and/or to the types of growth models and additional modules (e.g., modules for the simulation of forest management or disturbances) available in each country. For instance, in some countries projection systems emphasize future forest carbon stocks in response to Kyoto Protocol commitments (e.g. Iceland, Italy, Canada), whereas in other countries the tools were primarily developed to project wood availability for industry and were only later augmented to estimate biomass available for bio-energy and carbon stocks (e.g. Finland, Sweden, Portugal).

Another source of differences is climatic gradients that are evident from northern to southern Europe. One result is that northern climatic conditions inhibit the greater diversity in forest types and ecosystems that occur naturally in the south of Europe. Thus, whereas the Nordic countries have been able to develop advanced tools for simulating scenarios for their forests, the much greater complexity of southern European forests has often precluded development of similar tools. Therefore, approaches that include all forest ecosystem varieties are often lacking at the country level for these regions. For example, in Spain as many as 175 different tree species are assessed by the NFI, of which 30 cover a significant share of the total forest area (see Chap. 24). Such complexity has led to the development of stand level simulators for the most important tree species in specific regions of the country. If national estimates are needed, simpler approaches albeit with greater uncertainties are possible solutions.

Approaches to European forest management also differ with a general trend featuring more clear-cutting in the north and more coppice forests in the south. Superimposed on this trend are previous east-west differences in political systems that strongly affect approaches to both forest management and monitoring. Some eastern countries continue to use stand-wise forest inventories as the basis for forest management plans and as the source of data for input to projection systems that usually make projections based on NAI (Chap. 3).

The large range of environmental conditions for natural forest development and approaches to forest management throughout Europe has produced a tremendous variety of general approaches and methods for wood and biomass resource scenario analyses and interpretations of the results. The challenge of how to obtain the best pan-European estimates can be met in several ways (Chap. 4). Depending on the harmonisation strategy, approaches can range from a pure "bottom-up" approach, accommodating all differences in natural and socio-economic conditions and using only national tools, to a "top-down" approach, ignoring much of the natural and socio-economic variability, and using a centralised methodology based on NFI data and a common tool, producing comparable results.

These two approaches are linked through the trade-off between the most accurate descriptions of national and regional conditions but with little comparability among country results when using the "bottom-up" approach, or greater comparability of results but disregarding the topographic, climatic, vegetative and socio-economic conditions that are unique to countries and regions when using the "top-down" approach.

In North America, both Canada and the USA are large countries that span the continent and, therefore, have large numbers of biomes, each of which represents a different combination of topographic and climatic conditions and each of which is dominated by different forest types. Thus, construction of a single, all-inclusive projection system that would encompass the entire scope of these conditions would be nearly impossible. The two countries take somewhat different approaches to resolving this difficulty. In Canada, nearly all the commercial forest land is in public ownerships, and the provinces and territories have responsibility for managing these lands. One result is that management practices and policies and inventory programs vary considerably across the country, further compounding the difficulty of developing a common projection system. Canada's National Forest Carbon Monitoring Accounting and Reporting System (Kurz and Apps 2006), although national in scope, deals with biome differences by using provincial and territorial forest inventory data and growth and yield data to forecast future timber supply. In the USA, although most productive forest land is in private ownerships with potentially greatly varying management practices, all forest lands in all ownerships are covered by the country's nationally consistent inventory program. An important result is that nationally consistent data in a common format can be supplied to any projection system. The most widely used projection system, the Forest Vegetation Simulator (Crookston and Dixon 2005), uses these inventory data with a suite of biome-level growth and mortality models that are incorporated into a common software framework with a common user interface.

5.3 The Challenges

European countries use a large variety of approaches to model future woody biomass supply (Chap. 3). Even countries with similar ecological conditions have developed different approaches driven by differences in the economic importance of the forest sector, differences in the aims of the tools, and differences in inventory systems. As a basis, several steps could be taken towards harmonisation of national projection systems:

- Knowledge sharing to reduce the "noise" in different methodological approaches that are not based on different natural or management conditions
- Use of a common output format
- Harmonisation of the input and output variables for the national tools to the degree possible
- Use of a common general modeling structure and a common software structure
- Work towards common tools for countries with similar general conditions

In addition, all countries face similar challenges with respect to inclusion of climate change effects on increment and tree species composition, uncertain owner behaviour and modelling future demand of wood and wood products. Development of common approaches for these issues seems very useful. Such approaches need to be generic so they can be combined with different types of projection systems. Examples are the development of a method for integrating diverse owner behaviours into projection systems by Rinaldi et al. (2015), and the role of forestry mitigating the effects of climate change (Lundmark et al. 2014).

Despite attempts to better align the national simulators, there is a clear need for harmonised projections at the European scale. Several simulators have been developed that can produce such projections (Chap. 4). The challenge is to combine national data, expertise and modelling systems in a manner that maximises harmonisation. A prerequisite here is continued collaboration between national and European experts to exchange information on methods and assumptions that contribute to differences among approaches. A comparison between national and European simulators and projection studies would be a useful first step in this direction. Furthermore, projection systems in the USA and Canada may provide inspiration for solving some of the European issues with regard to the variety of ecological conditions that need to be covered.

Within the United Nations Economic Commission for Europe (UNECE), there is also the request to further harmonise projections between Europe and North America. It is unlikely that the same resource projection tools will be used for both continents, but current discussions focus on using the same global market models and harmonising economic scenarios between European and North American Forest Sector Outlook Studies (respectively EFSOS III and NAFSOS II). Another major challenge with respect to these studies is the inclusion of the Russian Federation in a comparable way.

Another challenge relates to "customization" of the results and information produced by different projection systems. There is often a lack of mutual understanding among supply side experts and demand side experts (i.e. wood industry) resulting in misunderstandings about availability of woody biomass (Chap. 2). The only way to overcome these communication problems is to work intensively together. The ITOC model (Inventory to Consumer biomass availability) (Mantau et al. 2016) is an example of a methodology for transforming forest inventory data to consumer biomass availability that facilitates the exchange of data among experts in different fields and increases mutual understanding.

Last but not least, estimation of the quality of information obtained from wood and biomass supply studies has not yet been rigorously or scientifically addressed. Global interrelationships among systems cause policy decisions to have all-embracing consequences. Policymakers decide on the policies to be implemented to meet stakeholders' objectives and preferences, while scientists are expected to assess the outcomes of alternative policies. In the case of woody biomass projection systems, the links, flows and relationships among all the elements comprising the system must be evaluated and the corresponding uncertainties must be assessed (Walker et al. 2003). Because simulation systems are composed of multiple modules (e.g. regeneration, growth, mortality) their complexities and the multiple error sources make uncertainty difficult to assess (Kangas 1999; McRoberts and Westfall 2014). One cause is that error propagates from trees to stands (e.g. Zhang et al. 1997; Cao 2006; Mäkinen 2010); a second cause is that uncertainty increases with greater prognosis length (Holm 1981; Kangas 1997); another cause is that deterministic projection systems force decision making to be based on mean predicted values but without appropriate knowledge of the risk (Fortin and Langevin 2012; Fortin et al. 2009; Antón-Fernández and Astrup 2012). Estimators based on likelihood or pseudo-likelihood functions (Lindstrom and Bates 1988; Wolfinger and O'Connell 1993; Pinheiro and Bates 1995, 2000; McCulloch and Searle 2001) facilitate accounting for the different error components, as well as the error structure within the sub-models that comprise individual-tree models. When coupled with a Monte Carlo approach, these estimators can simulate multiple error components and enable assessment of uncertainty related to the stochasticity of the processes. Multiple studies of the precision of growth projections have been conducted (e.g. Gertner and Dzialowy 1984; Mowrer and Frayer 1986; Gertner 1987; Mowrer 1991; Kangas 1997; Fortin et al. 2009; Fahlvik et al. 2014; Fortin et al. 2016). Unfortunately not all the projection systems described in this book have undergone similar analyses. However, the use of large-scale projection systems in decision-making processes, including the upcoming international negotiations on climate change, urge the importance of uncertainty assessment. In particular, the IPCC good practice guidance specify criteria related to both bias ("neither over- nor underestimates so far as can be judged") and precision ("uncertainties are reduced as far as practicable") (Penman et al. 2003). Sensitivity analyses have been performed for some tools, but a comprehensive approach that takes into account multiple sources of uncertainty is lacking. Such uncertainty analyses should start at the measurement level and include investigation of both sampling and modeling prediction errors. Finally, such analyses should also address whether the data and the models are adequate for the types of questions to be answered.

References

Alberdi I, Michalak R, Fischer C et al (2016) Towards harmonized assessment of European forest availability for wood supply in Europe. Forest Policy Econ 70:20–29. doi:10.1016/j.forpol.2016.05.014

Antón-Fernández C, Astrup R (2012) Empirical harvest models and their use in regional business-as-usual scenarios of timber supply and carbon stock development. Scand J Forest Res:1–14. doi:10.1080/02827581.2011.644576

Cao QV (2006) Predictions of individual-tree and whole-stand attributes for loblolly pine plantations. For Ecol Manag 236:342–347

Crookston NL, Dixon GE (2005) The forest vegetation simulator: a review of the structure, content, and applications. Comput Electron Agr 49:60–80. http://www.nrs.fs.fed.us/pubs/18474. Accessed 15 Nov 2016

Fahlvik N, Wikström P, Elfving B (2014) Evaluation of growth models used in the Swedish Forest Planning System Heureka. Silva Fenn 48. doi:10.14214/sf.1013

FOREST EUROPE (2015) State of European Forests 2015

Fortin M, Langevin L (2012) Stochastic or deterministic single-tree models: is there any difference in growth predictions? Ann For Sci 69:271–282. doi:10.1007/s13595-011-0112-0

Fortin M, Bédarda S, DeBloisa J, Meunierb S (2009) Assessing and testing prediction uncertainty for single tree-based models: a case study applied to northern hardwood stands in southern Québec, Canada. Ecol Model 220:2770–2781. doi:10.1016/j.ecolmodel.2009.06.035

Fortin M, Robert N, Manso R (2016) Uncertainty assessment of large-scale forest growth predictions based on a transition-matrix model in Catalonia. Ann For Sci 73:871. doi:10.1007/s13595-016-0538-5

Gertner G (1987) Approximating precision in simulation projections: an efficient alternative to Monte Carlo methods. For Sci 33:230–239

Gertner G, Dzialowy PJ (1984) Effects of measurement errors on an individual tree-based growth projection system. Can J For Res 14:311–316

Holm S (1981) Analys av metoder för tillväxtprognoser i samband med långsiktiga avverkningsberäkningar. Swedish University of Agricultural Sciences, Department of Biometry and Forest Management, Working Paper. 22 p. [In Swedish]

Kangas A (1997) On the prediction bias and variance of long-term growth predictions. For Ecol Manag 96:207–216

Kangas AS (1999) Methods for assessing uncertainty of growth and yield predictions. Can J For Res 29:1357–1364

Kurz WA, Apps MJ (2006) Developing Canada's national forest carbon monitoring, accounting and reporting system to meet the reporting requirements of the Kyoto Protocol. Mit Adapt Strat Glob Change 11:33–43

Lindstrom MJ, Bates D (1988) Newton–Raphson and EM algorithms for linear mixed-effects models for repeated-measures data. J Am Stat Assoc 83:1014–1022
Lundmark T, Bergh J, Hofer P et al (2014) Potential roles of Swedish forestry in the context of climate change mitigation. Forests 5:557–578
Mäkinen A (2010) Uncertainty in forest simulators and forest planning systems. Dissertationes Forestalis 97, 38p. doi:10.14214/df.97
Mantau U, Gschwantner T, Paletto A et al (2016) From inventory to consumer biomass availability – the ITOC-model. Ann For Sci. doi:10.1007/s13595-016-0582-1
McCulloch CE, Searle SR (2001) Generalized, linear, and mixed models. Wiley, New York
McRoberts RE, Westfall JA (2014) Effects of uncertainty in model predictions of individual tree volume on larger area volume estimates. For Sci 60:34–42
Mowrer HT (1991) Estimating components of propagated variance in growth simulation model projections. Can J For Res 21:379–386
Mowrer HT, Frayer WE (1986) Variance propagation in growth and yield projections. Can J For Res 16:1196–1200
Natural Resources Canada (2015) Forestry. http://www.nrcan.gc.ca/earth-sciences/geography/atlas-canada/selected-thematic-maps/16874. Accessed 15 Nov 2016
Oswalt SN, Smith WB, Mile PD, Pugh SA (2014) Forest Resources of the United States, 2012: a technical document supporting the Forest Service 2015 update of the RPA Assessment. General Technical Report WO-91. U.S. Department of Agriculture, Forest Service, Washington Office, Washington, DC, 218p. http://www.srs.fs.usda.gov/pubs/47322. Accessed 15 Nov 2016
Penman J, Gytarsky M, Hiraishi T, Krug T, Kruger D, Pipatti R, Buendia L, Miwa K, Ngara T, Tanabe K, Wagner F (eds.) (2003). Good practice guidance for land use, land-use change and forestry. Intergovernmental Panel on Climate Change. Institute for Global Environmental Strategies, Hayama
Pinheiro JC, Bates DM (1995) Approximations to the log-likelihood function in the nonlinear mixed-effects model. J Comp Graph Stat 4:12–35
Pinheiro JC, Bates DM (2000) Mixed-effects models in S and S-PLUS. Springer, New York
Rinaldi F, Jonsson R, Sallnäs O, Trubins R (2015) Behavioral modelling in a decision support system. Forests 6(2):311–327. doi:10.3390/f6020311
Vidal C, Sallnäs O, Redmond J, Alberdi I, Barreiro S, Hernández L, Schadauer K (2016) Introduction. In: Vidal C, Alberdi I, Hernández L, Redmond J (eds) National forest inventories: assessment of wood availability and use. Springer, Cham, pp 1–24. doi:10.1007/978-3-319-44015-6_1
Walker WE, Harremo€ees P, Rotmans J, Van Der Sluijs JP, Van Asselt MBA, Janssen P, Krayer Von Krauss MP (2003) Defining uncertainty – a conceptual basis for uncertainty management in model-based decision support. Integr Assess 4(1):5–17
Wolfinger R, O'Connell M (1993) Generalized linear mixed models: a pseudolikelihood approach. J Statist Comput Simul 48:233–243
Zhang S, Amateis RL, Burkhart HE (1997) Constraining individual tree diameter increment and survival models for loblolly pine plantations. For Sci 43:414–423

Part II
National Woody Biomass Projection Systems Based on Forest Inventory

Chapter 6
Austria

Thomas Ledermann, Georg Kindermann, and Thomas Gschwantner

6.1 Introduction

In Central Europe information about the available timber resources became necessary because of the increasing demand for wood used for mining, salt and charcoal production. Simple woodland surveys dating back to the Middle Ages were conducted for visual estimation of growing stock and costs of timber transport. The risk of over-utilization inevitably led to considerations of the sustainable use of forest resources (Carlowitz 1713; Gabler and Schadauer 2007). Thus, the estimation of increments, future stand development and yield of young forest stands gained importance. The first guidelines for constructing growth and yield tables were published in 1721 by Réaumur (Schwappach 1903; Pretzsch 2001). At the end of the eighteenth century Paulsen (1795) presented the first growth and yield table for Germany.

In Austria, the first yield tables were developed by Feistmantel (1854), Guttenberg (1896, 1915) and Schiffel (1904) and were used for several decades. However, experiences from the first country-wide forest survey and the first sample-based forest inventory indicated that the growth pattern of these yield tables did not apply to all regions of Austria (Sterba 1991). Consequently, Marschall (1975) analyzed existing yield tables and split Austria into four distinct yield table regions for Norway spruce. In forest enterprises, Marschall's (1975) collection of yield tables is

T. Ledermann (✉) • G. Kindermann
Department of Forest Growth and Silviculture, Federal Research and Training Centre for Forests, Natural Hazards and Landscape (BFW), Vienna, Austria
e-mail: thomas.ledermann@bfw.gv.at; georg.kindermann@bfw.gv.at

T. Gschwantner
Department of Forest Inventory, Federal Research and Training Centre for Forests, Natural Hazards and Landscape (BFW), Vienna, Austria
e-mail: thomas.gschwantner@bfw.gv.at

S. Barreiro et al. (eds.), *Forest Inventory-based Projection Systems for Wood and Biomass Availability*, Managing Forest Ecosystems 29,
DOI 10.1007/978-3-319-56201-8_6

still used in common practice. Yield tables apply best to even-aged, pure species stands whose treatment regimes are similar to those of the yield tables. These limitations, together with changing growth conditions in Austria (Neumann and Schadauer 1995), restrict the reliability of yield tables. To overcome this limitation, several forest growth simulators have been developed in Austria in recent decades (Huber et al. 2013). MOSES is a distance-dependent simulator based on data from permanent experimental plots (Hasenauer 2000). The distance-independent simulator PROGNAUS (Sterba et al. 1995; Ledermann 2006) was developed using sample tree and plot data from the Austrian National Forest Inventory (NFI). CALDIS (from Caldis vâtis = Celtic term for forest prophet) is a recent version of PROGNAUS and also based on Austrian NFI data. The modules, structure and growth models for both simulators are comparable, but CALDIS additionally integrates climate variables and uses longer time-series of data from the Austrian NFI covering the years 1981–2009. All of the mentioned methods, models and simulators (yield tables, plot- and tree-level simulators) have fields of application where they are superior. For country-wide estimation, PROGNAUS and CALDIS seem to be best because both can be applied directly to the Austrian NFI sample tree and plot data. Since the 1990s, PROGNAUS has been used for scenario simulations in wood supply and outlook studies for Austria based on NFI data (Sterba 1996; Neumann and Schadauer 2007; BFW 2009). CALDIS uses a larger data set from a longer time period, is climate sensitive, and shows some improvements in the model structure (Kindermann 2010a, b; Gschwantner et al. 2010b). It was used for the most recent study on green house gas emissions (Ledermann and Schadauer 2015) and to estimate the productivity of Austrian forests in terms of volume and biomass increment under various scenarios of tree species distributions and climate conditions. The methodological approaches and main results of these studies are presented in this report together with results from the NFIs conducted in Austria.

6.2 The Austrian NFI as Data and Information Source

The major source of up-to-date information on the state and change of forest resources are the observations and estimates produced by the Austrian NFI. It is the largest forest-related monitoring program at the national level and serves as the primary source of information for forestry, forest economy and forest ecology related issues. With regard to forest growth, the NFI is a tool to monitor the sustainability of wood resource utilization, and serves as the basis for growth simulations and outlook studies.

6.2.1 Overview on NFIs

The first assessment of forest resources that covered the entire area of Austria was conducted during 1952–1956. It was designed as a stand-wise inventory of single stands. Later, sample-based inventories gained more importance. The first NFI was conducted during 1961–1970 using a temporary, systematic sampling grid and individual-tree measurements on inventory plots. The second temporary inventory followed in the years 1971–1980. For the third NFI in 1981–1985, the sampling system was changed from temporary to permanent sample plot locations that were marked in the field. The plots have been re-measured in the subsequent years 1986–1990, 1992–1996, 2000–2002 and 2007–2009. Approximately 22,000 sample plots are systematically distributed on a 3.89 × 3.89 km sampling grid that covers the Austrian territory; of these plots, approximately 50% are located in forests. The sampling design features tracts, each of which comprises four plots. Each sample plot consists of two concentric circles with fixed radii of 9.77 and 2.60 m, respectively, and a Bitterlich sample plot (Bitterlich 1948, 1952). The large plot is used for forest area estimation and to assess stand- and site-specific information. Sample trees are selected within the small circular plot if their diameters at breast height (1.3 m, *dbh*) are in the range of 5.0 cm $\leq dbh \leq$ 10.4 cm. Trees with $dbh \geq$ 10.5 cm are selected on the Bitterlich plot using a basal area factor of 4 m^2/ha. Detailed descriptions of the Austrian NFI are available in Gabler and Schadauer (2008) and Gschwantner et al. (2010a).

6.2.2 Forest Resources in Austria

Since its establishment in the 1960s, the Austrian NFI has provided estimates on the state and change of forest resources at both national and sub-national levels. Table 6.1. gives an overview of the total forest area, the forest area available for wood supply, the minimum diameter, the standing stock, and removals and the increments in Austria for the last seven NFIs. The forest area available for wood supply equals the area of productive forests. Note that standing stock, harvest and increment correspond to the area available for wood supply and that the method for estimating increments changed between the second NFI in 1971–1980 and the third NFI in 1981–1985. Volume is reported as m^3 of stemwood over bark.

Age structure has an important influence on the standing stock, increment and removals. Table 6.2 indicates that the younger age classes account for a relatively large proportion of the Austrian forest area.

According to the most recent Austrian NFI in 2007–2009, Norway spruce (*Picea abies*) occupies 50.7% of forest area, followed by common beech (*Fagus sylvatica*) with 10.0%. European larch (*Larix decidua*) and Scots pine (*Pinus sylvestris*) are similarly represented occupying 4.6% and 4.5% of forest area, respectively. Species such as white fir (*Abies alba*), black pine (*Pinus nigra*), stone pine (*Pinus cembra*),

Table 6.1 Characterization of Austrian forest resources over time according to the national forest inventories

Period (year)	Area (1000 ha)	Area available for wood supply (1000 ha)	D_{min} (cm)	Standing stock (m^3/ha)	Removals (m^3/ha per year)	Increment (m^3/ha per year)
1961–1970	3691	3230	10.5	234	3.8	5.7
1971–1980	3754	3128	10.5	257		6.3
1981–1985	3857	3339	5.0	280	5.9	9.4
1986–1990	3878	3331	5.0	292	5.9	8.2
1992–1996	3924	3352	5.0	295	5.6	9.3
2000–2002	3960	3371	5.0	325	7.7	9.0
2007–2009	3991	3367	5.0	337		

Table 6.2 Percentage of area available for wood supply by age classes according to the national forest inventories

	Age (years)									
Period	0	1–20	21–40	41–60	61–80	81–100	101–120	>120	121–140	>140
1961–1970	1.9	18.8	20.2	16.3	15.1	11.3	8.4	8.0		
1970–1980	6.2	22.3	19.1	12.5	13.5	9.8	7.1	8.6	4.3	4.3
1986–1990	9.2	18.6	19.9	12.4	11.6	9.9	7.1	11.2	4.7	6.5
1992–1996	8.5	17.1	22.7	12.7	11.3	9.8	6.8	11.2	4.6	6.6
2000–2002	9.3	14.9	23.2	14	10.9	9.8	7.2	10.6	4.5	6.1
2007–2009	12.1	12.9	22.3	15.7	10.7	9.4	7.2	9.8	4.4	5.4

and oak (*Quercus* sp.) represent small forest areas (2.4%, 0.6%, 0.6%, and 2.0%). The remaining area is occupied by other broadleaved hardwoods (8.2%), other broadleaved softwoods (4.2%), and other coniferous tree species (0.2%).

6.2.3 NFI Data as the Basis for Growth Models

The availability of high quality inventory field data and the adequacy and interaction of the models in a growth simulator are crucial for accurately estimating wood and biomass resources. The forest growth simulators PROGNAUS and CALDIS consist of multiple sub-models that were parameterized using Austrian NFI data. For model parameterization, the data (species, diameter, height, per hectare expansion factor, height to crown base, temperature, precipitation, slope, exposition, relief, soil type, vegetation type, elevation, soil moisture, soil depth, depth of humus horizons and growth region) collected between 1981 and 2002 were used. During parameterization, the parameters of the models used for predicting individual-tree increments were estimated using regression techniques. Moreover, CALDIS was calibrated by adjusting the predicted individual-tree increments with a multiplier to eliminate average differences between the observed and estimated increments between the first and last observations. For model calibration data collected between 1981 and 2009 were used.

6.3 Description of the Growth Simulators

6.3.1 PROGNAUS

PROGNAUS is an individual-tree based, distance-independent growth simulator of the FVS-type (Forest Vegetation Simulator, Stage 1973; Wykoff 1990). The main components of PROGNAUS are a basal area increment model (Monserud and Sterba 1996), a height increment model (Nachtmann 2006), a mortality model (Monserud and Sterba 1999), and an ingrowth model (Ledermann 2002). Stand-level estimates are obtained by aggregating the tree-level predictions. All model components were developed using Austrian NFI data for time periods between 1981 and 2002. Thus, PROGNAUS is able to process any inventory data that provide the input variables according to the guidelines of the Austrian NFI (Gschwantner et al. 2010a). The increment models are independent of age and predict the increment directly, i.e. they do not follow the potential-modifier approach (Ek and Dudek 1980). The input data are limited to trees with *dbh* of at least 5 cm. The simulator can further be applied to even- and uneven-aged mixed stands of common Austrian tree species and for all types of commercial and pre-commercial thinning and harvesting strategies that are represented in the Austrian NFI data. Algorithms for simulating thinning and/or harvesting operations can be found in Söderbergh and Ledermann (2003). All components (sub-models) operate at the individual-tree level and can be applied either deterministically or stochastically for 5-year prediction intervals.

The simulator was implemented in Visual Basic Professional 6.0. Input data should be provided in text files. The simulator generates tree lists that include all trees prior to harvesting, harvested trees, trees that died naturally, and remaining trees (Ledermann 2006). Specifications regarding precision, bias and accuracy can be found in Monserud and Sterba (1996), Nachtmann (2006), Sterba and Monserud (1996, 1997), Sterba (1999), Monserud and Sterba (1999), and Sterba et al. (2001).

6.3.2 CALDIS

The forest growth simulator CALDIS consists of multiple modules and uses climate and soil variables as predictors. Therefore, the simulator is able to accommodate short term climate changes, at least to a certain extent. CALDIS is based on the same model concept as PROGNAUS (FVS-type). It is a distance-independent, individual-tree growth simulator whose sub-models were developed using Austrian NFI data. Model predictions can be aggregated to stand-level estimates. The prediction interval is 1-year, and the model can be used for both short-term (1-year) and long-term (several decades) predictions of forest development. The model is organized in modules. Growth predictions are typically deterministic, whereas harvest, regeneration and mortality estimates can be either deterministic or stochastic. The model can predict species composition, stand structure, stand treatment, and tree sizes for all site conditions observed in the Austrian NFI. For the main tree species, the model residual standard deviations for predictions of the annual *dbh* increment are in the range of 0.14–0.20 cm/year. Because the model is calibrated with Austrian NFI data, predictions are reliable for the 1981–2009 period.

CALDIS uses tree-level data (species, *dbh*, height, crown length) and plot-level data (temperature, precipitation, slope, exposition, relief, soil type, vegetation type, elevation, soil moisture, soil depth, and growth region). All data must be provided to the model except crown length which can be generated automatically. For missing information, estimates such as average soil type or vegetation type can be used. The type of thinning/harvesting can be defined as well as the species that should be used for regeneration. The felling decision can be made at the individual-tree level (e.g. remove this tree in the year 2015) or more generally for stands (e.g. clear cut when Lorey's height reaches a specified threshold).

Currently, two different versions of the growth simulator are available. CALDIS-CC v0.1 is the implementation of the growth model in C++. The increment module consists of sub-models for estimating basal area increment (Kindermann 2010a, b), height increment (Gschwantner et al. 2010b), and change in crown length (Ledermann 2011; Hasenauer and Monserud 1996). The implemented mortality model operates at the stand level and depends on stand density, i.e. trees are removed when the maximum stand density is reached (cf. Reineke 1933). Regeneration is done with planting. Thinning and harvesting can be done with a model which mimics the operations observed similar to the observations in the Austrian NFI or by defining stand density and/or tree size thresholds. The removals can be done stochastically using random numbers or deterministically by reducing a tree's per hectare expansion factor.

If a precompiled version of CALDIS-CC v0.1 is used, the only requirement is an operating system. For performance reasons it is possible to compile the source code on the machine where the simulations will be done. Therefore, a C++ compiler which fulfills the ISO/IEC 14882:2011 standard is needed.

Input data for CALDIS-CC v0.1 should be provided as a SQLite data base or as csv files. The simulator's output is a tree list for each year showing *dbh*, height, crown length, and per hectare expansion factor. In addition, trees that have been removed by thinning and harvesting are shown and can be used to estimate harvesting volume and harvesting costs. The values can be aggregated during a post process for any stratum by using software which is able to read and handle data of csv files. Special visualization features are not included in the simulator as the data outputs can easily be imported into Geographic Information System (GIS) or statistical software for state of the art visualization.

CALDIS-VB v0.1 is rather different from CALDIS-CC v0.1 because it is based on the source code and graphical user interface of PROGNAUS (Ledermann 2006). The increment module, which consists of the quoted sub-models for basal area increment (Kindermann 2010a, b), height increment (Gschwantner et al. 2010b), and change in crown length (Ledermann 2011; Hasenauer and Monserud 1996), is the only module common to both versions of CALDIS. While CALDIS-CC v0.1 uses a stand level mortality model, CALDIS-VB v0.1 resorts to a sub-model that estimates competition-induced mortality at the individual-tree level. CALDIS-VB v0.1 also contains a sub-model which estimates salvage cuts caused by storm, drought and bark-beetle attacks (Ledermann et al. 2010; Gschwantner et al. 2010b). For simulating forest regeneration CALDIS-VB v0.1 resorts to an ingrowth model (Ledermann 2002) that is also implemented in PROGNAUS. Tree volume and biomass models (Eckmüllner 2006; Gschwantner and Schadauer 2006; Ledermann and Neumann 2006; Offenthaler and Hochbichler 2006; Pollanschütz 1974; Rubatscher et al. 2006) are also incorporated in CALDIS-VB v0.1. These models are used to estimate stand volume as well as above and belowground biomass.

6.4 Outlook Studies and Simulation Scenarios

A study aimed at an overall assessment of the wood and biomass available from Austrian forests was carried out in the years 2007–2008 using Austrian NFI data from the period 2000–2002 (Neumann and Schadauer 2007; BFW 2009). The objective was to estimate harvest totals until 2020 that can be achieved on a sustainable basis using a comprehensive approach that covers all relevant aspects of using forests as wood and biomass resources. The forest growth predictions for the period 2000–2020 were obtained from the forest growth simulator PROGNAUS (Sect. 6.4.1). A subsequent study focused on the productivity of Austrian forest sites in the context of climate change. This study was done using Austrian NFI data from the period 2007–2009 and the climate sensitive forest growth simulator CALDIS-CC v0.1 (Sect. 6.4.2). The most recent study was carried out in the years 2013–2015. The objective of this study was to analyze the role of Austrian forests in the context of greenhouse gas emissions. For this study the Austrian NFI data from the period 2007/09 and the growth simulator CALDIS-VB v0.1 were used (Sect. 6.4.3).

6.4.1 Prediction of Available Wood and Biomass Amounts

6.4.1.1 Harvesting Scenarios

Three harvesting scenarios representing harvesting strategies based on different thinning intensities and final cut were developed and implemented in PROGNAUS. These scenarios accommodated all restrictions by the Austrian Forest Act. In

Table 6.3 Description of the three harvesting scenarios

	Main drivers	
Scenario	Thinning	Final cut
Constant standing stock	Reduce them to a minimum what is urgent	Keeps the overall standing stock constant (325 m^3/ha)
Silviculture	When silvicultural aspects suggest them	All stands with decreasing commercial value were cut
Adaption of standing stock	When silvicultural aspects suggest them	Reduces the overall standing stock to the level of the 1980s (280 m^3/ ha)

applying this harvesting model, the necessity and type of a harvesting intervention was determined. The scenario "Constant standing stock" keeps the standing stock at the current level; the scenario "Adaption of standing stock" reduces the standing stock to the same level as in the 1980s; and the scenario "Silviculture" thins and harvests as suggested by silvicultural or economic considerations. The main features of the harvesting scenarios are briefly described in Table 6.3.

6.4.1.2 Constraints in Harvesting and Wood Removal

Harvesting operations (pre-commercial thinning, thinning and final cuts) have only been done in forests without ecological, technical or economical restrictions or in forests that are not protected by natural conservation. For addressing ecological constraints, the sustainability of soil nutrients was assessed and was used to classify the inventory plots into two categories that prescribed the tree components that could be extracted during harvesting. Because branches, needles and leaves contain a considerable proportion of the nutrients stored in forest vegetation, these tree components are important for maintaining site productivity after harvesting, particularly on sites where the nutrient supply is limited. The harvesting and extraction of all above-ground tree components were considered to be ecologically possible only if all relevant nutrient elements were assessed to be sustainable. Otherwise, if one or more nutrient elements lacked sustainability, only the stems of trees could be harvested.

Technical and economic constraints influence decisions regarding the harvesting and logging technologies that can be used for different stand and site conditions. Particular factors include the slope, logging distance and tree diameters. Ecological constraints also influence harvest technology decisions. Machines that extract all above-ground tree biomass are not used for sites that lack nutrient sustainability because branches, needles and leaves must be left in the forest.

6.4.1.3 Costs and Revenue of Harvests

Depending on slope, distance to the nearest forest road and ecological restrictions, the costs associated with using various harvesting techniques were estimated using harvesting models. The harvesting costs were compared to the revenue that can be expected from the trees selected for harvesting. For estimating revenues, the selected trees first had to be partitioned into assortments that align with the actual wood timber trade guidelines (Kooperationsplattform Forst Holz Papier 2006). For this purpose, taper curve models, together with models that describe quality parameters, branch diameter, and length of wood decay, were applied in an assortment program to obtain quality and diameter classes for each tree (Eckmüllner et al. 2007). Several wood price scenarios were developed to represent the development of wood price in the years 2004–2006 and for the future. As a reference, the price of spruce for saw log quality in the diameter class 25–29 cm was set to 71 € per m^3 in *wood price scenario 1*, to 81 € per m^3 in *price scenario 2*, and to 100 € to *price scenario 3*. When applying the wood price scenarios to the assortments, the achievable revenues could be estimated. Harvesting interventions that did not reach the profit margin one (i.e., with costs greater than revenue) were not further considered in this study.

6.4.1.4 Achievable Amounts of Harvested Wood and Biomass

The study on the wood and biomass supply from Austrian forests (BFW 2009) produced results for several harvesting and price scenarios. The theoretically achievable amount of harvested timber was calculated for the whole Austrian forest area on the basis of the different harvesting scenarios (cf. Table 6.3). This potential was subsequently reduced by ecological, technical, and economic constraints or by restrictions from nature conservation objectives. Table 6.4 shows the available timber amount from Austrian forests on an annual basis for the period 2000–2020.

The scenario “Adaption of standing stock” gives the greatest estimates of annually available wood and biomass. However, a considerable reduction of the overall standing stock would require a broad discussion among environmental, energy and forest political interest groups. Thus, the scenarios “Constant standing stock” and “Silviculture” can be assumed to be more realistic from the current point of view. The ecological, technical and economic constraints have a considerable impact on the available wood and biomass amounts. Depending on the harvesting scenario and the price scenario, these reductions range between 19% and 28%. Constraints due to nature conservation decrease the available amounts of wood and biomass by additional 2%.

Table 6.4 Amounts of wood available according to the harvesting scenarios, price scenarios, and by taking into account the various constraints

	Wood available by harvest scenario (m^3/ha per year)		
Price scenarios and reduction of achievable amounts by constraints	Constant standing stock	Silviculture	Adaption of standing stock
Theoretically achievable	9.70	10.59	11.39
Wood price scenario 1			
Ecological, technical, economical constraints	7.30	7.62	8.34
Nature conservation constraints	7.09	7.42	8.13
Wood price scenario 2			
Ecological, technical, economical constraints	7.59	7.98	8.69
Nature conservation constraints	7.39	7.77	8.45
Wood price scenario 3			
Ecological, technical, economical constraints	7.83	8.28	8.99
Nature conservation constraints	7.62	8.07	8.75

6.4.2 *Productivity Estimates*

6.4.2.1 Model Calibration

CALDIS-CC v0.1 consists of multiple independently parameterized growth models. The core models implemented in CALDIS-CC v0.1 predict basal area increment, height increment, crown length, regeneration, mortality and removals. However, because the sub-model predictions with their residuals are interrelated, error propagation can cause overall growth predictions to deviate considerably in average from measurements after some prediction periods. To eliminate this effect, a *calibration* of the models is necessary. Data from the third Austrian NFI 1981–1985 (the first NFI using permanent plots) were used as a starting point. Increments were predicted annually until 2009, whereas harvests and ingrowth were simulated until 2009 using information provided by the Austrian NFIs. First, tree-level predictions of diameter, height and crown length were compared to the latest measurements. Then, the deviations were minimized using an iterative approach that simultaneously adjusted the model parameter estimates for each species individually until the total basal area, height increments, and crown length predictions closely approached the measured values.

6.4.2.2 Productivity Scenarios

CALDIS-CC v0.1 allows predictions of future stock, harvests and increments under various scenarios that reflect the effects of different types and intensities of thinning, rotation lengths, species selection and climate scenarios. However, all scenarios are influenced by the current age distribution, species combination and stand density. To circumvent this complexity, the growth model can start with a newly planted area and return the mean annual increment with an increment optimal rotation length (MAI_{max}). This feature is possible for any species for which the model was calibrated and for any location in Austria where the Austrian NFI provides information on site condition. Simply showing growth starting from the current forest situation masks the joint effects of climate change, age structure, species distribution, and stand density. Instead, showing MAI_{max} change as a descriptor of site productivity distinguishes the pure effect of climate change. However, the simulator does not account for hazard events such as wind throws, snow breakage, pests or diseases. To estimate productivity in terms of mean annual increments for the increment optimal rotation time, the following scenarios were investigated: (1) Current species distribution; (2) A mixed stand of the two best growing species, of 27 species, on the specific site; (3) the 1975–2007 climate; and (4) for a climate with 1.5 °C increased temperature.

The best growing species were selected by considering increment of merchantable volume and merchantable biomass as a means of determining the optimal species composition in terms of productivity.

6.4.2.3 Achievable Forest Productivity

Under the scenario of a current species distribution and the 1975–2007 climate, the estimated mean annual increment at increment optimal rotation time (MAI_{max}) for merchantable wood (diameter > 7 cm) is 12.6 m^3/ha per year which is equivalent to 5.37 t/ha per year dry matter biomass production. Under a climate scenario where the mean temperature is increased by 1.5 °C, the current species distribution would be able to produce, on average, 13.4 m^3/ha per year which is equivalent to a dry matter biomass production of 5.70 t/ha per year. Similarly, the MAI_{max} estimate for stem wood would be 11.9 m^3/ha per year (4.92 t/ha per year) with the 1975–2007 climate, and 12.8 m^3/ha per year (5.29 t/ha per year) with a 1.5 °C temperature increase. For comparison: the Austrian NFI reported a current annual increment (CAI) of stemwood of 9.0 m^3/ha per year for the period 2000–2009.

6.4.3 *Carbon Dynamics of the Austrian Forest Sector*

Since the UN climate summit in Durban in 2011, accounting of managed forests and harvested wood products is obligatory for the second Kyoto protocol commitment period (UN 2012). Therefore, it is important to analyze potentials and effects in the area of forest management and the whole wood product chain. The aim of this study was to estimate long-term effects of various assumptions concerning different pathways of wood utilization.

6.4.3.1 Coupling a Forest Growth Model with a Forest Sector Model and a Soil Carbon Simulator

To simulate the carbon dynamics of the Austrian forest sector, CALDIS-VB v0.1 was linked to the dynamic forest sector model FOHOW2 (Schwarzbauer and Stern 2010) and to the soil carbon model YASSO 07 (Liski et al. 2005). Using this model ensemble the whole Austrian wood product chain was simulated for five different scenarios of forest management and wood utilization. Climate change effects from RCP8.5 projections (IPCC 2013) were used for both forest stand and soil carbon modeling. The simulation runs were done until the year 2100.

6.4.3.2 Scenarios of Forest Management and Wood Utilization

In the reference scenario (R) the demand for wood as well as forest management followed trends of the recent years and were influenced by the same framework conditions as recently. Austria's National Renewable Energy Action Plan 2010 (NREAP) was implemented until 2020. Developments in the forestry sector followed the average of scenarios defined in the wood and biomass study (BFW 2009). Production of sawn wood, boards and paper followed resulting market conditions. In the *energy scenario* (1a) an increased allocation of fuel wood from national production was implemented. The NREAP was further developed leading to a strongly increased demand for fuel wood. In the *material use scenario* (1b) the material use of timber, especially the use of long-life harvested wood products, is assumed to be fostered through subsidies and the direct use of fuel wood is reduced. In the fourth scenario (1c) the framework conditions were similar to scenario 1b, but it was assumed that import possibilities could be extended through extending the import range (Lauri et al. 2013). In the fifth and last scenario (2) reduced timber utilization from managed forests was assumed due to policies towards increased conservation and growing carbon pools in forestry. Additional restrictions concerning logging were implemented (e.g. Natura 2000, EU biodiversity strategy), and an increase of natural forest reserves, biosphere parks and subsidies for measures was assumed.

6.4.3.3 Simulating Carbon Dynamics in Austrian Forests

For each of the defined scenarios the demand for wood from Austrian forests was estimated by FOHOW2 and transferred to CALDIS-VB v0.1. Using the Austrian NFI data from 2007 to 2009 CALDIS-VB v0.1 simulated the development of the Austrian forests by harvesting exactly the demanded amount of wood from the plots of the NFI. Economic, ecological and legal constraints were taken into account analogously to our first study on wood and biomass supply (Sect. 6.4.1). Stand carbon was estimated via volume and biomass models as well as species specific factors for wood shrinkage and dry matter. Soil carbon dynamics were simulated by

YASSO 07 soil carbon model. The required information on growing stock, litterfall, natural mortality, and harvesting residues was provided by CALDIS-VB v0.1.

The results of this simulation study revealed that the annual amount of utilized timber varied between 18.8 and 27.6 million m^3. The greatest values were simulated for the material use scenario (1b) and the least values for scenario 2, in which reduced timer utilization was envisaged. The simulations also revealed that under the assumptions of scenarios 1a, 1b, and 1c, Austrian forests serve as a carbon sink until the year 2040 and then turn into a source of carbon. In scenario 2, however, Austrian forests serve as carbon sink until the year 2100.

6.5 Discussion

Reliable countrywide estimates of wood and biomass supply require: (1) a large and representative data set, and (2) growth simulators which are able to handle these data.

The Austrian NFI is certainly such a data set that adequately represents the status of the Austrian forest resources. Seven sample-based field assessments have been completed within the time period from 1961 until 2007. The NFI covers all forest land within the Austrian territory and represents a solid data base for outlook studies and the development of forest growth models.

In the meantime, Austria has also gained a long history in the field of individual-tree growth modeling. Filla (1981) and Sterba (1983) were among the first researchers to introduce this modeling approach in Central Europe. In the 1990s, the individual-tree based growth simulators WASIM (Eckmüllner 1990), MOSES (Hasenauer 2000), and PROGNAUS (Sterba et al. 1995; Ledermann 2006) were released. The most recent version of such an individual-tree based growth simulator is CALDIS. However, forest growth models which rely on different modeling concepts have also been developed within the last two decades. BIOME-BGC is a mechanistic model that describes the physiology and biogeochemistry of forest ecosystem cycles (Pietsch et al. 2005). PICUS belongs to the group of gap models and allows spatially explicit predictions (Lexer and Hönninger 2001). The most recent version of PICUS is a hybrid model that combines the concepts of a gap model and a BGC-model.

Although many forest growth models (simulators) are available in Austria, only PROGNAUS and CALDIS have successfully been used for outlook studies on wood and biomass supply based on NFI data. The main reason is that both simulators have been developed from Austrian NFI data. This is an important issue because sampling properties of the model variables have an effect on the estimation of the model coefficients (Jaakkola 1967; Stage and Wykoff 1998). Consequently, using a sampling design that is different from the one used to develop a forest growth model can cause imprecise predictions (Hann and Zumrawi 1991). Another big advantage of PROGNAUS and CALDIS is that they do not require input variables other than those provided by the Austrian NFI. For example, Austrian NFI data are lacking the spatial information about the neighboring trees that are actually not included in the sample. The application of a distance-dependent growth simulator such as MOSES is therefore not straightforward. To overcome this problem, distance-dependent growth simulators often resort to artificially generated tree positions. However, the use of generated spatial information might allow the application of distance-dependent growth simulators but will certainly not provide additional information. The problem of missing input data applies to BIOME-BGC and PICUS as well. Comparing Austrian NFI data to the simulation results of BIOME-BGC, PICUS and PROGNAUS, Huber et al. (2013) pointed out that a large part of the NFI sample plots could not be used for model comparison because the detailed soil parameters necessary for BIOME-BGC and PICUS were not available. Furthermore, the application of BIOME-BGC was limited to more or less mono-specific sample plots, i.e., to plots for which a single tree species represents at least 80% of stand basal area. For 2224 Austrian NFI plots, BIOME-BGC and PICUS could handle only 51% and 59%, respectively. With PROGNAUS, however, it was possible to simulate all 2224 NFI plots. Huber et al. (2013) concluded that more complex models might produce better estimates, but their applicability is limited due to the need for detailed input data. A non-random reduction in sample size, which is perhaps driven by the availability of specific stand and/or soil parameters, could result in a non-representative sample of NFI plots. Such a non-representative sample as well as a simple reduction in sample size may have an impact on the projections, eventually leading to a loss in precision and/or accuracy. And this would be the opposite of what is expected from a more complex model.

The comparative study by Huber et al. (2013) clearly demonstrated the advantage of a forest growth simulator that has been developed from NFI data. All three outlook studies, which are presented in this book chapter, were done using Austrian NFI data and the PROGNAUS and CALDIS growth simulators. The application of these simulators was straightforward, and they could use the full information of the Austrian NFI (approximately 7000 sample plots). Thus, our experiences in estimating wood and biomass supply from NFI data and forest growth models are in full accordance with the findings by Huber et al. (2013).

References

BFW (2009) Holz- und Biomassenstudie. BFW Praxisinformation 18:24p

Bitterlich W (1948) Die Winkelzählprobe. Allgemeine Forst- und Holzwirtschaftliche Zeitung 59(1/2):4–5

Bitterlich W (1952) Die Winkelzählprobe. Forstwissenschaftliches Centralblatt 71(7/8):215–225

Carlowitz HC (1713) Sylvicultura oeconomica. Braun, Leipzig

Eckmüllner O (1990) Spielerisch durchforsten lernen: WASIM - Wachstumssimulation. Holzwirtschaft, Wald & Holz Rundschau 46:24–25

Eckmüllner O (2006) Allometric relations to estimate needle and branch mass of Norway spruce and Scots pine in Austria. Austrian J For Sci 123:7–16

Eckmüllner O, Schedl P, Sterba H (2007) Neue Schaftkurven für die Hauptbaumarten Österreichs und deren Ausformung in marktkonforme Sortimente. Austrian J For Sci 124:215–236

Ek AR, Dudek A (1980) Development of individual-tree based stand growth simulators: progress and applications. University of Minnesota, College of Forestry, Staff Paper 20, 25p

Feistmantel R (1854) Allgemeine Waldbestandestafeln oder übersichtliche Darstellung der vorzüglichsten Wachsthums- und Holzertrags-Verhältnisse der Forste.Wilhelm Braumüller, Wien, 110p

Filla K (1981) Die Parametrisierung von Einzelstammwachstumsmodellen über die bei Forstinventuren erhobenen Daten. Dissertation, University of Natural Resources and Life Sciences (BOKU), Vienna

Gabler K, Schadauer K (2007) Some approaches and designs of sample-based National Forest Inventories. Austrian J For Sci 124:105–133

Gabler K, Schadauer K (2008) Methods of the Austrian Forest Inventory 2000/02 – origins, approaches, design, sampling, data models, evaluation and calculation of standard error. BFW-Berichte 142:121p

Gschwantner T, Schadauer K (2006) Branch biomass functions for broadleaved tree species in Austria. Austrian J For Sci 123:17–34

Gschwantner T, Gabler K, Schadauer K, Weiss P (2010a) National Forest Inventory reports – Austria. In: Tomppo E, Gschwantner T, Lawrence M, McRoberts RE (eds) National Forest Inventories – pathways for common reporting. Springer, Berlin, pp 57–71

Gschwantner T, Kindermann G, Ledermann T (2010b) Weiterentwicklung des Wachstumssimulators PROGNAUS durch Einbindung klimarelevanter Parameter. In: Neumann M (ed) Auswirkungen des Klimawandels auf Österreichs Wälder – Entwicklung und vergleichende Evaluierung unterschiedlicher Prognosemodelle. Forschungsbericht A760631, Klima- und Energiefonds, Vienna, 150p

Guttenberg A (1896) Die Aufstellung von Holzmassen- und Geldertragstafeln auf Grundlage von Stammanalysen. Österreichische Vierteljahresschrift für Forstwesen. Österreichische Reichsforstvereine, pp 203–237 and 319–345

Guttenberg A (1915) Wachstum und Ertrag der Fichte im Hochgebirge. Deutike, Wien/Leipzig, 153p

Hann DW, Zumrawi AA (1991) Growth model predictions as affected by alternative sampling-unit designs. For Sci 37:1641–1655

Hasenauer H (2000) Die simultanen Eigenschaften von Waldwachstumsmodellen. Blackwell Wissenschafts-Verlag, Berlin/Wien, 131p

Hasenauer H, Monserud RA (1996) A crown ratio model for Austrian forests. For Ecol Manag 84:49–60

Huber MO, Eastaugh CS, Gschwantner T et al (2013) Comparing simulations of three conceptually different forest models with National Forest Inventory data. Environ Model Softw 40:88–97

IPCC (2013) Summary for policymakers. In: Stocker TF, Qin D, Plattner GK et al (eds) Climate change 2013: the physical science basis. Contribution of working Group I to the fifth assessment report of the intergovernmental panel on climate change. Cambridge University Press, Cambridge/New York

Jaakkola S (1967) On the use of variable size plots for increment research. In: DVFFA (ed) Proceedings of XIV. IUFRO-Kongress, Band VI, München, pp 371–378

Kindermann G (2010a) Weiterentwicklung eines Kreisflächenzuwachsmodells – Refining a basal area increment model. In: Nagel J (ed) Beiträge zur Jahrestagung 2010 in Körbecke/Möhnesee. Deutscher Verband Forstlicher Forschungsanstalten – Sektion Ertragskunde, Göttingen, pp. 82–95. ISSN: 1432–2609

Kindermann G (2010b) Eine klimasensitive Weiterentwicklung des Kreisflächenzuwachsmodells aus PROGNAUS – a climate sensitive refining of the basal area increment model in PROGNAUS. Austrian J For Sci 127:147–178

Kooperationsplattform Forst Holz Papier (2006) Österreichische Holzhandelsusancen 2006. Service-GmbH der Wirtschaftskammer Österreich, 310p

Lauri P, Kallio M, Schneider U (2013) The future development of the use of wood in Russia and its potential impacts on the EU forest sector. Scand J For Res 28:291–302

Ledermann T (2002) Ein Einwuchsmodell aus den Daten der Österreichischen Waldinventur 1981–1996. Austrian J For Sci 119:40–77

Ledermann T (2006) Description of PrognAus for Windows 2.2. In: Hasenauer H (ed) Sustainable forest management – growth models for Europe. Springer, Berlin/Heidelberg/New York, pp 71–78

Ledermann T (2011) A non-linear model to predict crown recession of Norway spruce (*Picea abies* [L.] Karst.) in Austria. Eur J Forest Res 130:521–553. doi:10.1007/s10342-010-0440-x

Ledermann T, Neumann M (2006) Biomass equations from data of old long-term experimental plots. Austrian J For Sci 123:47–64

Ledermann T, Schadauer K (2015) Treibhausgasbilanz im österreichischen Wald. Österreichische Forstzeitung 126(3):4–5

Ledermann T, Jandl R, Veselinovic B et al (2010) Ein Ansatz zur Abschätzung der sturminduzierten Ausfallwahrscheinlichkeit von Fichten- und Buchenbeständen des österreichischen Alpenvorlandes. In: Forstwissenschaften – Grundlage nachhaltiger Waldbewirtschaftung. Beiträge zur Forstwissenschaftlichen Tagung 2010 in Göttingen. Cuvillier Verlag, Göttingen, p 61

Lexer MJ, Hönninger K (2001) A modified 3D-patch model for spatially explicit simulation of vegetation composition in heterogeneous landscapes. For Ecol Manag 144:43–65

Liski J, Palosuo T, Peltoniemi M, Sievänen R (2005) Carbon and decomposition model Yasso for forest soils. Ecol Model 189:168–182

Marschall J (1975) Hilfstafeln für die Forsteinrichtung. Österreichischer Agrarverlag, Wien

Monserud RA, Sterba H (1996) A basal area increment model for individual trees growing in even- and uneven-aged forest stands in Austria. For Ecol Manag 80:57–80

Monserud RA, Sterba H (1999) Modeling individual tree mortality for Austrian forest species. For Ecol Manag 113:109–123

Nachtmann G (2006) Height increment models for individual trees in Austria depending on site and competition. Austrian J For Sci 123:199–222

Neumann M, Schadauer K (1995) Die Entwicklung des Zuwachses in Österreich anhand von Bohrkernanalysen. Allgemeine Forst- und Jagdzeitung 166(12):230–234

Neumann M, Schadauer K (2007) Holz- und Biomasseaufkommensstudie für Österreich – Hintergründe, Ausgangssituation und methodische Ansätze. In: Nagel J (ed) Beiträge zur Jahrestagung 2007 in Alsfeld. Deutscher Verband Forstlicher Forschungsanstalten – Sektion Ertragskunde, Göttingen, pp 193–199. ISSN: 1432–2609

Offenthaler I, Hochbichler E (2006) Estimation of root biomass of Austrian forest tree species. Austrian J For Sci 123:65–86

Paulsen JC (1795) Kurze praktische Anleitung zum Forstwesen. Verfasst von einem Forstmanne, Detmold, 152p

Pietsch SA, Hasenauer H, Thornton PE (2005) BGC-model parameters for tree species growing in central European forests. For Ecol Manag 211:264–295

Pollanschütz J (1974) Formzahlfunktionen der Hauptbaumarten Österreichs. Allgem Forstzeitung 85(12):341–343

Pretzsch H (2001) Modellierung des Waldwachstums. Blackwell Wissenschafts-Verlag, Berlin/Wien, 341p

Reineke LH (1933) Perfecting a stand density index for even-aged forests. J Agric Res 46:627–638

Rubatscher D, Munk K, Stöhr D et al (2006) Biomass expansion functions for *Larix decidua*: a contribution to the estimation of forest carbon stocks. Austrian J For Sci 123:87–101

Schiffel A (1904) Wuchsgesetze normaler Fichtenbestände. Mitteilungen aus dem forstlichen Versuchswesen Österreichs. K.K. forstliche Versuchsanstalt in Mariabrunn, Wilhelm Frick/Wien, 106p

Schwappach A (1903) Leitfaden der Holzmesskunde. Julius Springer, Berlin, p 173

Schwarzbauer P, Stern T (2010) Energy vs. material: economic impacts of a "wood-for-energy scenario" on the forest-based sector in Austria – a simulation approach. Forest Policy Econ 12:31–38. doi:10.1016/j.forpol.2009.09.004

Söderbergh I, Ledermann T (2003) Algorithms for simulating thinning and harvesting in five European individual-tree growth simulators: a review. Comput Electron Agric 39:115–140
Stage AR (1973) Prognosis model for stand development. USDA For. Serv. Gen. Tech. Rep. INT-137, Intermountain Research Station, Ogden, Utah, 32p
Stage AR, Wykoff WR (1998) Adapting distance-independent forest growth models to represent spatial variability: effects of sampling design on model coefficients. For Sci 44:224–238
Sterba H (1983) Single Stem Models from Inventory Data with Temporary Plots. In: Mitteilungen der Forstlichen Bundesversuchsanstalt Wien, Meeting on Forest Growth Modelling and Simulation, October 1982, Band 147, Österreichischer Agrarverlag, Wien, pp 87–101
Sterba H (1991) Forstliche Ertragslehre, Vorlesungsunterlagen. Universität für Bodenkultur, Wien, 160p
Sterba H (1996) Holzaufkommen in Österreich. Die Sägeindustrie Österreichs (CD und Begleittext). Wirtschaftskammer Österreich, Wien
Sterba H (1999) PROGNAUS – ein Validierungsbeispiel. In: Kenk G (ed) Beiträge zur Jahrestagung der Sektion Ertragskunde d. DVFFA, pp 24–32
Sterba H, Monserud RA (1996) Validation of the Single Tree Stand Growth Simulator PROGNAUS with Permanent Plot Data. In: Köhl M, Gertner GZ (eds) Caring for the forest: research in a changing world. Statistics, mathematics and computers. Proceedings of the meeting of IUFRO S. 4. 11-00 held at IUFRO XX World Congress, 6–12 Aug 1995, Tampere Finland. Birmensdorf, Swiss Federal Institute for Forest Snow and Landscape Research (WSL/FNP), pp 36–49
Sterba H, Monserud RA (1997) Applicability of the forest stand growth simulator PROGNAUS for the Austrian part of the Bohemian Massif. Ecol Model 98:23–34
Sterba H, Moser M, Monserud RA (1995) PROGNAUS – Ein Waldwachstumssimulator für Rein- und Mischbestände. Österreichische Forstzeitung 106(5):19–20
Sterba H, Korol N, Rössler G (2001) Ein Ansatz zur Evaluierung eines Einzelbaumwachstumssimulators für Fichtenreinbestände. Forstwissenschaftliches Centralblatt 120:406–421
UN (2012) Report of the conference of the parties serving as the meeting of the Parties to the Kyoto Protocol on its seventh session, held in Durban from 28 Nov to 11 Dec 2011, Addendum
Wykoff WR (1990) A basal area increment model for individual conifers in the northern Rocky Mountains. For Sci 36:1077–1104

Chapter 7
Bulgaria

Nickola Stoyanov, Maria Stoyanova, Angel Ferezliev, and Radoslav Milchev

7.1 Introduction

7.1.1 Bulgarian Forest Resources

In the period 1990–2015, the total forest area in Bulgaria has increased by 19,800 ha per year, to 3.83 million ha in 2015 (35.2% of total land area) (Forest Europe 2015). However, only 57.8% (2.2 million ha) is considered to be forest area available for wood supply. In 1997, the process of restoring forest areas to their former owners began. Consequently, the share owned by the State decreased from 84.9% in 2000 to 74.1% in 2010. About 12.2% is owned by municipalities and 10.2% by private owners. Growing stock increased from 122 m^3/ha in 1990 to 182 m^3/ha in 2015, with the total growing stock in 2015 of 699 million m^3. The total annual increment amounts to 14.4 million m^3 (3.7 m^3/ha per year), while fellings are 7.0 million m^3 (1.8 m^3/ha per year, 49% of annual increment) (FOREST EUROPE 2015).

Nearly half of the forest in Bulgaria consists of coppice (Fig. 7.1). Oak is the main species, but there are considerable shares of other species such as beech, hornbeam and Turkey oak. About 23% of the forest consists of high-stem broadleaved forest, mainly beech and oak. Conifers are present on about 30% of the area, of which half is Scots pine. Norway spruce and Black pine are the most important other conifers.

Pure forests in which a single tree species dominates occupy 45.6% of the area of all forests. Forests with 2-3 tree species constitute 9.9%, and mixed forests with

N. Stoyanov (✉) • R. Milchev
University of Forestry, Sofia, Bulgaria
e-mail: nickst@ltu.bg; stoyanovnick@abv.bg; the_mentor@mail.bg

M. Stoyanova • A. Ferezliev
Forest Research Institute – BAS, Sofia, Bulgaria
e-mail: mst@mail.orbitel.bg; obig@abv.bg

S. Barreiro et al. (eds.), *Forest Inventory-based Projection Systems for Wood and Biomass Availability*, Managing Forest Ecosystems 29,
DOI 10.1007/978-3-319-56201-8_7

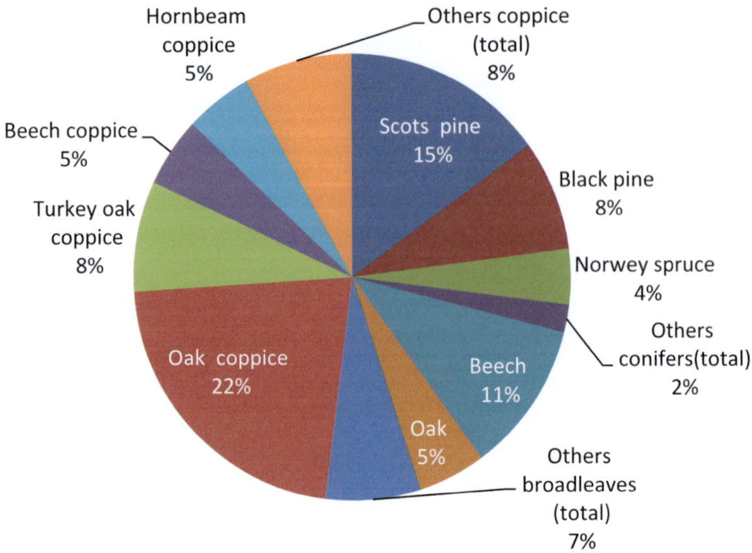

Fig. 7.1 Species distribution of Bulgarian forests (Kostov and Rafailova 2009)

4-5 tree species represent 44.5%. Compared to 2005, the area of pure forests increased by 182,949 ha and reduced the area of mixed forests with 4-5 tree species by 164,299 ha. There was no significant change in the area of forests with 2–3 tree species.

Statistical harvesting data reported by Eurostat for Bulgaria during the period 2006–2011 is as follows. The share of industrial wood over the years varied between 48 and 55% of the volume of harvested round wood. Harvested firewood amounted to 2669 thousand m^3 on average for the period (47%). The amount of firewood sold for the period 2005–2011 was less impacted by the economic recession than by fluctuations in the amounts of industrial wood sold in the same period. The volume of harvested round wood in 2011 was 531 thousand m^3 more than reported in 2005, and the volumes of industrial wood and firewood were 327 thousand m^3 and 204 thousand m^3 greater, respectively, than reported in 2005.

During the period 2006–2011, the percentage distribution of harvested wood by categories – large-sized, medium, small and firewood was, respectively 20%, 18%, 4%, and 57%. For softwood, the percentages were 37%, 32%, 7%, and 25%, and for hardwood the percentages were 11%, 11%, 2%, and 75%. Firewood from coniferous and deciduous tree species forms about 57% of the total amount of harvested timber. Deciduous extractions were mainly for firewood and represent 75% of all deciduous wood, while extractions from coniferous species produced mainly medium- and large-sized timber.

7.1.2 *Forest Inventory History*

Bulgaria has more than 110 years of history of forest inventory and elaboration of Forest Management Plans (FMPs). The first State Service on Measuring and Organization of Bulgarian Forests was created in 1901. In the period 1901–1919, it inventoried 225 thousand ha and elaborated the first three FMPs. By 1944, only one third of the Bulgarian forests were managed according to an FMP, although FMPs are required for all forests by the forest law. In the period 1950–1954, the staff and the number of forest inventory teams were increased considerably. In these four years, they inventoried forests and elaborated FMPs covering 2890 thousand ha. Subsequently, the entire forest was inventoried, and all forestry enterprises were working according to FMPs. After 1997, private companies began implementing forest inventories, elaborating FMPs, and competing with state companies. At present, all forests in Bulgaria from all owners have FMPs which are renewed every 10 years. Thus, in every forest management unit we have up-to-date information about forest stands and management plans for 10-year periods.

Apart from the inventories connected to the FMPs, Bulgaria experimented with a statistical inventory design in the 1970s. This inventory featured a stratified sampling design with the number of plots based on the expected coefficient of variation. Although the results produced the desired accuracy, it was only applied in two regions (Belyakov 1971, 1977; Krastanov 1977).

7.2 The Standwise Forest Inventory

Sub-compartments are the smallest territorial units of forests and remain relatively constant because they are the basis for forest inventory and management. The main variables that define a sub-compartment are species and management system.

During the inventory, forests are described using quantitative and qualitative variables that describe the set of activities and operations within the compartments and sub-compartments. The range of variables and evaluation indicators is mandatory for all forests, as determined according to the needs of forest management practices and the requirements of international agreements and documents, to which Bulgaria is a party.

During field data collection, information is collected and analysed to ascertain the results of forestry activities in the measured area since the last forest measurement. Forestry and harvesting activities planned for the next period are based on the results and findings of current economic activity.

Description of attributes assessed:

1. General Information: number of compartment, number of sub-compartments, rotation age felling, type of forest;

2. Characteristics of the trees: tree species composition, tree species crown cover area, tree species age, stand density index, average diameter, average height, productivity class, stock, use (removals);
3. Description of the forest cover: origin, form, structure, condition, type of mixture, trees above the forest, Landscape Assessment;
4. Description of the habitat: exposure, slope, altitude, relief, bedrock, soil nature, habitat, optimal future species composition, litter, coverage of grass, shrubs, technical valuable medicinal plants;
5. Reforestation:- composition, age, height and percentage of cover;
6. Health Record;
7. Biodiversity;
8. Transportation conditions: distances in meters to forest roads, road category, distance to the closest settlement and category of the cutting area, and other variables specified of whole deciduous wood in the methodology for economic evaluation of forest lands. Distances are measured from the base map of forestry for planned activities: cutting, thinning, schedule for reforestation, and land preparation
9. Other information.

Sample tree volume is predicted with general volume models of whole deciduous wood using diameter at breast height (*dbh*), form factor (F), and height (h) as predictor variables. Volume is expressed in m^3.

Data and results from the FMPs and Programmes are used for national and regional forest statistics and for international reporting (Global FRA, Forest Europe, Natura 2000 Network directive reporting, etc.). The data are also used for the management of forest units, for planning forest and harvesting activities, for estimation of future production possibilities of timber and forest bioenergy, and for research.

7.3 Projecting Woody Biomass Availability in Bulgaria

7.3.1 History of Projections

The first experiences in analyzing the state of the forest and projecting their long-term development in Bulgaria date from 1965 (Biolchev et al. 1965). This first projection was for the period 1965–2000. In the 1970s, two more studies followed (Iliev et al. 1973; Kostov et al. 1976). The main source of information was the FMPs. A study by Vachovski (1980) analysed developments in the forest sector from 1950 to 1975 and gave a projection until 2000. Dakov (1987) updated this projection for the period 1990–2040. This projection is the last study under the old regime and aimed to assist forest policy development. A new projection for the demand and production of wood products and wood harvest for the period 2005–2025 was published in 2007 (Vachovski 2007). The last projection for the dynamics of forest resources in Bulgaria was elaborated in the period 2005–2008 and financed by the National Board of Forests (Kostov and Rafailova 2009). This projection used the European Forest Information Scenario Model (EFISCEN).

7.3.2 Models in Use

7.3.2.1 Bulgarian Model for the Dynamics of Forest Resources Proposed by the University of Forestry in Sofia, Bulgaria in 1989

For the model developed by the University of Forestry, Sofia, the assessment units are tree species (by forest types) with the characteristics of their age classes. The model for the dynamics of forest resources is an automated procedure for processing diverse forest information for purposes of characterizing its future status and, depending on its current state, the natural processes occurring in it and the politics of its future management. The essence of the model is to identify the main characteristics of forests – areas, densities and stocks of tree species. The value of the harvesting is calculated as a percentage of the available table stock at a given age and forest appraisal index, reduced by the value of the current density. To determine the state of the forest at the beginning of the forecast, available data are used in the forest reporting forms.

The projection results include changes of values of the large number of indicators during the forecast period, variously characterizing forest on three levels of integration of information: by tree species and age classes; by tree species; and by indexes.

Kostov (1993a) presents the results of considered dynamics and gave the model its name: FRAM (Forest Resources Assessment Model). The FRAM model has multiple advantages: it directly uses management data; it has a simple structure and is divided into sub-models; it is open for changes in the management regimes; and there is no need for a powerful computer application. The disadvantages of the model are that it uses only data from FMPs which are not uniform over years; it does not use sub-models to accommodate the different sensitivities of species to environmental changes; it does not include job opportunities at the change of ownership; and most importantly, it is not comparable with other European studies.

The matrix model that was developed for the projection of forest resources in Sweden (Sallnäs 1990) was applied to Bulgarian forests and resulted in two scenarios for the period 1980–2040 (Kostov 1993b). The state of the forest is represented as a distribution of forest area over a matrix defined by age and volume classes. Area transitions between cells represent aging (moving to a higher age class), growth (moving to a higher volume class), thinning (moving to a lower volume class) and final felling (moving area to a separate class with zero age and volume). This model forms the core of the EFISCEN simulator (Nabuurs et al. 2006) that was applied by Kostov and Rafailova (2009).

7.3.3 Main Results from the Projections for Dynamics of Forest Resources for the Period 2005–2050 in Bulgaria

The latest projection for Bulgarian forests was done by Kostov and Rafailova (2009) using the EFISCEN model. The model was initialized using Bulgarian official forest statistics for 2005 which were based on inventory data from the latest FMP for each forest enterprise. Although these data are not fully harmonized among the forest enterprises, they are well-suited for use with the EFISCEN model. All forests are represented as even-aged and as distributions by tree species and owners.

Four scenarios were elaborated for the period 2005–2050: reference scenario, scenario for maximum sustainable use, optimistic scenario and pessimistic scenario. All scenarios were elaborated on the basis of points of view for possible development of forest sector by accepted international and national experts.

Based on the projections, the potential use of wood from coniferous forests can be increased from the present 4.7 million m^3 to 8 million m^3 per year, mainly by an increase in thinning activity. There is no potential for increase of final felling in the case of Norway spruce and European fir. Most of the unrealized potential is therefore in Scots pine stands and other coniferous plantations. The potential use of wood from high-stem broadleaved forests may be increased from the present 2.4 million m^3 of standing volume to 4.2 million m^3. The increase is both in thinnings and in final felling. In the latter, however, there is almost no resource to increase the harvest from felling in the case of the main species such as oaks and beech.

The potential for increase of harvest of wood from coppice forests is significant, regardless of the general tendency of decreases in their areas because of conversion to high-stem stands. The potential harvest increase is mainly based on a felling increase of about 1 million m^3 per year.

As a whole, it is possible to increase the timber harvest in the country from the present 7.2 million m^3 to about 11 million m^3 per year (Fig. 7.2). Unrealized potential is mainly in thinnings from most species and final felling in coppice forests. These are likely to yield lower-quality assortments and biomass for energy purposes. The prospect of increasing the volume of wood harvest for sawing, veneer and veneered plywood is low. The main species, such as Norway spruce, European fir, high-stem oaks and beech, have no resource to increase the harvest of large-sized round wood from old (mature) stands. The average annual increment will remain around 4.5 m^3/ha per year in the moderate scenarios, but range from 3.6 m^3/ha per year for the pessimistic scenario to 5.3 m^3/ha per year for the optimistic scenario (Fig. 7.2). In the basic scenario, the average growing stock will increase further to 243 m^3/ha in 2050. The pessimistic scenario projects a decrease to 135 m^3/ha due to lower increment rates, while the optimistic scenario ends at 199 m^3/ha, mainly due a higher felling level early in the projection period.

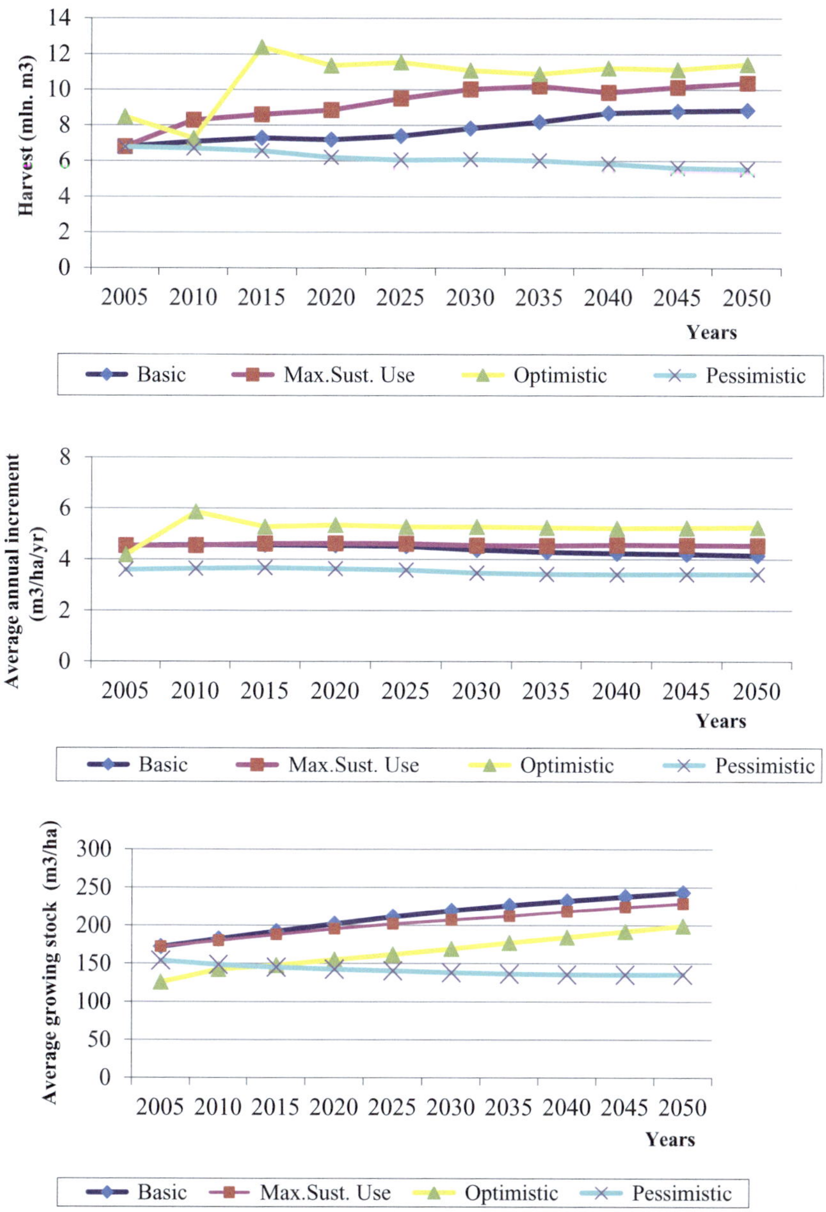

Fig. 7.2 Development of annual harvest, increment and growing stock in the scenarios

7.4 Discussion and Conclusions

Bulgaria began preparing FMPs by 1901. Since 1955, all forests have been covered by FMPs and their underlying stand-wise inventories. FMPs focus mainly on timber resources and do not provide sufficient information for multifunctional forest management planning and evaluation. Other disadvantages of the FMP system are that the information is only renewed every 10 years; that some forest enterprises are late in updating their FMPs; and that there are differences in reporting among owners. An NFI based on a statistical design is believed to solve these problems. Policy processes and documents also stress the need to begin developing a Bulgarian NFI, but so far it has not been implemented. Inventory research institutes and units, as well as specialized design organizations, should be involved in this effort. Experiments conducted in the 1970s show the feasibility of a statistical approach.

From 1965 until now, seven projections for the development of forestry and forest resources were elaborated. Earlier projections were directly based on the FMPs using different methods. In the last projection the EFISCEN model was applied. It showed that the positive development of Bulgarian forests as reported in the period 2000–2010 can continue into the future. It also showed that there is considerable potential to increase the harvest level, mainly in thinning operations and in coppice stands. A continuous harvest of 11–12 million m^3 per year until 2050 should be possible, which is greater than previous estimates. However, to reach this level, the development of regional programs for evaluation, rational utilization, and use of forest resources is a necessary step to improve the efficiency of forest policy in the country.

References

Belyakov P (1971) Edroploshna inventarizacia na gorskite resursi v gorite na Strandja planina [Large-scale inventory of forest resources in the forests of Strandzha Mountain]. Gorskostopanska nauka 3:45–63

Belyakov P (1977) Varhu deistvitelnata tochnost na edroploshnata inventarizacia v gorite na Strandja [On the actual accuracy of the Large-scale inventory in the forests of Strandzha]. Gorskostopanska nauka:55–74

Biolchev A, Vlasev V, Vachovski H, et al (1965) Sastoyanie na gorite u nas I perspektivi za povishavane na tiahnata proizvoditelnost. V: Potrebnosti na narodnoto stopanstvo ot darvesina I vazmojnosti za tiahnoto zadovoliavane [Status of forests in the country and prospects for increasing their productivity. In: Needs of the national economy from timber and possibilities for their satisfaction], Sofia, pp 5–21

Dakov M (1987) Programa za uskoreno razshireno vazproizvodstvo i nai-efektivno kompleksno izpolzvane na gorskite resursi na N. R. Balgaria za perioda 1990 – 2040 g., Otchet na tema po NIS pri VLTI s rakovoditel akad. Mako Dakov [Programme for accelerated expanded reproduction and most effective integrated use of forest resources of PR Bulgaria for the period 1990–2040, Report on the theme at the Scientific Research Sector at the University of Forestry, headed by Acad. Mako Dakov]

FOREST EUROPE (2015) State of Europe's Forests 2015

Iliev A, Andonov A, Dimitrov St et al (1973) Prognoza za dinamikata na gorskite resursi v Narodna Republika Balgaria za perioda 1971–2000 g [Prognosis for dynamics of forest resources in People's Republic of Bulgaria 1971–2000)], Sofia, Zemizdat

Kostov G (1993a) The Bulgarian Forest Resources Assessment Model – some base data and results, SUAS, Department of Operational Efficiency, College of Forestry, Garpenberg, Sweden

Kostov G (1993b) Management of Forest Stands in Bulgaria (A Base Data for TAM), SUAS, Department of Operational Efficiency, College of Forestry, Garpenberg, Sweden

Kostov G, Rafailova E (2009) Dinamika na gorskite resursi v Balgaria pri razlichni rejimi na stopanisvane [Dynamics of forest resources in Bulgaria at different regimes of management], Sofia, Avangard Prima

Kostov P, Zhelev I, Belyakov P et al (1976) Sastoyanie na gorskite resursi na NRB i saobrajenia za tiahnoto kompleksno izpolzwane [Status of forest resources in Bulgaria and considerations for their complex use] Zemizdat 11

Krastanov K (1977) Vazmojnosti za ustanoviavane na dyrvesnia zapas v izborni gori s pomoshta na matematiko- staticheski metod [Establishment of the tree stock of elective forests with the help of mathematical and statistical method]. Gorskostopanska nauka 4:21–33

Nabuurs GJ, Van Brusselen J, Pussinen A, Schelhaas MJ (2006) Future harvesting pressure on European forests. Eur J For Res 126:391–400

Sallnäs O (1990) A Matrix Growth Model of the Swedish Forest. SUAS, Faculty of Forestry, Uppsala, Studia Forestalia Suedica, № 183

Vachovski Chr (1980) Tendencii I perspektivi za potreblenie I proizvodstvo na darvesina prez perioda 1950–2000 g [Tendencies and perspectives for use and production of wood in the period 1950–2000], Sofia, Zemizdat

Vachovski Chr (2007) Prognoza za potreblnie, proizvodstvo I dobiv na darvesina I darvesni produkti prez perioda 2005–2025 godina [Prognosis for use, production and harvesting of wood and wood products in the period 2005–2025], Sofia, GTZ

Chapter 8
Canada

Juha M. Metsaranta, Carolyn E. Smyth, and Werner A. Kurz

8.1 Introduction

Forecasting the quantity and distribution of forest biomass has been of interest in Canada for at least the last 80 years (Jenkins and Guernsey 1936; Nautiyal 1979; Hall and Richardson 2001; Paré et al. 2011). This includes questions about supply (how much, where, and in what form), discussions of trade-offs with other social, economic, and environmental values, including other potential uses wood, and issues of economic and environmental feasibility of biomass use. Existing forecasts have quantified live biomass, residues from forest management (Cambero et al. 2015), underutilized species, trees killed by disturbance (Dymond et al. 2010a; Barette et al. 2013), bioenergy plantations (Allen et al. 2013; Amichev et al. 2014), or several possible sources (Bedarul Alam et al. 2012; Ter-Mikaelian et al. 2015). These studies have often evaluated the climate change mitigation potential of bioenergy and the increased use of long-lived wood products to substitute carbon-intensive alternatives (Lemprière et al. 2013; Smyth et al. 2014).

Annual State of the Forest reports produced since 1990 by the Canadian Forest Service of Natural Resources Canada (e.g. Natural Resources Canada 2015) describe Canada's forest sector, including species and forest type, the impact of natural disturbances, and the importance of the forest sector to Canada's national economy. Canada's provinces and territories have constitutional responsibility for

J.M. Metsaranta (✉)
Natural Resources Canada, Canadian Forest Service, Northern Forestry Centre, Edmonton, AB, Canada
e-mail: juha.metsaranta@canada.ca

C.E. Smyth • W.A. Kurz
Natural Resources Canada, Canadian Forest Service, Pacific Forestry Centre, Victoria, BC, Canada
e-mail: carolyn.smyth@canada.ca; werner.kurz@canada.ca

S. Barreiro et al. (eds.), *Forest Inventory-based Projection Systems for Wood and Biomass Availability*, Managing Forest Ecosystems 29,
DOI 10.1007/978-3-319-56201-8_8

forest management, and most (94%) forest in Canada is on publically owned lands. Jurisdictions develop legislation, regulations, policies, and practices to support their forest management obligations. Forest inventory data (Leckie and Gillis 1995), forest legislation (National Council for Air and Stream Improvement 2014), and harvest policy (Roach and Berch 2014) therefore vary across Canada. National-level forest monitoring involves collaboration between the federal government and Canada's provinces and territories. Canada's National Forest Inventory (NFI, Stinson et al. 2016) and Canada's National Forest Carbon Monitoring Accounting and Reporting System (NFCMARS, Kurz and Apps 2006; Kurz et al. 2013; Stinson et al. 2011) are examples of programs that provide the best available strategic data across all jurisdictions in Canada and can be used to conduct analyses such as forecasting future woody biomass availability. The forest inventories used by NFCMARS are those maintained by Canadian provinces and territories rather than the NFI. The NFI is a statistical survey program (Stinson et al. 2016), whereas the provincial and territorial inventories provide complete coverage of the forests managed for timber supply. These provincial and territorial inventories are used in combination with growth and yield data by the provinces and territories to forecast future timber supply and determine Allowable Annual Cut (AAC) levels. The NFCMARS uses these same data, already prepared for use in timber supply modelling by the provinces and territories, as its principle inputs, standardized as far as possible for differences in the design of forest inventories in each jurisdiction. This strong linkage between provincial and territorial timber supply modelling and national forest C modelling is advantageous because it creates opportunities to link tools and assure consistency between management-unit scale and national-scale modelling and analysis.

Methods available for projecting future stocks of woody biomass are generally already well described in the literature. A unique consideration in Canada is that forests face increasing risk from natural disturbance by wildfire, insects, and disease (Kurz et al. 2008a; Balshi et al. 2009; Weed et al. 2013) and may be experiencing changes in production and mortality rates due to CO_2 fertilization, nitrogen deposition, and drought (Girardin et al. 2011; Michaelian et al. 2011; Hember et al. 2012). Increases in forest productivity are occurring in some parts of Canada (Hember et al. 2012), but not consistently (Girardin et al. 2011) and are unlikely to be large enough to offset expected losses from increases in disturbances (Kurz et al. 2008b; Metsaranta et al. 2010). Disturbance risk has not always been accounted for in existing forecasts of future biomass availability, but is known to increase the vulnerability of the future timber supply (Gauthier et al. 2015). We have therefore chosen to focus on methods for quantifying such risks, which are likely to be of interest in all countries experiencing a changing climate, even those with relatively intensive forest management regimes that have not traditionally been concerned with disturbance risk. Canada's NFCMARS is used for science, policy analysis, and reporting of C dynamics for the 230 of 348 million ha of Canada's forest included in the system (Fig. 8.1).[1] In this chapter, we briefly describe Canada's NFCMARS, a

[1] Not all of Canada's forest are included in the NFCMARS because large areas are not part of the "managed forest" according to Canada's definition of forest for reporting under the United Nations Convention on Climate Change.

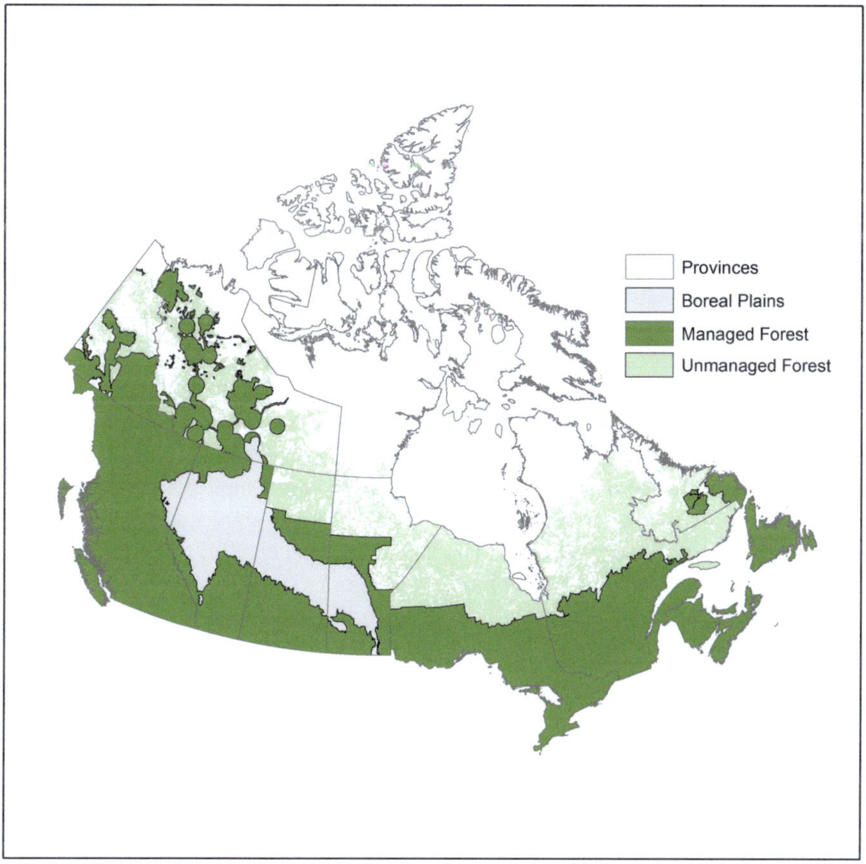

Fig. 8.1 The spatial distribution of managed and unmanaged forest lands in Canada. The 230 million hectares of managed forest land are included in Canada's National Forest Carbon Monitoring, Accounting, and Reporting System. The boreal plains ecozone, to which the examples given for forecasting future disturbances apply, is highlighted in *grey*

conceptual framework for assessing potential future disturbance risk (Kurz et al. 2008a; Kurz et al. 2008c; Dymond et al. 2010a, b) and give examples of how this framework can quantify natural disturbance risks to forest biomass availability in Canada (Metsaranta et al. 2010; Smyth et al. 2014).

8.2 Methods

8.2.1 National Forest Carbon Monitoring Accounting and Reporting System (NFCMARS)

Canada's NFCMARS is built around the Carbon Budget Model of the Canadian Forest Sector (CBM-CFS3; Kurz et al. 2009; Kull et al. 2011). It brings together the inputs required to run CBM-CFS3 nationally, including: forest inventory data, empirical wood volume yield models, statistics on forest management activities, and remote sensing to estimate area, type, and location of natural disturbance by wildfire and pests. The various forest data, information, and models used by Canada's NFCMARS and how they fit together is shown in Fig. 8.2. Briefly, the CBM-CFS3 is a stand- and landscape-level model simulating above- and below-ground biomass, detrital and soil carbon (C) dynamics. Provincial forest inventory data (age, species composition, site index, etc.) are used to describe forest characteristics and empirical yield models to predict productivity. Stand-level allometric models predict above- and below-ground biomass from wood volume (Li et al. 2003; Boudewyn et al. 2007). The model tracks 11 detrital and soil C pools and five biomass categories (foliage, merchantable wood, other wood, coarse and fine roots) by species group (softwoods and hardwoods). Disturbances cause transfer of biomass out of the ecosystem through combustion

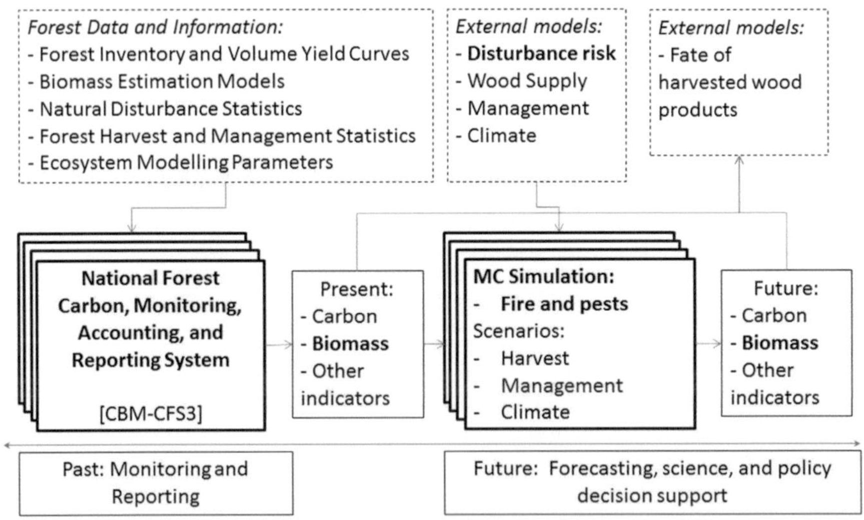

Fig. 8.2 Canada's National Forest Carbon Monitoring, Accounting and Reporting System including its core model, the Carbon Budget Model of the Canadian Forest Sector, CBM-CFS3), the data and information used as input to the system, the external models used to derive model input scenarios, including Monte Carlo simulation for assessing future disturbance risk, and the external models that use information derived from the system. The system is used for reporting and forecasting. The parts described in this paper are in *bold*

(fire) or transfers to forest products (harvest), as well as within ecosystem transfers from living to dead pools, where it subsequently decays. Impacts vary spatially by disturbance type to reflect differences in disturbance characteristics (Kurz et al. 2009). They may be stand replacing (clear-cut harvest and wild fire, which also reset the age of the affected stand to zero), cause partial mortality and growth reduction (pests), or transition stands to alternative forest type (including in the case of regeneration failure to non-forest ecosystems). Disturbance impacts are simulated using matrices defining proportional C transfers between pools and fluxes to the atmosphere or forest products in an event-driven framework. The NFCMARS operates in accounting mode for reporting (e.g. Environment Canada 2015), and in projection mode for evaluating science and policy scenarios, including future disturbances risk (Dymond et al. 2010a; Kurz et al. 2008a, c), potential climate change effects on C and GHG balances (Metsaranta et al. 2010, 2011), and the climate mitigation potential of forest management (Smyth et al. 2014). Details on model assumptions, structure, and scenario development are in the cited papers.

8.2.2 Conceptual Framework to Assess Disturbance Risk

We describe a conceptual basis for forecasting future disturbance risk, and use examples from two existing analyses to show how these forecasts can examine biomass supply indicators (Metsaranta et al. 2010; Smyth et al. 2014). Area burned by wildfire in Canada has a large inter-annual variation between high and low disturbance years (see for example the 1959–2013 annual area burned in the boreal plains ecozone of Western Canada Fig. 8.3a). Similarly, records of insect outbreak derived from surveys or other proxies show years with negligible impact interspersed with periods of pest occurrence. Future natural disturbances can only be forecast with probabilities estimated from a combination of observations from the recent past and expert judgement. Monte Carlo simulation is then used to estimate the combined effects of different disturbances on model outcomes, which provides both a point (mean or median) prediction and estimate of uncertainty (confidence intervals). The main natural disturbances types affecting forests in Canada, fires and insects, can respectively be thought of as annual or cyclical risk agents. Annual risk agents have an annual probability of occurrence that can depend on previous years or other regions, with impacts typically occurring only in the event year. Future area burned can be forecast as an independent draw from a probability distribution derived from historical area burned records (Metsaranta 2010), and future risk modelled by adjusting the parameters of these distributions (Metsaranta et al. 2010). Cyclic risk agents like insects are more complex, with periods when occurrence probability is low (outbreak interval), and periods when it is high (outbreak length), the sum of which correspond to the outbreak cycle. If the insect is not known to occur but may spread to a geographic region, further assessment is required to estimate range expansion probability. Conversely, there may also be the possibility of shifts in climatic suitability that cause range contraction. A common group of pests that reduces productivity and can cases causes mortality are the aspen defoliators, the forest tent caterpillar (*Malacosoma disstria*) and large aspen tortrix (*Choristoneura conflictana*). Several recent studies have documented patterns of aspen defoliator outbreaks in Canada (Cooke and Lorenzetti 2006; Cooke et al. 2012). Based on these historical data, we show in Fig. 8.3b–d examples of the parameters and modelling rules that would be used to forecast future outbreaks of aspen defoliators in the boreal plains ecozone of western Canada, including outbreak interval (Fig. 8.3b), length (Fig. 8.3c), and the distribution of impact types over the course of an outbreak (Fig. 8.3d). Parameterizations currently exist for seven major pests (Kull et al. 2011): two aspen defoliators, mountain pine beetle (*Dendroctonus ponderosae*), spruce budworm (*Choristoneura fumiferana,* derived from MacLean et al. 2001), jack pine budworm (*Choristoneura pinus*), hemlock looper (*Lambdina fiscellaria*), and spruce beetle (*Dendroctonus rufipennis*). A software tool used to forecast future outbreak timing and extents has been applied nationally for all both forest fire and insects (Kurz et al. 2008c), and regionally for mountain pine beetle (Kurz et al. 2008a) and spruce budworm (Dymond et al. 2010b).

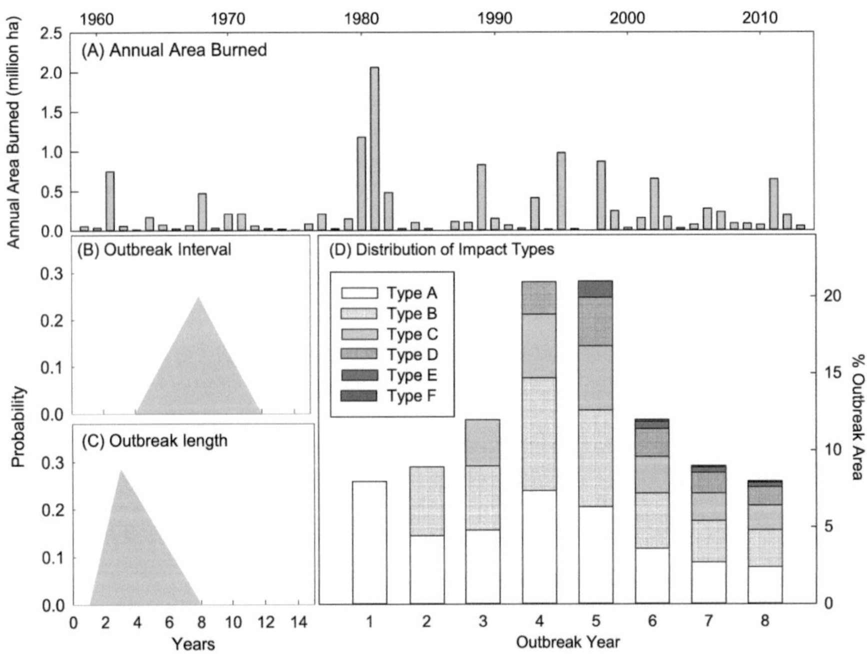

Fig. 8.3 Fire statistics and parameters for forecasting aspen defoliators in the boreal plains ecozone of western Canada, which would be the basis for assessing the risk of these particular disturbance types to future biomass supply in that region. Probability distribution for forecasting annual area burned would be developed from the 1959–2013 area burned statistics in (**a**). The modeling rules for future outbreaks of aspen defoliators include the probability distribution for outbreak intervals when defoliation does not occur (**b**), outbreak lengths when defoliation does occur (**c**), and the distribution of different severities of impact (1–6 years of consecutive defoliation, Types A through F) for outbreaks of different length (**d**)

8.2.3 *Future Harvest Projections*

Analysis of future biomass availability also requires projections of future harvest, typically estimated from wood supply analyses that, in turn, estimate amounts of wood volume or biomass that can sustainably be harvested in the long-term. In Canada, these analyses, or Annual Allowable Cut (AAC) determinations, are the responsibility of provincial management agencies and are governed by a myriad of policies and regulations (National Council for Air and Stream Improvement 2014) and are set out in forest management plans. These projections account for both the supply of wood at present and in the future, as well as potential future demand for both biomass and traditional forest products like lumber and pulp in the whole forest value and supply chains (Mansuy et al. 2015). In addition, they account for various regulatory constraints related to the requirement that forests be managed for multiple social, economic, and environmental benefits, of which biomass is only one potential value of interest. They are formulated as an optimization problem under a series of constraints. Although CBM-CFS3 does not solve such problems, it can use results from such analyses to derive scenarios to evaluate biomass supply consequences of wood harvest projections (Kull et al. 2011).

8.3 Example Results

A preliminary assessment of the risk of climate change-induced increases in annual area burned was extracted from Metsaranta et al. (2010), where a simplified version of NFCMARS was used to project the C dynamics of Canada's managed forest to 2100 under two scenarios, (1) future annual area burned is similar to late twentieth century observations; and (2) future annual area burned increases so that the mean in 2100 is either two or four times greater than historical observations, depending on region. Forecasts of total aboveground biomass (including wood, bark, branches, and foliage), and total harvested biomass (wood and bark only) were predicted for each scenario. Relative to the median result for the historical scenario, available above ground biomass in 2100 was reduced by 10–13% in the increased area burned scenario (Fig. 8.4a). However, total biomass harvested was not affected (Fig. 8.4b), indicating sufficient biomass was still available. Further considerations are discussed in Metsaranta et al. (2010). A second relevant study examined fire risk in a climate change mitigation potential context (Smyth et al. 2014). Sophisticated harvest forecasts to 2050 were derived from provincial and territorial AAC calculations, while future area burned was forecast as an average only, with an additional scenario with a 20% increase. Similar to Metsaranta et al. (2010), increased burned area had negligible impact on mitigation potential. Future pests or other risk agents and changes in future productivity and mortality rates were not thoroughly assessed in these studies. Improvements to data and models used by NFCMARS to account for these considerations are ongoing.

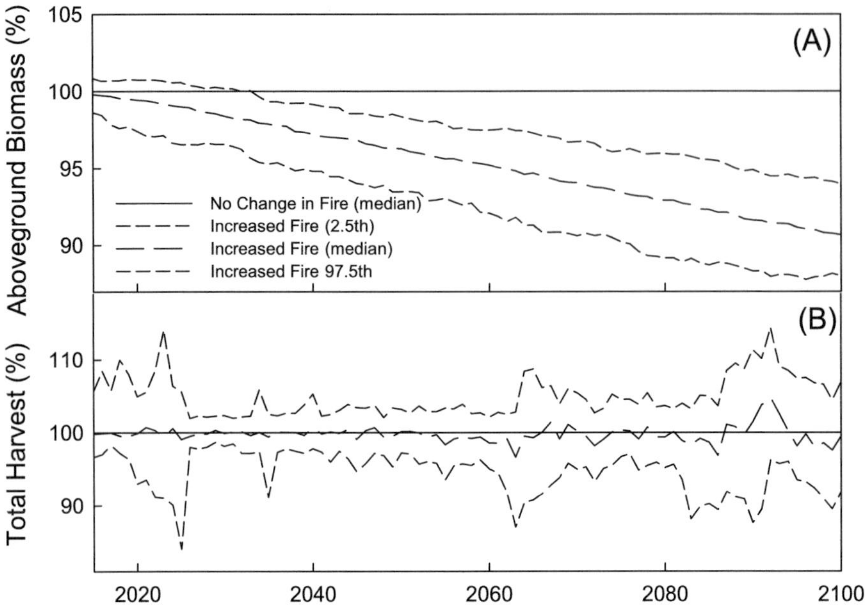

Fig. 8.4 Projections of (**a**) above ground biomass and (**b**) total biomass harvest in Canada's managed forest (2015–2100) for two future scenarios of future fire in Canada: (1) similar to late twentieth century observations; and (2) increases of two or four times depending on region. The median value for scenario (1, *solid line*) represents the annually normalized baseline value. Values for scenario (2) are calculated relative to this normalized baseline value, and represent the 97.5th, 50th, and 2.5th of 100 Monte Carlo simulations of future area burned

8.4 Discussion and Conclusions

We presented results from two existing assessments that used Canada's NFMCARS to examine predicted climate change impacts on future C dynamics (Metsaranta et al. 2010) and the climate change mitigation potential (Smyth et al. 2014) of Canada's managed forest to demonstrate that the system can be used to forecast natural disturbance risk to biomass supply. In our results, future harvests could be achieved regardless of increased future area burned, because harvest targets were applied to larger geographic regions where the model can shift harvest to other eligible stands. In reality, forest management occurs on smaller land units with less flexibility due to local constraints on required product assortments, harvest costs, and transportation distances. Thus, 10–13% reductions in available biomass would likely result in localized wood shortages, for both biomass and traditional wood products such as lumber or pulp, causing reduced harvest, a shift to salvage operations, or both. Improved efficiencies in harvest methods through 'better utilization' (that increases the stemwood utilization rate for harvest and increases the proportion of salvage harvest) could offset these product shortages by decreasing the waste in harvesting operations (Smyth et al. 2014). Fire risk increases the vulnerability of the timber supply (Gauthier et al. 2015), but this risk has not typically been well accounted for by wood supply analysis (Savage et al. 2010). Thus, we expect that the forecast increase in the severity of the fire regime should also influence the ability of forests to continue to meet society's demands for biomass.

Bioenergy is often presented as a sustainable climate change mitigation option. Increasingly, uncertainties about actual emission reductions and the timeframe over which these reductions can actually occur are challenging this notion (Bernier and Paré 2013; Klopp and Fredeen 2014; Smyth et al. 2017a; Ter-Mikaelian et al. 2015), particularly since the timing of the emissions of the alternative use of biomass (e.g. long-lived wood products) must now be accounted for in the analysis and cannot be assumed to be instantly oxidized. Increasing harvest rates for bioenergy supply have not been found to reduce GHG emissions because the forgone sequestration of the forest cannot be balanced by the forest regrowth or from substituted fossil fuel emissions, thereby resulting in a net increase in GHG emissions to the atmosphere. Dead feedstocks such as harvest residues have higher mitigation potential because the alternate fate of this fibre is combustion in slashpiles or progressive release of C through decay. In addition to harvest residues, salvage of dead wood can be a source of raw material for forest-based bioenergy, with a potentially shorter time to a climate breakeven point (McKechnie et al. 2011; Zanchi et al. 2012). The net result of natural disturbance is typically not an immediate and total loss of biomass (apart from what is combusted by fire), but a transfer of live biomass to dead wood (Kurz et al. 2008a; Dymond et al. 2010a), degrading over time until no longer salvageable (Barrette et al. 2015). The stochastic nature of disturbance makes predicting quantities of dead wood available for salvage at a particular point in time difficult, posing supply management challenges (Shabani et al. 2014). In addition to the feedstock supply, the effectiveness with which fossil C is displaced has a major influence on bioenergy emission reductions, as well as the conversion efficiency of woody biomass combustion (Richter et al. 2009; Smyth et al. 2017b).

Potential improvements to forest biomass supply forecasts could be realized through remote sensing (Wulder et al. 2012; Beaudoin et al. 2014), particularly for forest area not currently included in Canada's NFCMARS. Evaluating the accuracy of available estimates relative to ground-based observations remains challenging. Underlying all biomass estimates are basic allometric models used to predict biomass by component (wood, bark, branches, and foliage) from mensurational properties of trees (breast height diameter, tree height, and species). Owing to the labour-intensive nature of its collection, few additional data have been collected since intensive efforts to develop such models across Canada in the 1980s (Aldred and Alemdag 1989). The many historically available models have recently been harmonized into a single set of models estimated from the historical data (Lambert et al. 2005; Ung et al. 2008) that are now the standard model used in biomass estimates in Canada. Canada's NFCMARS can also produce estimates of root biomass availability (Smyth et al. 2013), though even fewer data exist for estimating below ground biomass (Li et al. 2003). However, harvest of additional biomass from stumps and coarse roots (Berch et al. 2012) may deplete soil nutrients, undesirably reducing forest productivity (Thiffault et al. 2011).

At present, technical and scientific improvements to Canada's NFCMARS are ongoing. Several recent papers describe model and system performance, uncertainty, and improvements to parameters and process representation (Smyth et al. 2011; Hilger et al. 2012; Bona et al. 2013; Shaw et al. 2014, 2015). Additional details on structural uncertainties and proposed system improvements that are the subject of research and development are in Kurz et al. (2013) and Bernier et al. (2012). These are aimed at making increased use of high-resolution spatially-explicit data, not just for estimation of biomass, but also the spatial distribution of harvest and natural disturbances as well as incorporating the unmanaged forest that is currently not accounted for in any assessments undertaken by the system, improved representation of the impact of climate change on ecosystem processes, and better representation of management activities and the fate of harvested wood products. As a result of these improvements, the system will continue to provide a platform to support science and policy analysis that integrates knowledge on ecosystem processes, as well as economic information, to allow projections of many characteristics of Canada's managed forest, including the potential supply of forest derived biomass and how this is affected by natural disturbance associated risks.

References

Aldred AH, Alemdag IS (1989) Guidelines for forest biomass inventory. Forestry Canada, Petawawa National Forestry Institute, Chalk River

Allen D, Mckenney DW, Yemshanov D, Fraleigh S (2013) The economic attractiveness of short rotation coppice biomass plantations for bioenergy in Northern Ontario. For Chron 89:66–78

Amichev BY, Hangs RD, Konecsni SM et al (2014) Willow short-rotation production systems in Canada and Northern United States: a review. Soil Sci Soc Am J 78:S168–S182

Balshi MS, Mcguire AD, Duffy P et al (2009) Vulnerability of carbon storage in North American boreal forests to wildfires during the 21st century. Glob Chang Biol 15:1491–1510

Barette J, Thiffault E, Paré D (2013) Salvage harvesting of fire-killed stands in northern Quebec: analysis of bioenergy and ecological potentials and constraints. J Sci Tech For Prod Proc 3:16–25

Barrette J, Thiffault E, Saint-Pierre F et al (2015) Dynamics of dead tree degradation and shelf-life following natural disturbances: can salvaged trees from boreal forests 'fuel' the forestry and bioenergy sectors? Forestry 88:275–290

Beaudoin A, Bernier PY, Guindon L et al (2014) Mapping attributes of Canada's forests at moderate resolution through kNN and MODIS imagery. Can J For Res 44:521–532

Bedarul Alam M, Pulkki R, Shahi C (2012) Woody biomass availability for bioenergy production using forest depletion spatial data in northwestern Ontario. Can J For Res 42:506–516

Berch SM, Curran M, Dymond C et al (2012) Criteria and guidance considerations for sustainable tree stump harvesting in British Columbia. Scand J For Res 27:709–723

Bernier P, Paré D (2013) Using ecosystem CO_2 measurements to estimate the timing and magnitude of greenhouse gas mitigation potential of forest bioenergy. GCB Bioenergy 5:67–72

Bernier PY, Kurz WA, Lempriere T, Ste Marie C (2012) A blueprint for forest carbon science in Canada 2012–2020. Natural Resources Canada, Canadian Forest Service, Ottawa

Bona KA, Fyles JW, Shaw C, Kurz WA (2013) Are mosses required to accurately predict upland black spruce forest soil carbon in national-scale forest C accounting models? Ecosystems 16:1071–1086

Boudewyn P, Song X, Magnussen S, Gillis MD (2007) Model-based, volume-to-biomass conversion for forested and vegetated land in Canada. Information Report BC-X-411. Canadian Forest Service, Pacific Forestry Centre, Victoria
Cambero C, Hans Alexandre M, Sowlati T (2015) Life cycle greenhouse gas analysis of bioenergy generation alternatives using forest and wood residues in remote locations: a case study in British Columbia, Canada. Res Cons Recycling 105A:59–72
Cooke BJ, Lorenzetti F (2006) The dynamics of forest tent caterpillar outbreaks in Québec, Canada. For Ecol Man 226:110–121
Cooke BJ, Macquarrie CJK, Lorenzetti F (2012) The dynamics of forest tent caterpillar outbreaks across east-central Canada. Ecography 35:422–435
Dymond CC, Titus BD, Stinson G, Kurz WA (2010a) Future quantities and spatial distribution of harvesting residue and dead wood from natural disturbances in Canada. For Ecol Man 260:181–192
Dymond CC, Neilson ET, Stinson G et al (2010b) Future spruce budworm outbreak may create a carbon source in Eastern Canadian forests. Ecosystems 13:917–931
Environment Canada (2015) National inventory report (1990–2013) of greenhouse gas sources and sinks in Canada: the Canadian government's submission to the UN framework convention on climate change. Environment Canada, Gatineau
Gauthier S, Berner PY, Boulanger Y et al (2015) Vulnerability of timber supply to projected changees in fire regime in Canada's managed forests. Can J For Res 45:1439–1447
Girardin MP, Bernier PY, Raulier F et al (2011) Testing for a CO_2 fertilization effect on growth of Canadian boreal forests. J Geophys Res 116:G01012
Hall JP, Richardson J (2001) ENFOR – Energy from the forest. For Chron 77:831–835
Hember RA, Kurz WA, Metsaranta JM et al (2012) Accelerating regrowth of temperate-maritime forests due to environmental change. Glob Chang Biol 18:2026–2040
Hilger AB, Shaw CH, Metsaranta JM, Kurz WA (2012) Estimation of snag carbon transfer rates by ecozone and lead species for forests in Canada. Ecol Appl 22:2078–2090
Jenkins JH, Guernsey FW (1936) Wood and charcoals as motor fuel. Department of Mines and Resources, Canadian Forest Service, Ottawa
Klopp WS, Fredeen AL (2014) Harvesting the dead and decaying forests: Potential carbon storage in harvested wood products. For Chron 90:614–619
Kull SJ, Rampley GJ, Morken S et al (2011) Operational-scale carbon budget model of the Canadian forest sector (CBM-CFS3) version 1.2: user's guide. Canadian Forest Service, Edmonton
Kurz WA, Apps MJ (2006) Developing Canada's national forest carbon monitoring, accounting and reporting system to meet the reporting requirements of the Kyoto Protocol. Mit Adapt Strat Glob Change 11:33–43
Kurz WA, Dymond CC, Stinson G et al (2008a) Mountain pine beetle and forest carbon feedback to climate change. Nature 452:987–990
Kurz WA, Stinson G, Rampley G (2008b) Could increased boreal forest ecosystem productivity offset carbon losses from increased disturbances? Phil Trans Roy Soc B 363:2259–2268
Kurz WA, Stinson G, Rampley GJ et al (2008c) Risk of natural disturbances makes future contribution of Canada's forests to the global carbon cycle highly uncertain. PNAS 105:1551–1555
Kurz WA, Dymond CC, White TM et al (2009) CBM-CFS3: a model of carbon-dynamics in forestry and land-use change implementing IPCC standards. Ecol Model 220:480–504
Kurz WA, Shaw CH, Boisvenue C et al (2013) Carbon in Canada's boreal forest – a synthesis. Environ Rev 21:260–292
Lambert MC, Ung CH, Raulier F (2005) Canadian national tree aboveground biomass equations. Can J For Res 35:1996–2018
Leckie DG, Gillis MD (1995) Forest inventory in Canada with emphasis on map production. For Chron 71:74–88
Lemprière TC, Kurz WA, Hogg EH et al (2013) Canadian boreal forests and climate change mitigation. Environ Rev 21:293–321

Li Z, Kurz WA, Apps MJ, Beukema SJ (2003) Belowground biomass dynamics in the Carbon Budget Model of the Canadian Forest Sector: recent improvements and implications for the estimation of NPP and NEP. Can J For Res 33:126–136

Maclean DA, Erdle TA, Mackinnon WE et al (2001) The spruce budworm decision support system: forest protection planning to sustain long-term wood supply. Can J For Res 31:1742–1757

Mansuy N, Thiffault E, Lemieux S et al (2015) Sustainable biomass supply chains from salvage logging of fire-killed stands: a case study for wood pellet production in eastern Canada. Appl Energy 154:62–73

Mckechnie J, Colombo S, Chen J et al (2011) Forest bioenergy or forest Carbon? Assessing trade-offs in greenhouse gas mitigation with wood-based fuels. Environ Sci Technol 45:789–795

Metsaranta JM (2010) Potentially limited detectability of short-term changes in boreal fire regimes: a simulation study. Int J Wildland Fire 19:1140–1146

Metsaranta JM, Kurz WA, Neilson ET, Stinson G (2010) Implications of future disturbance regimes on the carbon balance of Canada's managed forest (2010–2100). Tellus B 62:719–728

Metsaranta JM, Dymond CC, Kurz WA, Spittlehouse D (2011) Uncertainty of 21st century growing stocks and GHG balance of forests in British Columbia resulting from potential climate change impacts on ecosystem processes. For Ecol Man 262:827–837

Michaelian M, Hogg EH, Hall RJ, Arsenault E (2011) Massive mortality of aspen following severe drought along the southern edge of the Canadian boreal forest. Glob Chang Biol 17:2084–2094

National Council for Air and Stream Improvement (2014) Compilation of Canadian provincial and federal regulations relevant to forest management activities. Special Report No. 14-03. National Council for Air and Stream Improvement, Research Triangle Park

Natural Resources Canada (2015) The State of Canada's forests: annual report 2015. Natural Resourcs Canada, Canadian Forest Service, Ottawa

Nautiyal JC (1979) The place of forestry in the energy question. Can J For Res 9:68–75

Paré D, Bernier P, Thiffault E, Titus BD (2011) The potential of forest biomass as an energy supply for Canada. For Chron 87:71–76

Richter D, Jenkins DH, Karakash JT et al (2009) Wood energy in America. Science 323:1432–1433

Roach J, Berch SM (2014) A compilation of forest biomass harvesting and related policy in Canada. Technical Report 81. Ministry of Lands, Forests, and Natural Resource Operations, Victoria

Savage DW, Martell DL, Wotton BM (2010) Evaluation of two risk mitigation strategies for dealing with fire-related uncertainty in timber supply modelling. Can J For Res 40:1136–1154

Shabani N, Sowlati T, Ouhimmou M, Rönnqvist M (2014) Tactical supply chain planning for a forest biomass power plant under supply uncertainty. Energy 78:346–355

Shaw CH, Hilger AB, Metsaranta J et al (2014) Evaluation of simulated estimates of forest ecosystem carbon stocks using ground plot data from Canada's National Forest Inventory. Ecol Model 272:323–347

Shaw CH, Bona KA, Kurz WA, Fyles JW (2015) The importance of tree species and soil taxonomy to modeling forest soil carbon stocks in Canada. Geoderma Reg 4:114–125

Smyth CE, Kurz WA, Trofymow JA (2011) Including the effects of water stress on decomposition in the Carbon Budget Model of the Canadian Forest Sector CBM-CFS3. Ecol Model 222:1080–1091

Smyth CE, Kurz WA, Neilson ET, Stinson G (2013) National-scale estimates of forest root biomass carbon stocks and associated carbon fluxes in Canada. Glob Biogeochem Cycles 27:1262–1273

Smyth CE, Stinson G, Neilson E et al (2014) Quantifying the biophysical climate change mitigation potential of Canada's forest sector. Biogeosciences 11:3515–3529

Smyth CE, Kurz WA, Rampley G, Lemprière TC, Schwab O (2017a) Climate change mitigation potential of local use of harvest residues for bioenergy in Canada. Glob Change Biol Bioenergy 9:817–832

Smyth CE, Rampley G, Lemprière TC, Schwab O, Kurz WA (2017b) Estimating product and energy substitution benefits in national-scale mitigation analyses for Canada. Glob Change Biol Bioenergy 9:1071–1084

Stinson G, Kurz WA, Smyth CE et al (2011) An inventory-based analysis of Canada's managed forest carbon dynamics 1990 to 2008. Glob Chang Biol 17:2227–2244

Stinson G, Magnussen S, Boudewyn P et al (2016) Chapter 12 Canada: National resource availability reports. In: Vidal C, Alberdi I, Hernández L, Redmond J (eds) National forest inventories – assessment of wood availability and use. Springer, Cham

Ter-Mikaelian MT, Colombo SJ, Chen J (2015) The burning question: does forest bioenergy reduce carbon emissions? A review of common misconceptions about forest carbon accounting. J For 113:57–68

Thiffault E, Hannam KD, Paré D et al (2011) Effects of forest biomass harvesting on soil productivity in boreal and temperate forests – a review. Environ Rev 19:278–309

Ung CH, Bernier P, Guo XJ (2008) Canadian national biomass equations: new parameter estimates that include British Columbia data. Can J For Res 38:1123–1132

Weed AS, Ayres MP, Hicke JA (2013) Consequences of climate change for biotic disturbances in North American forests. Ecol Monogr 83:441–470

Wulder MA, White JC, Nelson RF et al (2012) Lidar sampling for large-area forest characterization: a review. Remote Sens Environ 121:196–209

Zanchi G, Pena N, Bird N (2012) Is woody bioenergy carbon neutral? A comparative assessment of emissions from consumption of woody bioenergy and fossil fuel. Glob Change Biol Bioenergy 4:761–772

Chapter 9
Czech Republic

Miloš Kučera

9.1 Introduction

According to the first National Forest Inventory (NFI) carried out between 2001 and 2004, the total area of forests in the Czech Republic is 2,751,586 ha of which 2,396,705 ha (87%) are classified as Forest Available for Wood Supply (FAWS). The remaining 354,881 ha (12.9%) include protection forests and protected forest areas. Protection forests represent 2.8% of the forest area and include forest areas in extremely unfavorable sites, high altitude forests and *Pinus mugo* stands. Forest parks and forest reserves, the first zones considered as protected landscape areas, represent 10.1% of the forest area in the Czech Republic. The area of forest not available for wood supply is estimated using auxiliary information from other sources, such as: Forest Management Plans (FMP), Forest Development Regional Plans, and Forest Typology Classifications. All forests under FMP (90% of forests) are classified according to their prevalent primary functions. This classification includes all possible legal and site restrictions regarding forest management.

Czech forests are mainly owned by the state which holds 61%, followed by private forest owners with 18.4% and municipalities with 12.8%.

Coniferous forests dominate in the Czech Republic covering 67% of the forest area. The main tree species in Czech forests are *Picea abies* (48%), *Pinus sylvestris* (14%), *Fagus Sylvatica* (7%), and *Quercus* (7%). These four species cover 76.2% of the forest area and represent 83.8% of the growing stock. The Czech wood industry is based on processing spruce and pine timber. These two species together represent 71% of the growing stock. Total growing stock of Czech forests varies according to the data source: National Forest Inventory (NFI) or FMP. In 2003, the total growing stock was 900 million m^3 according to the NFI, but only 650 million m^3 were

M. Kučera (✉)
Forest Management Institute, Brandýs nad Labem-Stará Boleslav, Czech Republic
e-mail: kucera.milos@uhul.cz

S. Barreiro et al. (eds.), *Forest Inventory-based Projection Systems for Wood and Biomass Availability*, Managing Forest Ecosystems 29,
DOI 10.1007/978-3-319-56201-8_9

reported based on information obtained from the FMPs. For the same period, differences were also observed for the mean growing stock per hectare which varied from an NFI estimate of 333 m^3/ha to an FMP estimate of 251 m^3/ha.

In 2012, according to the FMPs, the total annual increment in Czech forests was 17.9 million m^3 and the mean annual increment per hectare was 6.9 m^3/ha, whereas the total volume of timber harvested was 15.1 million m^3, of which 13.1 million m^3 was from conifers and 2 million m^3 from broadleaf species. The mean annual harvest per hectare was 5.7 m^3/ha in 2012.

The Czech Republic has two main data sources for forest and forestry information: the NFI and the FMPs. These two sources have different aims and different content, but they both provide data about forests that can be used in forest policy and decision making processes. However, for different reasons, the use of long-term predictions in policy decision making processes is not very common in the Czech Republic. Nevertheless, predictions based on FMPs data have been obtained regularly since 1983 and are provided to the Ministry of Agriculture.

Short-term estimates of wood available for harvesting were first carried out in the 1950s using inventory data collected to prepare FMP. A decade later, a new inventory was carried to prepare or renew existing forest management plans. Inventory units corresponded to forest enterprises for which harvestable wood estimates were produced for a 5-years period (1961–1965, Anonymous 1960). In 1978, new legislation for more comprehensive FMP included information on logging volume with the result that logging forecasts were made until 2020 (Nymburský 1983). Ten years later a new logging forecast until 2010 was published, and in 2006 the Forest Management Institute Brandýs nad Labem (FMI) prepared and published a study on logging perspectives in the Czech Republic (Vašíček et al. 2006). For this study, the FMP database was used and included some tests of different management approaches: namely, the effect of reduced rotation periods in certain management sets, the positive effects of thinning on volume 40 years later, as well as a combination of both. In addition to the FMP database, data on the growing stock from the first cycle of the whole-scale statistical NFI were also used (Synek et al. 2014).

In the course of the European Forest Sector Outlook Study II (EFSOS II) (UN-ECE/FAO 2011) which aimed to forecast possible or probable wood availability, the EFISCEN model was used to produce results by individual countries. Again, a combination of FMP and NFI data was used, and FMI provided NFI data from the first cycle on forest areas and standing volumes by administrative regions, ownership, ecological series, groups of tree species and age classes. However, the FMP database had to be used to provide data on increment which would only be available after the second cycle of the NFI is completed. Hypothetical logging perspectives for three main scenarios in the Czech Republic are presented in Synek et al. (2014).

9.2 Data

9.2.1 National Forest Inventory

The aim of the NFI is to establish comprehensive information about the state and development of forests in the Czech Republic from the point of view of both sustainable environment and economic use. The main tasks of the Czech NFI are to provide information about forests for state organization purposes, for the evaluation of forestry activities and for assessing fulfilment of forest management goals. NFI data are also used for national and international reporting. The NFI was first implemented in the Czech Republic in 2001, was carried out between 2001 and 2004, and is described in Cerny et al. (2010). The second NFI was launched in 2011 and was terminated in 2015. In the course of this NFI, the COST E43 land use classification (Tomppo et al. 2010) was adopted instead of the national land use classification used during NFI1 Štěrba and Jankovská (2007).

The third NFI is planned to start in 2016. Based on analyses of the first inventory grid and all the technology used in the first NFI, it was decided to change the sampling design. The new sampling design allows new inventory technology to be incorporated, and the transition to a continuous inventory with plot re-measurement every five years. The continuous inventory will enable the production of results annually.

9.2.2 Forest Management Plans

A FMP is prepared every 10 years for each forest owner and includes management guidelines such as binding provisions for maximum cumulative volume of felled timber, minimum area of thinning activities in stands less than 40 years of age, and species-specific planting prescriptions to improve the soil, stabilize stands against weather hazards, and produce desired species and age distributions. FMPs also provide basic information about the stand, the stand layer and the forest region for which the FMP is intended. After the stand is described, it is classified with respect to forest type, the prevailing function of the forest, the forest management type, the duration of the plan and the area it covers, and the rotation and regeneration periods. Additionally, the stand storey is classified in terms of species composition and structure, making reference to the share of species and stand age, respectively, as well as to the site class. Records of the stand density, the mean height and mean diameter at breast height (*dbh*), the volume of the mean stem, the growing stock per species and the volume of standing deadwood, and abiotic and biotic damages are also recorded.

9.3 Timber Harvest Predictions According to Forest Management Plans

The Timber Harvest Prediction (THP) tool is a projection tool used in the FMI for simulating growing stock and harvest in Czech forests. The tool is applied to the area of forest covered by the FMP which represents about 90% of the total forest area in the Czech Republic. The THP tool assesses the current state of Czech forests and predicts the future state and development of the forest using data from FMP as input. Estimates for different area units can be produced. Estimates are regularly produced for levels two and three of the national subdivisions defined by the Nomenclature of Territorial Units for Statistics (NUTS II and NUTS III). Estimates are based on growth tables for the main tree species in the Czech Republic (Cerny et al. 1996) and on the harvesting percentage defined according to ordinance No. 84/1996 on forest management planning.

9.3.1 Input Data

The inputs consist of data at storey level, and the each storey represents a different vertical layer within the stand. The stand storey is the lowest unit of space distribution in the forest according to FMPs and is characterized by the following variables:

- Storey level – area, age, management type, rotation period, stocking, regeneration period;
- Species level – tree species, share of each tree species, area of each tree species, growing stock per species based on FMP, absolute and relative site class per species.

In addition to FMP data, other types of inputs are also required such as the final harvest and thinning percentages which are defined by law.

9.3.2 Description of the Algorithm

The THP tool is applied within each stand storey separately for each species and produce tabular outputs. The outputs are structured by management types, age classes, broadleaf and coniferous species. The tool runs on 10-year time-steps meaning that THP predicts the state of forest for t_0+10, $t_{10}+10,\ldots$, t_n+10 where t_0 is the current state of the forest according to the FMP, and n is the total number of years considered in the prediction. Each simulation runs in eight steps applied within each time-step. These eight steps are responsible for calculating the removals and updating the values for the standing forest.

Final Felled Volume

The final felled volume is calculated separately for each species in the storey according to the FMP prescribed the final felled volume. The felled volume of a given species is obtained by multiplying the species growing stock in the storey according to the FMP for the time t, by the harvesting percentage/100 which is set by law and is defined by rotation period and regeneration period as prescribed in the FMP. The total felled volume in the storey for time t_0 +10 is the sum of the volumes of all the species felled in the storey.

Area of Final Felling's

The area of final fellings is calculated similarly to the felled volume by multiplying the area of each species in the storey as described in the FMP for the time t, by the harvesting percentage. The total area of final fellings per storey in time t +10 is the sum of areas of the felled species in the storey.

Volume of Thinning

The volume resulting from thinnings, including both pre-commercial and management thinnings, is calculated separately for each species in the storey according to the percentage of pre-commercial and management thinnings prescribed in the FMP. The total volume of thinnings per storey is the sum of the thinned volumes of all the species in the storey. The thinned volume by species is the product of the growing stock defined in the FMP for the time t by the thinning percentage divided by 100 which is set by law.

Age

The age of the species in the storey is increased after 10 years and the storey is shifted to the next age class

Storey Area

Storey area at time *t+10* is calculated as the storey area at time *t* minus the area subjected to final felling.

Storey After Final Felling

For the storey or for the part of the storey that was final felled according to the FMP prescription, the age is shifted to the youngest age class and a new storey is established. The area of the new storey equals the area that was under final felling. The tree composition assigned to the new storey is the same as in the previous storey. The volume of the established stand is assigned based on the youngest age tabulated in the yield tables (Cerny et al. 1996) which are also used to predict the growth of the stand for subsequent periods.

Growing Stock

The growing stock at time t+10 is calculated in two steps. First, the growing stock after final harvest is updated by discounting the felled volume resulting from the final felling from the growing stock at time *t*. Second the growing stock resulting from the previous step is increased based on the increment coefficient defined in the growth tables.

9.3.3 *Outputs of the Timber Harvest Prediction Tool*

The end user of the THP outputs is the Ministry of Agriculture, and FMI is responsible for providing predictions. Predictions are prepared every year for periods of 30 years after the FMPs have been updated (one tenth of the FMPs is updated every year). Outputs have a standardized structure according to the ministry's request. Predictions results are produced for NUTS I and NUTS III levels and are structured by age classes, forest type (coniferous and broadleaf) and by rotation and regeneration period. The evolution of stands is presented by 17 age classes (class 1: ages from 1 to 10, class 17: ages from 171 to 180) and forest area (ha), growing stock (m^3) and total harvested volume (m^3) are possible outputs.

9.4 Discussion and Conclusions

Timber Harvest Predictions are official predictions which use forest data at national level. These predictions use only FMP data and reflect some disadvantages that derive from the nature of the FMPs. FMPs are prepared according to their main purpose which is forest management support and provision of guidelines to the forest owner. FMPs mainly include information about forest management and a description of the stands. Moreover, FMPs do not cover all forests in the Czech Republic: according to the second NFI 10% of forests are not covered. Another limiting fact is that some forest owners do not provide FMP data to the central database resulting in information gaps. Further, this prediction method is only possible for stands of a certain age and rotation period. Finally, some FMPs that describe selection forests or forests in National parks, do not use age as a descriptive variable which leaves these forests out of the prediction.

The accuracy of the FMP data and the uncertainty involved in THP predictions is not known. FMPs do not provide direct information about harvest and increment. The volume of harvest for national statistics is obtained by the Czech statistical office by sampling forest owner questionnaires. Increment is determined by the growth tables using summary FMP data. Differences in total and mean growing stock estimates depending on the FMP or NFI data source produce completely different starting points for predictions and therefore also different predictions.

For future predictions it will be very useful to replace FMP data with NFI data. The NFI provides reliable data with known accuracy, and its data structure is suitable for use with prediction tools. Also, fundamental variables such as harvest and increment can be calculated using inventory data. Moreover, the use of inventory data from a continuous inventory (to be implemented in 2016) will enable development of forest growth models. Once the growth models are developed, the next step will be to develop an appropriate national simulation tool that can use NFI data.

References

Anonymous (1960) Inventarizace lesů 1960 [Forest Inventory 1960] Praha, Ministerstvo zemědělství, lesního a vodního hospodářství

Cerny M, Parez J, Malik Z (1996) Růstové a taxační tabulky hlavních dřevin České republiky [Growth tables of the main tree species of the Czech Republic]. Institute for Forest Research IFER

Cerny M, Kucera M, Cienciala E, Beranova J (2010) The Czech Republic. In: Tomppo E, Th G, Lawrence M, RE MR (eds) National forest inventories – pathways for common reporting. Springer, Cham, pp 311–331. ISBN:978-90-481-3232-4

Nymburský B (1983) Úkoly hospodářské úpravy lesů v rozvoji lesního hospodářství [Tasks of Forest Management Planning in the Development of the National Economy] Praha, Ministerstvo zemědělství, lesního a vodního hospodářství

Synek M, Vašíček J, Zeman M (2014) Outlook of logging perspectives in the Czech Republic for the period 2013–2032. J For Sci 60(9):372–381

Štěrba P, Jankovská Z (eds) (2007) National Forest Inventory in the Czech Republic 2001–2004: introduction, methods, results. Ústav pro Hospodářskou Úpravu Lesů Brandýs nad Labem (ÚHÚL)

Tomppo E, Gschwantner T, Lawrence M, McRoberts RE (eds) (2010) National forest inventories – pathways for common reporting. Springer, Heidelberg/Dordrecht/London/New York

UN-ECE/FAO (2011) The European Forest Sector Outlook Study II (EFSOS II) 2010–2030. UN-ECE/FAO

Vašíček J, Hána J, Kraus M, et al. (2006) Těžební možnosti na území lesů ČR [Logging perspectives of forests in the territory of the Czech Republic]. Lesnická práce 85:240–242

Chapter 10
Denmark

Vivian Kvist Johannsen, Thomas Nord-Larsen, Torben Riis-Nielsen, Lars Graudal, and Erik Schou

10.1 Introduction

The Danish land area is 43,098 km^2 and consists of the Jutland peninsula and more than 400 islands of varying sizes. The country is mostly flat with low elevations. The climate is temperate maritime with cool summers and mild winters. During the latest ice age (Weichsel), which ended 11–12,000 years ago, the Scandinavian ice shield covered the islands and part of Jutland, resulting in generally sandy soils on the peri-glacial, alluvial plains in western Jutland, and more clay rich moraine soils in the rest of Denmark. Today, forest land covers 615,000 ha or 14.3% of the total land area (Nord-Larsen et al. 2014). Coniferous forest is found in all parts of the country, but is most common on the sandy soils in the western parts, while deciduous trees are mainly found on less sandy soils in eastern Denmark. Coniferous forests cover 39% of the forest area, while deciduous forests cover 41%. Mixed forests cover 11%, Christmas tree plantations cover 5%, and the remainder is temporarily un-stocked and auxiliary areas (Nord-Larsen et al. 2014). According to national forest inventory (NFI) data, the annual Danish forest harvest is 3.3 million m^3 per year, with total removals, including dead trees and lost volume, of 4.6 million m^3 year (Nord-Larsen et al. 2014).

The forest area is owned mainly by private persons (60%), private companies (10%) and state forests (18%). The remaining parts of the forest area are owned by other public bodies or foundations. In total, there are 28,000 individual owners, of which approximately 600 own more than 100 ha, indicating a much skewed distribution of forest ownership. Less than half the area is covered by updated forest

V.K. Johannsen (✉) • T. Nord-Larsen • T. Riis-Nielsen • L. Graudal • E. Schou
Department of Geosciences and Natural Resource Management, Section for Forest, Nature and Biomass, University of Copenhagen, Copenhagen, Denmark
e-mail: vkj@ign.ku.dk; tnl@ign.ku.dk; trni@ign.ku.dk; lgr@ign.ku.dk; eschou@ign.ku.dk

S. Barreiro et al. (eds.), *Forest Inventory-based Projection Systems for Wood and Biomass Availability*, Managing Forest Ecosystems 29,
DOI 10.1007/978-3-319-56201-8_10

maps and management plans, with a clear tendency for the larger forest owners to have more information and management plans (Nord-Larsen et al. 2014).

Increasing focus on biomass for bioenergy and maintenance of forest carbon pools has created a need to know more about potential instruments and incentives that may be applied in forest policies and forest management. In response, a number of studies were performed in 2000-2013 to assess future development of Danish forests and their wood production (Nord-Larsen and Heding 2002; Johannsen et al. 2010; Nord-Larsen and Suadicani 2010; Johannsen et al. 2011; Graudal et al. 2013). The most recent analyses were aimed at assessing the possibilities for increasing production and optimizing the use of Danish forest wood resources over the next 100–500 years on a sustainable basis, with due consideration to other forest functions. The initiative for the analyses came from both private forest owners and the National Nature Agency which has responsibility for developing forest policies.

Most previous studies of the effects of forest management on wood production have been limited to the analysis of one or a few forest management interventions, while the effects of a larger number of interventions that can be combined in different ways have received less attention. In the present study we aimed to study the effect of multiple silvicultural measures and how they both individually and collectively influence both potential harvest and resources in the forest in terms of living biomass and carbon pools.

10.2 Data

Assessment of forest biomass production potentials started with the current forest area and its aggregated distribution to species, age-class, regions and soil types provided by the Danish NFI. The Danish NFI was initiated in 2002 and is a continuous, sample-based inventory with partial replacement of sample plots based on a 2 × 2 km grid covering the Danish land surface (Johannsen et al. 2013a). The sample of permanent and temporary field plots has been systematically divided into five non-overlapping, interpenetrating panels; each of which is measured in a single year and constitutes a systematic sample of the entire country. Hence all the plots are measured in a 5-year cycle. Approximately one-third of the plots are permanent and are re-measured in every 5-year cycle, whereas two-thirds are temporary and are moved randomly within the particular 2 × 2 km grid cells in subsequent cycles of measurements.

In each square 2 × 2 km grid cell, a cluster of four circular plots (Primary Sampling Unit, PSU) is placed at the corners of a square with 200 m side length. Each circular plot (Secondary Sampling Unit, SSU) has a radius of 15 m. When plots include different land-use classes or different forest stands, the individual plot is divided into Tertiary Sampling Units (TSU).

Based on an analysis of aerial photos, each sample plot (SSU) is assigned one of three categories reflecting the likelihood of plot-level forest or Other Wooded Land (OWL) cover: (0) unlikely to contain forest or other wooded land cover, (1) likely to contain forest, and (2) likely to contain other wooded land. All plots in the last two categories are inventoried in the field. In the 2008-2012 NFI-cycle, 9425 plots covering 4138 clusters were classified as forest or OWL based on aerial photos and were thus selected for field inventory; only three plots were not inventoried in the field.

Each plot is composed of three concentric circles with radius of 3.5, 10 and 15 m. In the 3.5-m circle, a single caliper measurement of diameter is made at breast height (1.3 m from ground, *dbh*) for all trees taller than 1.3 m. Trees with diameters larger than 10 cm are measured in the 10 m circle, and only trees with diameters larger than 40 cm are measured in the 15 m circle. Measurements of total height are obtained for a random sample of 2–6 trees. Further on this subsample, crown height, age, and diameter at stump height are measured, and the presence of defoliation, discoloration, mast, mosses, and lichens are recorded. The presence of regeneration on the plots is registered as well as the species, age and height of the young trees.

Species are grouped into 13 main species groups based on similarity in growth and silvicultural practices such as thinning practices and rotation length. For each group, the area and the average carbon pools are estimated based on the NFI sample plots for one full cycle of measurements which, for the scenario modelling, was taken to be the period 2008–2012. Johannsen et al. (2013a) provide a full description of procedures for estimation of biomass and carbon pools.

For economic analyses, the current average prices of main assortments were collected based on private and public accounts, as well as their estimates of costs of reg eneration/afforestation (Schou and Thorsen 2013). Interest rates of 1.5% were based on the general recommendation from the National Bank of Denmark for long-term investments.

10.3 Methods

The scenario modelling and prognosis tool was applied to the entire Danish forest area based on the NFI input data, and was programmed to run in the Statistical Analysis System (SAS) environment. The models are deterministic and are based on stand level growth models for thinning harvests estimates and on regional modelling of regeneration probability. For geographical regions a simplification was applied, and only two regions were considered. The methods have been developed and used over a period of more than 10 years, and have been implemented in a number of publications, but do not have a specific name (Larsen and Johannsen 2002; Nord-Larsen and Heding 2002; Nord-Larsen and Suadicani 2010; Johannsen et al. 2011; Graudal et al. 2013).

10.3.1 Area Composition

Using the current distribution of forest area by species, age-classes, and geographical regions, the forest area is projected for one period (1-year time-step) using the probabilities of regeneration and annual afforestation given by the scenario. Regenerated areas are assumed to be established with the same species that previously occupied the area, unless otherwise stated in the silvicultural measures applied (see later paragraph on this issue).

Rotation age depends largely on the tree species and growing conditions. For spruce on fertile, clay soils in the eastern part of the country, the desired rotation age is 40–50 years. On gravelly soils in eastern Jutland and northern Zealand, the desired rotation age is 60–70 years, and on sandy soils in western and northern Jutland, rotation age may be 80–90 years. However, frequent wind throws, attacks by bark beetles, and debilitation by root rot often significantly shorten the actual rotation age.

The age-class distribution by species is projected assuming that the forest area in each age-class that has not been harvested progresses into the subsequent age-class after each year. The area harvested each year is re-assigned to the first age-class of the same species or another species in cases where species-change policies are modelled. The probability that the forest area is transferred to the subsequent age-class after a year is termed the *transition probability* whereas the net flow to or from the species classes is termed the *conversion probability*.

Transition probabilities are derived from an analysis of the two successive forest censuses in 1990 and 2000 (Nord-Larsen and Heding 2002). For each species class, the aggregated transition probability at any given point in time was modelled from the observed transition possibilities, and the area weighted production class in each county, using a logistic model of the form:

$$p(regeneration) = \frac{1}{1 + \left[\beta_0 + \beta_1 \cdot (1/PK)\right] \cdot e^{-\beta_2 \cdot T}} + \varepsilon, \qquad (10.1)$$

where PK is production class expressed as total volume production per hectare for a full rotation, T is age and β_0 to β_2 are species-specific parameters. Accumulated transition probabilities are illustrated in Fig. 10.1. By basing the estimation of the transition probability models on two successive forest inventories, the effects of windthrow (especially occurring in conifers such as Norway spruce) are included directly in the model which results in short rotation ages for most conifers (see lower graphs in Fig. 10.1). For further details on the estimation procedure see Nord-Larsen and Heding (2002).

10.3.2 Harvest from Thinning and Clear-Cut Volumes

Historically, Danish forest wood production has been mainly from homogeneous, even-aged forest stands. Regeneration of forest stands is done by clear felling and subsequent replanting. Between planting and clear felling, the stand is usually

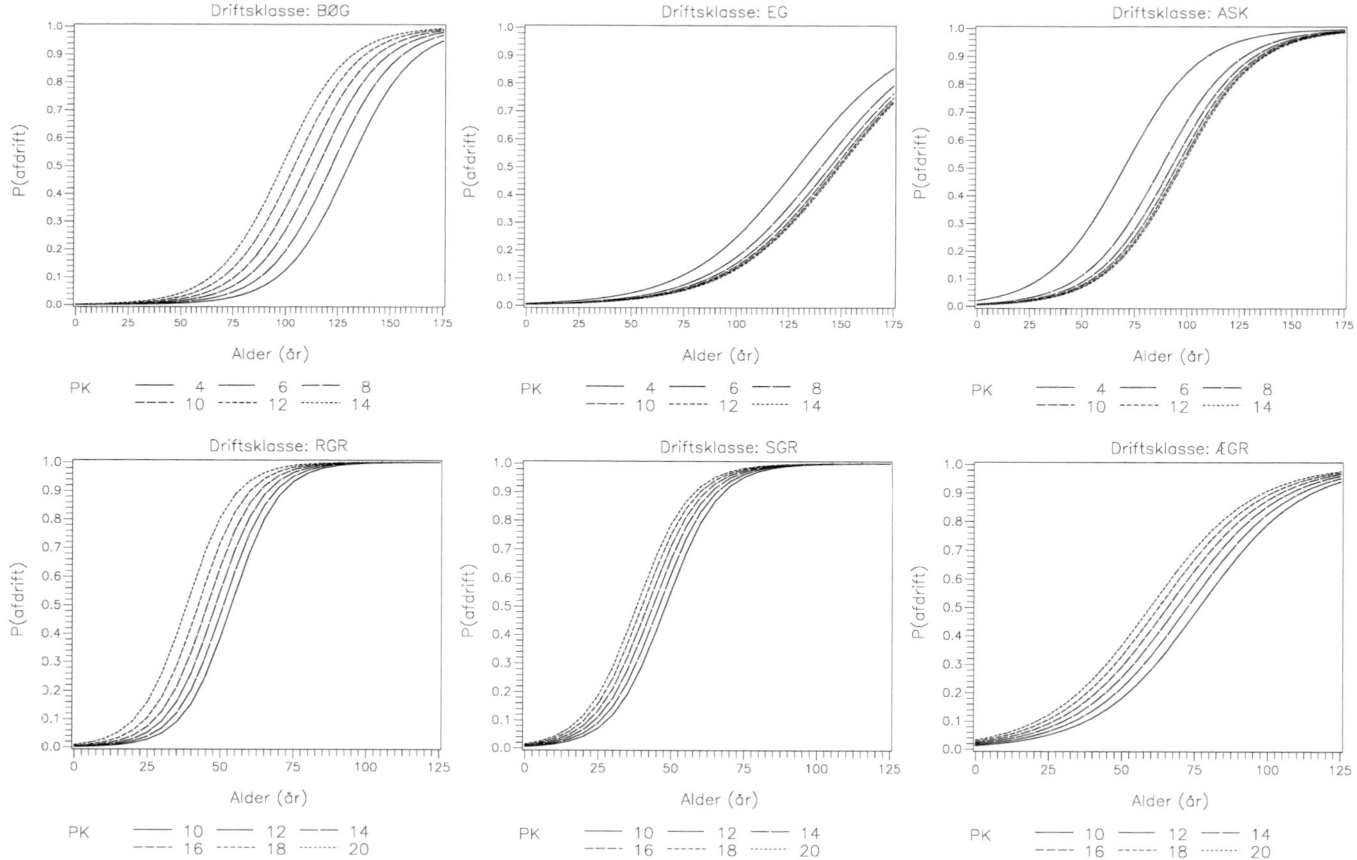

Fig. 10.1 Accumulated clear cut probability (*P(afdrift)*) for different species and production classes (Nord-Larsen and Heding 2002). From *top left* to *lower right*: beech (BØG), oak (EG), ash (ASK), Norway spruce (RGR), Sitka spruce (SGR), and silver fir (ÆGR). The x-axes denote stand age and *PK* refer to different production classes, i.e. average annual volume production at the optimal rotation age

thinned. As a rule of thumb, thinnings are carried out with a frequency corresponding to about a tenth of stand age. In coniferous species, the market for wood chips and the risk of windthrow have created a silvicultural practice of frequent, heavy thinnings in stands less than 14–15 m tall and subsequently management without thinnings until clear-felling.

Using the current distribution of forest area by species, age-classes and the species-specific, regional average production class, volume growth is estimated using mathematical formulations of species-specific yield tables (Møller 1933; Møller and Nielsen 1959; Kjølby 1958; West-Nielsen 1950; Magnussen 1983; Henriksen 1957, 1958, Elingård-Larsen and Jensen 1985; Morville 1948). The species-specific, regional average production class is a measure of site quality corresponding to the potential average volume production, expressed in cubic meters, at the rotation age optimizing volume growth. Newer models have been developed for general growth simulations, but because these models require the definition of specific thinning regimes, the older, well-known models were applied. Thinning volume outside bark is estimated as the difference between estimated volume per hectare at the beginning and at the end of the projection period and an estimate of total production during the period. The estimates are calculated as averages over 10 years. These models are only used to estimate harvested volume from intermediate thinnings.

Final harvest volume is estimated from the predicted area of regeneration and NFI estimated volume per hectare for each combination of species, age-class and geographical region. This provides estimats of the harvested volume from final fellings.

10.3.3 Scenarios – Silvicultural Measures

Specific silvicultural measures can be analyzed either one at a time or in combination. Each measure causes the prognosis to be adjusted according to the expected impacts on the species and age-class distribution, standing volume and resulting products in different geographical regions.

The silvicultural measures included the following:

- Afforestation, establishment of new forest: How much forest is planted on former agricultural land per year in hectares?
- Species choice in afforestation: Which tree species are used for these new forests?
- Rotation age: What is the expected rotation age of the forests?
- Species choice in regeneration: Which species are used for regeneration of existing forests?
- Intensity of regeneration: How is forest regenerated? More intensive regeneration: higher planting density, use of fast growing cover crops, providing early and higher biomass production
- Level of forest set-a-side: How large areas are not available for forest production to serve purposes such as biodiversity

- Utilization degree: How great is the utilization rate of the harvest?
- Assortment choice: What is the wood used for purposes such as firewood or timber?
- Breeding: How good is the planting material in planted forest in terms of breeding intensity (Hansen et al. 2013)

For each of the silvicultural measures, adjustments to the prognosis models were implemented so that the treatment change due to the measure had a direct effect on each step of the estimation process. This allowed detailed responses to have effects in the simulations. The calculations related to different scenarios are described in Graudal et al. (2013) and in several background reports (Hansen et al. 2013; Johannsen et al. 2013b; Schou and Jellesmark 2013).

Other silvicultural measures such as the use of fertilizer and pesticides could have been considered. However, because they are not commonly used in the Denmark and because the energy requirements for producing fertilizers and pesticides and their environmental impacts would complicate the overall assessment, they have been left out of the analysis.

10.3.4 Output for Further Analysis

Summary data for each year in the prognosis are collected and stored in datasets for further analysis and dissemination to private and public stakeholders and publications.

For each year, summary data for area, species and age-class distributions are provided, as well as summary data for biomass (in volume, dry matter and carbon equivalents) for the entire country. The biomass estimates are directly based on the NFI estimates and the methods applied therein.

Economic data were applied to the results to facilitate estimation of the direct economic effects of the simulations, without making the economic factors a part of the simulations directly.

10.3.5 Scenario Modelling

The overall effects on development of the forest resources were analyzed individually for each of the nine silvicultural measures, including all the different levels they could be assigned (Table 10.1). Furthermore, four different combinations (scenarios) of the silvicultural measures were modelled and the results analyzed. Other scenarios, consisting of combinations of levels for the different silvicultural measures, could have been analyzed as well.

The four scenarios are shown in Table 10.2 and are as follows: (1) a scenario reflecting current forest management (business-as-usual, BAU), (2) a scenario

Table 10.1 Silvicultural measures and different levels used in the analyses

Silvicultural measures	Level 0	Level 1	Level 2	Level 3
Afforestation	1900	2280	4560	0
Species choice, afforestation	As current	Only conifers	Only broadleaved	
Rotation age	As current	Lower rotation age	Higher rotation age	
Species choice in regeneration	As current	Transition towards conifers	Transition towards broadleaved	
Intensity of regeneration	As current	Higher stem number and nurse trees	Higher stem number, nurse trees and breeding	
Level of forest set-a-side	0	20% of broadleaved, approx. 10% of forest area	50% of broadleaved, approx. 25% of forest area	100% of broadleaved, approx. 50% of forest area
Utilisation degree	As current – 80%	Higher – 100%	Lower – 70%	
Assortment choice	As current	High firewood/energy wood proportion	Only high energy wood proportion in conifers	
Breeding (Hansen et al. 2013)	As current (low improvement)	High improvement	High speed improvement	

Table 10.2 Levels of the silvicultural measures listed in Table 10.1 for the four scenarios BAU, BIO, ECO, and KOMBI

Silvicultural measure number	Silvicultural measure description	Scenario			
		BAU	BIO	ECO	KOMBI
SK1	Afforestation	0	0	2	2
SK2	Species choice, afforestation	0	1	2	0
SK3	Rotation age	0	1	2	0
SK4	Species choice in regeneration	0	1	2	0
SK5	Intensity of regeneration	0	1	0	1
SK6	Level of forest set-a-side	0	0	1	1
SK7	Utilisation degree	0	1	2	0
SK8	Assortment choice	0	1	2	1
SK9	Breeding	0	1	1	2

where focus is on increasing the production of biomass for bioenergy (BIO), (3) a scenario where focus is on the production of biomass through afforestation, but with emphasis on domestic (broadleaved) species in the regeneration and in the afforestation (ECO), and (4) a scenario that aims to optimize biomass production, while also considering a more broad concern for protection and domestic species (KOMBI).

The summary data were analyzed in terms of multiple factors including: expected volume in harvest, amount of biomass for energy and industry, as well as increment and carbon pool in standing living volume.

10.4 Results

Because the primary purpose of these analyses was to assess the likely effects of multiple silvicultural measures, they are all assessed in relation to the current practice, i.e., business as usual. Therefore, the summary output tables were presented as relative values for a number of summary data for the analyzed area (Tables 10.3 and 10.4).

The effects of different silvicultural practices influenced potential biomass harvest both in the short and long-term. In the short term, the level of set aside land (SK6) and the assortment distribution had the most pronounced effects on biomass for energy. In the long-term, also the level of afforestation and intensive regeneration had a substantial effect on biomass production. The effect of different silvicultural measures also had to pronounced effects on the overall biomass production in the four scenarios (Fig. 10.2). Projections showed that in 2050, the BIO and KOMBI scenarios led to a 50% increase in the production of biomass for energy.

10.5 Discussion and Conclusions

The methods described for the prognosis tools for the Danish forest area and resources related to it have been developed over time and have been adjusted for different purposes. However, the current tools are flexible and the results are consistent with findings from long-term (50–150 years) experimental forest trials on growth and other data sources on the development of the forest area. The tools have already gained more use for derived analyses, and further developments are expected.

Table 10.3 Results of simulations with focus on harvest

	All in%. of BAU	Overall harvest			Biomass for energy			Wood for industry		
	Parameter/year	2020	2050	2100	2020	2050	2100	2020	2050	2100
	BAU	100	100	100	100	100	100	100	100	100
SK1	Afforestation 2280 ha/year	100	101	104	100	102	104	100	101	104
	Afforestation 4560 ha/year	100	108	127	100	113	131	100	105	125
	Afforestation 0 ha/year	100	94	80	100	91	78	100	96	82
SK2	Afforestation decidous	100	100	98	100	102	102	100	98	96
	Afforestation coniferous	100	101	104	100	98	99	100	104	108
SK3	Rotation shorter	103	98	100	103	100	101	103	97	99
	Rotation longer	97	98	98	98	98	100	96	98	96
SK4	Regeneration more conifers	100	101	103	100	99	101	100	102	105
	Regeneration more broadleaved	100	101	100	100	108	107	100	95	96
SK5	Intensive regeneration/ nurse trees	100	115	113	100	135	132	100	100	100
SK6	Forest set-a-side Level 1	82	91	93	84	88	93	80	93	94
	Forest set-a-side Level 2	67	75	82	72	75	80	65	75	84
	Forest set-a-side Level 3	49	51	57	43	50	57	52	51	58
SK7	Utilisation degree Higher	111	112	112	112	112	112	111	111	111
	Utilisation degree Lower	89	88	88	88	88	88	89	89	89
SK8	Assortment bioenergy	115	118	117	183	188	185	67	66	67
	Assortment bioenergy/ minor	112	114	113	157	162	159	79	79	80
	Assortment optimised	103	106	104	121	133	128	89	86	87
SK9	Breeding	100	101	113	100	102	115	100	101	112
	Breeding intensive	100	103	119	100	104	121	100	102	118

Table 10.4 Results of simulations with focus on increment, utilization degree and carbon

	All in pct. of BAU	Increment			Utilisation degree			Carbon in living biomass		
	Silvicultural measure/year	2020	2050	2100	2020	2050	2100	2020	2050	2100
	BAU	100	100	100	83	76	79	100	100	100
SK1	Afforestation 2280 ha/year	100	102	105	82	75	78	100	101	104
	Afforestation 4560 ha/year	102	117	134	81	70	75	101	109	129
	Afforestation 0 ha/year	99	88	76	83	82	84	100	94	79
SK2	Afforestation decidous	100	98	96	83	77	78	100	100	102
	Afforestation coniferous	101	106	112	82	74	77	100	100	94
SK3	Rotation shorter	97	100	99	87	74	79	97	96	97
	Rotation longer	99	100	100	82	75	77	100	103	100
SK4	Regeneration more conifers	100	103	107	82	75	79	100	100	94
	Regeneration more broadleaved	99	98	97	83	75	74	100	100	102
SK5	Intensive regeneration/ nurse trees	103	116	115	80	77	79	101	108	107
SK6	Forest set-a-side Level 1	101	97	98	74	76	79	105	111	106
	Forest set-a-side Level 2	101	96	95	60	63	71	106	119	115
	Forest set-a-side Level 3	101	96	92	48	47	56	106	123	132
SK7	Utilisation degree Higher	100	100	100	92	85	88	96	90	90
	Utilisation degree Lower	100	100	100	73	67	69	105	110	110
SK8	Assortment bioenergy	100	100	100	95	90	92	100	100	100
	Assortment bioenergy/minor	100	100	100	93	88	89	100	100	100
	Assortment optimised	100	100	100	84	80	82	100	100	100
SK9	Breeding	100	105	120	83	74	75	100	101	110
	Breeding intensive	100	108	129	83	73	74	100	102	114

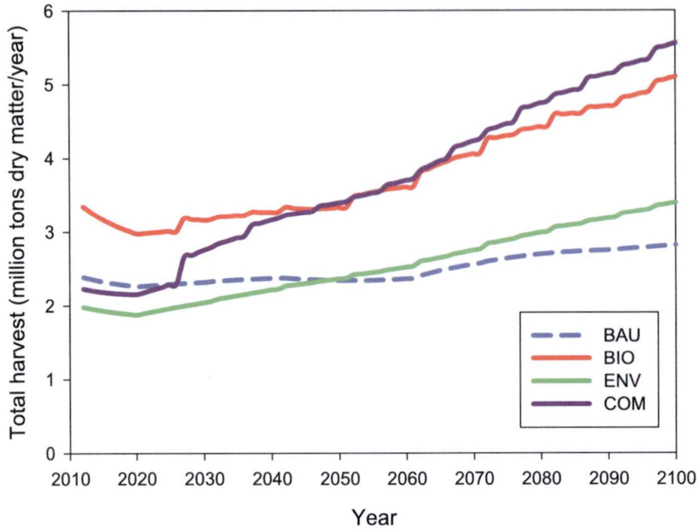

Fig. 10.2 Visualization of total harvest (million tons dry matter per year) for the four scenarios

10.5.1 Uncertainties

Technically the program and tools can handle a more refined geographical resolution, but then with a greater uncertainty because the NFI sample size for each species and age-class becomes smaller. On the other hand, because the tools have primarily been developed to assess the likely effects of silvicultural measures, the simplification in two regions can be justified.

The actual effects of the silvicultural measures at a national scale will, if implemented in operational forest policy, depend on the degree to which they are implemented on the entire forest area by all forest owners.

Changes in growing conditions or risk of forest damage caused by climate change have not been included in the models, and this uncertainty is not addressed in the models and the scenarios. Likely effects include increased growth due to higher temperatures and increased rainfall, but changes in seasonal patterns of precipitation and increased risk of windthrow may have the opposite effect.

10.5.2 Further Developments

The tools developed have not been fully analyzed with respect to their sensitivity towards changed preconditions or random variation in the individual factors or model parameter estimates. Such analyses could be considered in the further developments.

Use of the tools for prognoses of more than 500 years indicated steady states, which were consistent with results obtained from long-term field experiments. Additional such analyses will be pursued further at a later date.

References

Elingård-Larsen E, Jensen NPD (1985) Tilvækstoversigt for nobilis (Yield table for noble fir). Dansk Skovforenings Tidsskrift 70:249–272

Graudal L, Nielsen UB, Schou E et al (2013) Muligheder for bæredygtig udvidelse af dansk produceret vedmasse 2010–2100 (The possibilities for a sustainable increase in the Danish wood production 2010–2100). University of Copenhagen (IGN), Frederiksberg

Hansen JK, Nielsen UB, Graudal L (2013) Analyse af muligheder for at øge biomasseproduktionen fra de danske skove gennem forædling (Analysis of the possibilities to increase biomass production from Danish forests through forest tree breeding). University of Copenhagen (IGN), Frederiksberg

Henriksen HA (1957) Forsøgsvæsenets prøveflader i Abies-arter (Sample plots of Abies species). Det forstlige Forsøgsvæsen i Danmark 23:281–344

Henriksen HA (1958) Sitkagranens vækst og sundhedstilstand i Danmark (Growth and health of Sitka spruce in Denmark). Det forstlige Forsøgsvæsen i Danmark 24:1–372

Johannsen VK, Nord-Larsen T, Riis-Nielsen T et al (2010) Re-vised: Acquiring and updating Danish forest data for use in UNFCCC. Forest & Landscape Denmark, Frederiksberg

Johannsen VK, Nord-Larsen T, Suadicani KM (2011) Submission of information on forest management reference levels by Denmark. Forest & Landscape Denmark, Hørsholm

Johannsen VK, Nord-Larsen T, Riis-Nielsen T et al (2013a) Skove og plantager 2012 (Forests and plantations 2012). Skov & Landskab, Frederiksberg

Johannsen VK, Nord-Larsen T, Bentsen NS et al (2013b) Scenarieberegning for biomasseproduktion i skov – virkemidler og forudsætninger (Calculation of scenarios for biomass production in forests – means and prerequisites). University of Copenhagen (IGN), Frederiksberg

Kjølby V (1958) Ær. Naturhistorie, tilvækst og hugst (Sycamore. Natural history, growth and thinning). Dansk Skovforening, Frederiksberg

Larsen PH, Johannsen VK (2002) Skove og plantager 2000 (Forests and plantations 2002). Danmarks Statistik, Skov & Landskab and Skov- og Naturstyrelsen, Copenhagen

Magnussen S (1983) En tilvækstoversigt for rødgran på Østersønære lerede morænejorder (A yield table for Norway spruce on clayey morainic tills close to the Baltic sea). Dansk Skovforenings Tidsskrift 68:215–246

Møller CM (1933) Boniteringstabeller og Bonitetsvise Tilvækstoversigter for Bøg, Eg og Rødgran i Danmark (Yield tables for beech, oak and Norway spruce in Denmark). Dansk Skovforenings Tidsskrift 18(457–513):537–623

Møller CM, Nielsen C (1959) Bonitetsvise tilvækstoversigter for ask i Danmark (Yield tables for ash in Denmark). Dansk Skovforenings Tidsskrift 44:340–402

Morville K (1948) Skovfyrrens Vækst og Form (Growth and form of Scots pine). Dansk Skovforenings Tidsskrift 33:545–560

Nord-Larsen T, Heding N (2002) Træbrændselsressourcer fra danske skove over ½ ha – opgørelse og prognose 2002 (Wood fuel resources from Danish forests larger than ½ ha – assessment and prognosis 2002). Skov & Landskab, Hørsholm

Nord-Larsen T, Suadicani KM (2010) Træbrændselsressourcer fra danske skove over ½ ha – opgørelse og prognose 2010 (Wood fuel resources from Danish forests larger than ½ ha – assessment and prognosis 2010). University of Copenhagen (Skov & Landskab), Frederiksberg

Nord-Larsen T, Johannsen VK, Riis-Nielsen T et al (2014) Skove og plantager 2013 (Forests and plantations 2013). Skov & Landskab, Frederiksberg

Schou E, Thorsen BJ (2013) Økonomisk baggrundsnotat til Perspektiver for skovenes bidrag til grøn omstilling mod en biobaseret økonomi. Muligheder for bæredygtig udvidelse af dansk produceret vedmasse 2010–2110 (Economical background note for the report: The possibilities for a sustainable increase in the Danish wood production 2010–2100). University of Copenhagen (IGN), Frederiksberg

West-Nielsen G (1950) Rødgranens produktionsforhold på den midtjyske hede (The growth of Norway spruce on former Atlantic heathland in central Jutland). Hedeselskabets Tidsskrift 71:118–135

Chapter 11
Estonia

Allan Sims

11.1 Introduction

11.1.1 Descriptive Statistics

The Republic of Estonia is located in northern Europe on the eastern coast of the Baltic Sea. The total area of Estonia is 45,339 km^2, with a north-south distance of 240 km and an east-west distance of 360 km. According to the National Forest Inventory (NFI) of 2012, Estonia has 2,2 million ha of forest land which represents approximately half of the country's area (Raudsaar et al. 2014). According to the Forest Act, the definition of forest includes all forest land listed in the land register and any land of at least 0.1 ha with woody plants with a height of at least 1.3 m and a crown cover of at least 30% grow (RT I 2006).

Based on the geographical division of plants, Estonia primarily belongs to the northern area of the nemoral-coniferous or mixed forest belt of the temperate zone of the northern hemisphere. The main tree species is Scots pine which represents 34% of the forest area and 30% of growing stock with a mean volume per ha of 229 m^3. The second most important tree species is birch (Silver and Downy) which covers 31% of the forest area and represents 23% of the growing stock. The third most important species is Norway spruce, having a share of 16% of the forest area and 23% of the growing stock. Other species are aspen and alder (Grey and Black) (Raudsaar et al. 2014). Most Estonian forests are mixed forests, only 16% of stands are pure and even-aged (Adermann 2012).

In 1958, the total forest area in Estonia was 1.42 million ha, the total growing stock was 131.18 million m^3 and the average growing stock was 103 m^3 per ha.

A. Sims (✉)
Estonian University of Life Sciences, Tartu, Estonia

Estonian Environmental Agency, Tallinn, Estonia
e-mail: allan.sims@eesti.ee; Allan.Sims@Envir.ee

S. Barreiro et al. (eds.), *Forest Inventory-based Projection Systems for Wood and Biomass Availability*, Managing Forest Ecosystems 29,
DOI 10.1007/978-3-319-56201-8_11

From 1958 until 2010, the total forest area increased by 55% to 2.2 million ha, the total growing stock increased by 249% to 458.49 million m^3, and the average growing stock increased by 113% to 219 m^3 per ha (Raudsaar et al. 2014).

11.1.2 Forest Inventory History

In Estonia, two types of national forest inventories are conducted: (1) Standwise Forest Inventory (SFI), and (2) sample plot-based National Forest Inventory (NFI). The NFI started in 1999, whereas the SFI is almost 100 years old. The SFI is carried out by licensed forest inventory companies, whereas the NFI is carried out by the Estonian Environment Agency, which is a state agency administered by the Ministry of the Environment.

In Estonia, forest management requires forest inventory data for every stand, and the data must have been gathered within the last 10 years. Forest inventory is a licensed job in Estonia; thus, only companies that have a forest inventory license can undertake it. At the moment, 14 organizations are licensed, including the Estonian University of Life Sciences and the State Forest Management Centre. The SFI is carried out according to the *Forest Inventory Instruction* (FII) (RTL 2009), which was enacted on the basis of the Forest Act (RT I 2006). The instructions include definitions, methods and mathematical models to assess stand characteristics.

SFI data must be entered into the Forest Register (FR) database which is managed by the Estonian Environment Agency. The data in the FR are partially available for public use in a web-based information system (http://register.metsad.ee/avalik/). About 80% of Estonian forests are inventoried by SFI and are represented with data in the FR.

Criteria and instructions for management planning are fixed by law and described in the *Instruction for Forest Management* (IFM) (RTL 2006). This document defines the minimum density before and after thinning, the minimum clear-felling age by species, the minimum number of seedlings for planting, etc. The document was compiled by a working group that consisted of representatives from forest inventory companies, the Private Forest Centre, governmental organizations, and the Estonian University of Life Sciences.

11.2 Forest Inventory Data

11.2.1 Standwise Forest Inventory

SFI data are collected for forest management and are usually re-acquired every 10 years, because forest owners cannot carry out forest management with older data. Traditionally, forest inventories were carried out every 10 years, but nowadays active forest managers are more interested in updated forest inventory data; thus, their forest inventories are conducted more frequently.

SFI data include two levels of information: stand and stand-element level data which represents a group of trees of the same species in the same storey. At stand level, characteristics such as site type, site index, stand area and volume increment are collected, whereas at stand-element level, tree species are identified and mean height, mean diameter, basal area or number of trees, growing volume, damage characteristics at tree level and age are collected.

According to the FR, in 2013 there were 1.52 million forest stands with a total area of 1.90 million ha and a stand average area of 1.25 ha (Raudsaar et al. 2014).

11.2.2 National Forest Inventory

The first NFI was initiated in 1999 and has been carried out in 5-year cycles (Adermann 2012). The NFI sampling intensity is one sample plot per 1000 ha. The NFI sample plot network includes more than 4600 permanent plots which are measured every five years with one-fifth of the plots re-measured each year. The network contains two types of sample plots: (1) permanent plots measured for growing and standing stock estimation, and (2) temporary plots monitored for regeneration and thinning. Plots have a radius of 7 m (temporary) or 10 m (permanent).

11.3 Data and Methods for Projecting Woody Biomass

The methodology for growth projections, forest management and the models for stand simulation are published in the law (VVM 2008) as well as several algorithms for software development. There is no forest simulator available for public use, although private companies have constructed their own software based on the algorithms in the law.

Species-specific difference equations are used to project the growth in diameter and height and the evolution of the number of trees (Kiviste 1997). Timber assortments are calculated for all tree species based on Ozolins (2002) taper curve models.

11.3.1 Input Data

SFI data are used for growth simulation. The stand-element level input variables are tree species, age, mean diameter, mean height, number of trees. The IFM provides additional information on management thresholds such as the maximum stand density, the minimum density after thinning, the felling age.

11.3.2 Simulation Software

Stand growth is projected in 1-year time-steps using the difference equations until stand cutting age is achieved. Projections of height, diameter and number of trees allow other variables such as volume, basal area and stand sparsity (average distance between trees in meters) to be predicted. Stand growth is projected until stand age equals the final felling age. For each simulation year, stand density is compared with the maximum stand density and if the stand is denser than 95% of the maximum density, thinning is applied. Thinning intensity is based on the minimum density after thinning as specified in the forest management rules. Only intermediate thinning is considered and is implemented by decreasing the number of trees. Thinned volume is calculated by multiplying the volume of the average tree by the number of trees removed. For most tools developed by the companies, when the final felling age is reached, the stand is harvested and stops being projected. However some companies have implemented a stand initialization module based on the previous stand characteristics.

For every thinning and final felling timber assortments are calculated. The costs of silvicultural operations and the current year assortment prices are used to calculate the net present value for each stand.

11.3.3 Outputs

The outputs vary from one tool to another because each company has its own software, and there is no standard output.

11.4 Discussion and Conclusions

Estonia has almost 100 years of history in forest inventory and, therefore, the directive for unifying data collected by companies is well-documented. Inventory and stand variable calculations are carried out according to the FII, and management operations are selected if needed according to the IFM.

For long-term projections of available forest resources using the traditional management regime, a development manual is available for constructing a stand level growth simulator based on SFI data. The advantage of using SFI data is that 80% of Estonian forests are covered by this inventory system; however, the disadvantage is that these data can be outdated because forest management has as long as 10 years to re-acquire data and because data are not rapidly updated after cutting. On the other hand, the advantage of the NFI data is that the sample plots that are measured in the same year form a systematic sample of the entire country and, therefore, are more reliable for country level statistics of forest resources. Nevertheless, only one sample plot per 1000 ha is measured which makes these data unsuitable for small, single-owner, parcel-level simulations.

No long-term forest resources projection has been made for the whole of Estonia, primarily because the total growing stock has increased by 249% over the last 50 years due to forest area increase and average growing volume per ha increase. Therefore, any projections based on today's data could give results that are in error by more than 100%.

The simulation is used mainly for long-term cash flow prediction for private forest owners about their forest land. In forest science, the simulations are also used in different analyses for selecting final felling age, using management rules, etc. The Copernicus Project (European Space Agency) increases the opportunity to combine remote sensing (radar and multi-spectral imagery) and NFI data for analyses of issues such as land change and forest mosaic which will hopefully result in the development of a new wood availability simulation tool capable of combining NFI and remote sensing data.

References

Adermann V (2012) Eesti Metsad 2010 [Estonian Forests 2010]

Kiviste A (1997). Eesti riigimetsa puistute kõrguse, diameetri ja tagavara vanuseridade diferentsmudel 1984–1993. a. metsakorralduse takseerkirjelduste andmeil [Height, diameter and volume difference equations based on Estonian state forest standwise data at 1984–1993]. EPMÜ collection of scientific papers (63–75). Estonian Agricultural University, Tartu

Ozolins R (2002) Forest stand assortment structure analysis using mathematical modelling. – Metsanduslikud uurimused XXXVII. ISSN 1406-9954, pp 33–42

Raudsaar M, Merenäkk M, Valgepea M (2014) Yearbook Forest 2013. Estonian Environment Agency, Tartu

RT I (2006) Metsaseadus [Forest Act]. RTI 30:323. https://www.riigiteataja.ee/akt/123032015210. Accessed 1 May 2015

RTL (2006) Metsamajandamise eeskiri [Instruction for Forest Management]. RTL 2:16. https://www.riigiteataja.ee/akt/126022014017. Accessed 1 May 2015

RTL (2009) Metsa korraldamise juhend [Forest Inventory Instruction]. RTL 9:104. https://www.riigiteataja.ee/akt/13124148. Accessed 1 May 2015

VVM (2008) Kaitstavat loodusobjekti sisaldava kinnisasja riigi poolt omandamise ja ettepanekute menetlemise kord ning kriteeriumid, mille alusel loetakse ala kaitsekord kinnisasja sihtotstarbelist kasutamist oluliselt piiravaks ning kinnisasja väärtuse määramise kord ja alused. https://www.riigiteataja.ee/akt/103092013006. Accessed 1 May 2015

Chapter 12
Finland

Tuula Packalen, Kari T. Korhonen, and Olli Salminen

12.1 Introduction

In Finland, forests cover 22 million hectares of which more than 10% are reserved as conservation and wilderness areas (Finnish Statistical Year Book of Forestry 2013). The volume of growing stock is more than 2300 million m^3, and consists of Scots pine (50%), Norway spruce (30%), and deciduous trees (20%), mainly birch (Finnish Statistical Year Book of Forestry 2013). Net annual increment of growing stock is estimated to be more than 100 million m^3 per year (Finnish Statistical Year Book of Forestry 2013). In 2012, total drain was 70 million m^3 per year, of which 52 million m^3 per year were industrial roundwood removals (Finnish Statistical Year Book of Forestry 2013). In Finland, trees sequester more CO_2 than is emitted from fellings and natural mortality. The net sink in 2012 was 36 Mt CO_2 (Finnish Statistical Year Book of Forestry 2013; Statistics Finland 2014). In 2012, the consumption of energy in Finland was 1374 PJ of which approximately 25% was wood-based (Finnish Statistical Year Book of Forestry 2013). Approximately half of wood-based energy was consumed as industrial black liquor (a by-product of the pulp manufacturing process), other waste and by-products, and the rest as solid wood fuels in heating and power plants and fuel wood consumed by small-sized dwellings such as private houses or farms. Specifically, almost 8 million m^3 of forest chips were used in heating and power plants, and by small-sized dwellings. Wood fuels account for approximately 80% of all renewable energy consumed. To comply with Europe 2020 target (Europe 2020 Targets 2011), Finland has agreed to increase

T. Packalen (✉) • K.T. Korhonen
Natural Resources Institute Finland, Joensuu, Finland
e-mail: tuula.packalen@luke.fi; kari.t.korhonen@luke.fi

O. Salminen
Natural Resources Institute Finland, Vantaa, Finland
e-mail: olli.salminen@luke.fi

S. Barreiro et al. (eds.), *Forest Inventory-based Projection Systems for Wood and Biomass Availability*, Managing Forest Ecosystems 29,
DOI 10.1007/978-3-319-56201-8_12

its share of renewable energy sources from 28.5% (in 2005) to 38% by the year 2020. The national objective (Finland's National Forest Programme 2008) is that forest chips will account for half of the expected increase by 2020 when 12–13 million m^3 of forest chips will be consumed annually.

In Finland, industrial roundwood is harvested mainly using a cut-to-length system, i.e., as saw logs and pulpwood. The potential sources of energy wood include whole felled trees including crown, felled stems without branches or components of trees that do not satisfy the requirements for industrial use. Felled trees may be rejected for industrial use because of poor quality or small size of trees. Tree components rejected for industrial use include tops of stems, living and dead branches, foliage, off-cuts of stems, stumps and roots. Biomass harvested for energy use is usually converted to chips in the forest, at the roadside or at the site of end-use. The amount of residues left in the forest after cutting depends mainly on tree species, volume, size and branchiness of felled trees, and the amount of decayed wood. In Finland, the average recovery of residues from a logging site is 65–75% of the total residue or approximately 20–30% of the wood harvested for industrial use. Profitability is greatest for logging residue at sites where mature spruce-dominated forests have undergone final fellings.

Approximately 52% of forestry land is owned by non-industrial private forest owners (NIPF). There are 345,000 NIPF holdings, whose average size is 30 ha (holdings less than 2 ha excluded, Finnish Statistical Yearbook of Forestry 2013). State forests (35%) are managed by Metsähallitus, a state enterprise. Industrial private forests (8%) consist of forest land owned by wood processing companies. Other public forests (5%) consist of municipal, foundational, and church forests. In Finland, forest owners make their own decisions concerning cutting and other management activities that consequently affect the supply of timber, and future forest conditions for all citizens. Forest management decision-making for different forest ownerships is framed by the Forest Act (1093/1996), governed by Ministry of Agriculture and Forestry (MAF). (Finnish Statistical Yearbook of Forestry 2013).

Societal objectives for the use of forests are defined in the participatory processes of national and regional forest programmes covering forests in all ownership categories including non-industrial private, company, state and other ownerships. The focus of the programmes, as well as the role of the National Forest Inventory (NFI) has changed over decades. The first Finnish forest programmes referred to as HKLN, Teho and MERA I-III were designed after the Second World War to support intensive work in forest management and improvement. The timescale of these programmes was several decades. Simple projections of growing stock development under different financing programmes were calculated manually based on NFI results. Since the 1980s, the Forest 2000 Programme and its successors such as the National Forest Programmes (NFPs) in 1999 and 2008 (Finland's National Forest Programme 2008) have had wider interests in forests and forestry than solely timber production. Parallel to forest programmes, conservation programmes and climate strategies have been designed and their success monitored. All of them have been supported by the MELA model calculations based on NFI sample plot and tree data (see e.g. Auvinen et al. 2007; Kärkkäinen et al. 2008; Matala et al. 2009; Nuutinen et al. 2009; Sievänen et al. 2014; Kallio et al. 2013). For the most recent NFP (Finland's National Forest Programme 2008), a partial equilibrium model, SF-GTM, was used (Uusivuori et al. 2008) to seek market balance between industrial and energy demand and supply of wood.

The name MELA originates from the Finnish word MEtsäLAskelma (forestry analysis). The MELA software is written in FORTRAN and consists of two executable components, MELASIM and MELAOPT (Redsven et al. 2013). In addition to national level analyses, MELA has been applied at regional and enterprise levels for strategic analysis based on NFI data. For operational planning in state, company and private (estate) forest stand data are used instead of NFI data. The most recent advances in calculation methods include the integration of multisource NFI (MS-NFI), administrative and land-form data sets for landscape level forest scenarios used in the preparation of a strategic forestry programme at the local (village) level (Mäkelä et al. 2011; Kärkkäinen et al. 2011; Nuutinen et al. 2011; Kärkkäinen et al. 2013).

In the following, the NFI-MELA framework for studying the simultaneous recovery of industrial wood and energy wood from cuttings and consequent development of growing stock during next decades is presented.

12.2 Data

The MELA input data are based on inventory sample plots from the Finnish NFI. In Finland, the first NFI inventory was conducted by the 1920s (NFI1 1921–1924) and since then, NFIs have been conducted regularly at 5–10 years cycles. The latest field measurements were carried out for NFI11 (2009–2013). Based on forest sample plots located in temporary clusters, reliable forest statistics can be estimated for the entire country and for large areas of more than 200,000 ha such as regional Forestry Centres (Korhonen 2016). Cluster sampling features such as the distance between clusters, cluster shape, the number of field plots per cluster and the distance between plots within a cluster vary by region according to the spatial variation of the forests and the road network density (Korhonen 2016).

For each NFI sample plot, tally tree data such as tree species and diameter at breast height are augmented using sub-model predictions to compensate for missing MELA sample-tree variables such as height and age, and MELA sample-plot variables are calculated using the sample plot data (Fig. 12.1). For the simulation, modelling units are constructed and classified into three forest management categories based on ecological and legal (e.g. conservation areas) constraints: (1) no restrictions on wood production, (2) restrictions on wood production exist, but wood production is not totally forbidden, (3) no wood production is allowed. In the first category, all typical forest management measures such as thinning and regeneration cuttings are allowed; in the second category, clear cuttings are forbidden; and in the third category, all forest management measures were forbidden.

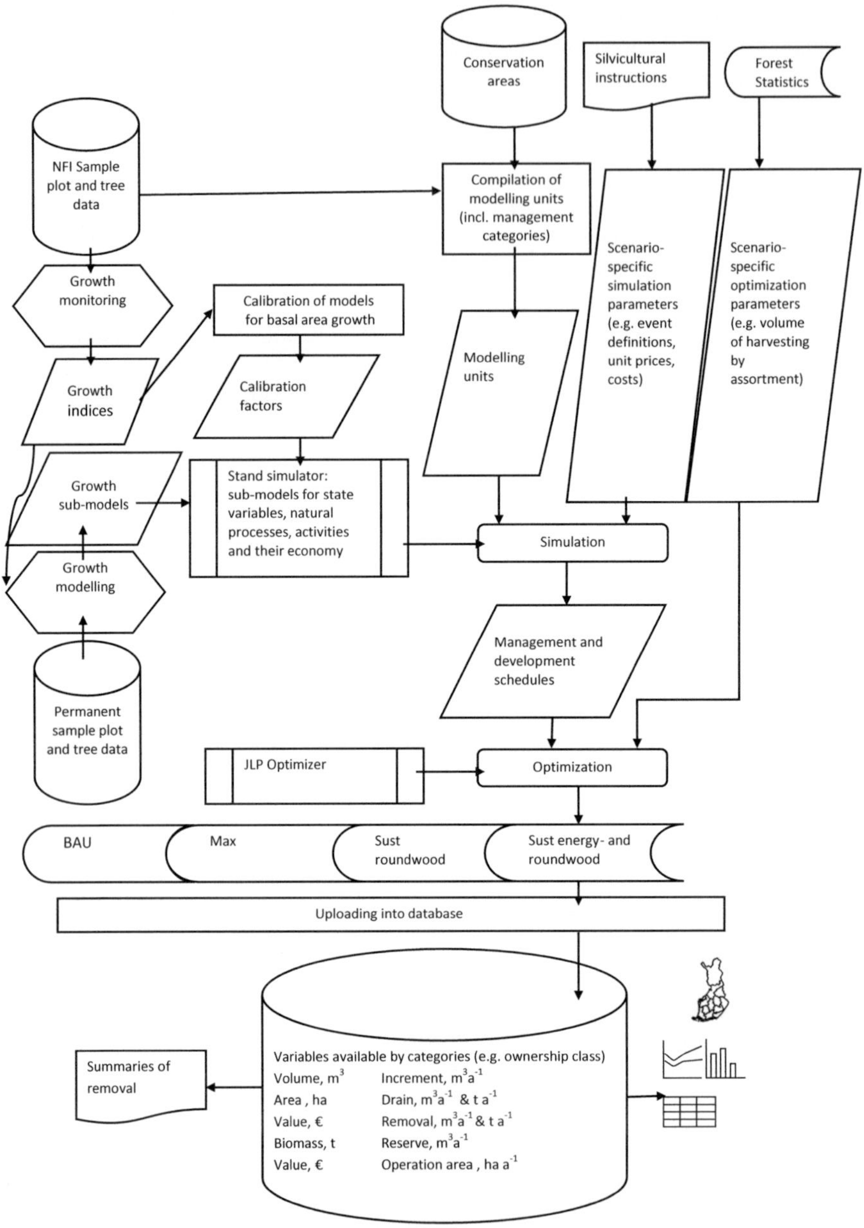

Fig. 12.1 The flowchart of MELA analysis based on the NFI sample plot and tree data

Other data sources in the MELA analyses include spatially referenced data on conservation areas, statistical information on harvesting for business-as-usual scenarios and economic information on unit prices and costs (Metinfo 2013).

12.3 Methodology

In this study the national woody biomass resource projections were carried out using the Finnish forestry dynamics model MELA. The approach adopted in the MELA model (Siitonen et al. 1996; Redsven et al. 2013) is based on integrated stand-level simulation and forest area-level optimization. The stand simulator automatically generates several alternative feasible management schedules for the modelling units, such as forest stands or sample plots representing the stands. The stand simulator uses non-spatial empirical sub-models for individual trees (Hynynen et al. 2002). The events in the simulation consist of natural processes such as in-growth, growth and tree mortality and management activities defined by the built-in basic event routines (e.g. artificial regeneration with selection of tree species, clearing of regeneration area, soil preparation, tending of young stands, cuttings, ditching, fertilisation, pruning of pine). The event parameter (see above the list of event routines) of the MELA model enables the definition of a set of optional events for each analysis within the built-in event routines and their arguments. For example, thinning events can be constrained to remove only industrial roundwood, both industrial roundwood and energy wood or only energy wood. Final felling events can be constrained so that only industrial roundwood, or industrial roundwood and waste wood, or industrial roundwood, waste wood and stumps are removed. The feasibility of management activities can be defined, for example, by management categories. In the model, a set of calibrated tree-level sub-models is used to develop the growing stock predictions for the MELA description (sample) trees of the sample plots in a 5-year time-step (Hynynen et al. 2002). In collaboration with the NFI and growth and yield modellers, growth indices based on climate data and NFI growth measurements (Henttonen 2000) are used when calibrating sub-models before running the MELA simulations. The MELA stand simulator reads in the results to calibrate the sub-models. Forest state variables such as volume (Laasasenaho 1982) or dry weight (Repola 2009a, b) are estimated using specific models for different tree components. The components available for energy wood include roundwood (also stems not fulfilling the requirements for industrial wood because of small size or poor quality), foliage, living and dead branches, the tops of stems as well as stumps and roots. The tops of stems and stems not satisfying the requirements for industrial wood are grouped together as waste wood. Minimum top diameters for pulpwood are 6.5 cm for spruce and deciduous trees and 6.3 cm for pine and other conifers.

After the simulation, an optimization package based on linear programming, called JLP (Lappi 1992), is used to select an optimal combination of management schedules. The optimization task is given as an objective function and a set of constraints. Thousands of variables describing the forest state are available for the definition of the optimization task.

Because the amount and type of potential cuttings for roundwood are of great interest in the Finnish forest sector, four different cutting scenarios are typically defined to map the production potentials. Scenarios are generated through optimization tasks: (1) to maximize the net present value of timber production by using a 4% interest rate with the same flow of saw logs, pulpwood and energy wood as during the past 5 years (business-as-usual scenario), (2) to maximize the net present value of timber production by using a 5% interest rate (Max scenario), (3–4) to maximize the net present value of timber production by using a 4% interest rate with non-decreasing flow of wood and net income over a 50-year period and net present value after the 50-year period greater than or equal to the beginning (two variants of Sust scenario, both referring to the sustainable flow of wood and income). In Finland, there are three ways to record statistics on harvesting: (1) area estimates based on harvesting plans submitted to forest administration (Forest Centers), (2) area estimates based on the NFI, and (3) volume estimates based on wood sales data. For the MELA business-as-usual (BAU) scenario, the sales volume estimates from statistics are used. Therefore, we call the BAU scenario demand-based, i.e., the scenario is driven by harvesting of wood as recorded by timber assortment for the past few years. In the traditional maximum sustainable (Sust) scenario, the flow of wood includes sawlogs and pulpwood (Sust industrial roundwood) and the modern maximum sustainable scenario also includes energy wood (Sust energy- and industrial roundwood). Both Sust scenarios are supply-based, i.e., scenarios are driven by harvesting potential determined based on NFI and auxiliary land-use data to map forests and wood available for supply.

The unit wood prices (€/m^3) and costs (€/ha) of felling, silvicultural and forest improvement work are based on the deflated average realised prices and costs during the last 10-year period (Metinfo 2013). Forest management activities are simulated based on the current silvicultural instructions given as event parameters. For example, thinning rules are based on dominant height and basal area. Correspondingly, the rotation period is determined by the basal area weighted by breast height diameter and/or stand age. The rules are defined separately for each tree species and site type.

12.4 Results

The most recent results are available at MELA Analysis Service (http://mela2.metla.fi/mela/tupa/tupaindex-en.htm, retrieved 7 April 2014). The web-based dissemination is designed for easy and fast knowledge transfer. The MELA Analysis Service is updated regularly by Metla, based on the most recent NFI data and synchronized with the publication of the NFI outputs.

The variables cover growing stock (in 2010, 2020, 2030 and 2040) and harvest (in the periods of 2010–2019, 2020–2029, 2030–2039) by assortments in different strata.

During the period 2008–2012, the amount of extracted industrial roundwood was 50 million m^3 per year and the amount of energy wood was 12 million m^3 per year (Metinfo 2013). If the fellings follow this BAU scenario, the volume of growing stock is projected to increase from the initial 2310 million m^3 to 3428 million m^3 by 2040.

The projected industrial roundwood removal in the Max scenario during the period 2010–2019 is 100 million m^3 per year, while the projection of potential energy wood is 27 million m^3 per year. During the period 2030–2039, the projected potential recovery of industrial wood is 72 million m^3 per year, and energy wood is 21 million m^3 per year. In the Max scenario, the volume of growing stock is projected to decrease by the end of the year 2019. After 2020, volume is projected to begin gradually increasing.

In the traditional Sust cutting scenario (Sust industrial roundwood), the potential recovery of industrial roundwood during the period 2010–2019 is projected to be 77 million m^3 per year. During the period 2030–2039, the potential recovery of industrial roundwood is projected to reach 86 million m^3 per year. In the modern Sust cutting scenario (Sust energy- and industrial roundwood), the projected potential recovery of industrial roundwood during the period 2010–2019 was 73 million m^3 per year, while the amount of potential energy wood was 21 million m^3 per year. During the period 2030–2039, the potential recovery of industrial roundwood was projected to be 78 million m^3 per year and energy wood 22 million m^3 per year. In the both Sust scenarios, the volume of growing stock is projected to increase slightly and to be larger than in the Max scenario.

The proportion of saw logs in cuttings is projected to decrease and pulpwood to increase, especially in the Max-cutting scenario. The proportion of different biomass components is also estimated to change in the future, reflecting the decrease in the amount of final fellings, and consequently a decrease in amount of wastewood.

12.5 Discussion

In all scenarios, due to changes in the structure of the forests during the next 50 years, the proportion of saw logs is projected to decrease and the amount of pulpwood from cuttings is projected to increase. In the future, the area of forest where thinning is feasible will increase. On average, the biomass of living branches and foliage in relation to the biomass of the stem is greater in young trees than in mature trees. Therefore, in the future the proportion of residues will be greater relative to that of harvestable stems for industrial use. During the next decades, the amount of spruce-dominated mature forest will decrease; consequently, the total amount of residues from the final cuttings of spruce-dominated forests will decrease. The actual harvest of industrial wood and raw material for energy wood will depend, on the one hand, on the capability of industry to pay for wood for different end-use purposes and, on the other hand, on the interests of forest owners. Satisfaction of the national wood energy target will create new challenges to develop cost-efficient harvesting systems for collecting forest chips from other types of cutting as well. Because the volume of growing stock is increasing in all scenarios, carbon pools in standing trees are increasing in spite of cuttings. Clearly, the mitigation potential of trees is considerable; especially when the carbon in removed trees is sequestered in wood-based products and the role of wood energy as substitute for carbon-based energy is considered.

The MELA scenarios outline the production potential of the Finnish forestry in terms of industry and energy wood, and remaining carbon in growing stock. The MELA approach makes it possible to track any wood, biomass (energy) or carbon component of a tree, either a standing tree or a cut tree, up to the stand or regional level. In addition, we can model the development conditional on different types of management and cutting operations. When using NFI sample plot and tree data, we can accommodate the variety of different types of stands, their geographic and site conditions as well as their current growing stock in initial state. Tree-level growth models accommodate competitive interactions within a stand. Therefore, the effects of stand density, species mixture, various size classes and age can be accommodated. The effects of forest management on the growth of trees and on the amount of different timber assortments and biomass components can be estimated quite accurately with tree-level models as shown by sub-model validations (e.g. Matala et al. 2003). However, the basic models as such are not applicable for Continuous Cover Forestry (CCF). In addition, the detailed descriptions of forest conditions makes it possible to address profitability of operations for each site depending on technology available.

When comparing the national supply analysis with the European Forest Sector Outlook Studies (EFSOS) (UN-ECE and FAO 2011), the European study seems to predict the development of growing stock in Finland in right magnitude according to specified fellings. However, in Finland the cutting potential is considerably greater than sustainable removal due to the age structure of the forest. Therefore, the specified wood demand ("real supply") can be harvested in many ways, each of them resulting in different development of forest resources. This applies specifically for BAU which for decades has been much less than harvesting potential. Obviously, wood sales and consequent harvesting reflect market demand for different timber assortments. According to the EFSOS results, the EFISCEN model favors final fellings to thinnings. In national scenarios, thinnings are more common than final fellings in future decades. Concerning wood demand, EFSOS forecasts an increase in the Finnish production in sawn wood, newsprint as well as printing and writing papers, and consequently also in wood demand. Because Finland is a large exporter and the role of the Finnish producers is based on the global market, the future magnitude of production is subject to uncertainty. In addition, the role of imported wood may increase due to differences in profitability among countries and, therefore, affect the domestic harvest. Therefore, use of a partial equilibrium model such as

SF-GTM (Kallio et al. 2013) is necessary when analyzing the balance between wood demand and supply.

Scenarios are always conditional on underlying assumptions. As opposed to demand-based BAU, supply-based BAU (e.g. Antón-Fernández and Astrup 2011) could be used. The NFI sample plot data includes information on forest ownership group (private, state, industrial, others). However, the data have not yet been used in modeling the effects of forest owner interest despite the existence of national forest owner surveys. There are also gaps of knowledge related to complex interactions between climate change and other natural processes, including forest damage and carbon and nutrient cycles in soils. Consequently, there are uncertainties related to the optimal allocation of wood, with changing energy demand and the valuation of the different wood-based and forest-based products and services. Uncertainties in this type of modelling are related to the use of data and sub-models, the incorporation of the models into the simulator and the scenario assumptions. To date, the main focus has been the validation of sub-models (Hynynen et al. 2002; Matala et al. 2003). More research is needed to validate the MELA analysis and provide uncertainty estimates.

References

Antón-Fernández C, Astrup R (2011) Empirical harvest models and their use in regional business-as-usual scenarios of timber supply and carbon stock development. Scand J For Res 27:379–392

Auvinen A-P, Hildén M, Toivonen H et al (2007) Evaluation of the Finnish National Biodiversity Action Plan 1997–2005. Monogr Boreal Env Res 29:54

Europe 2020 targets (2011) Available at: http://ec.europa.eu/europe2020/pdf/targets_en.pdf. Retrieved 4 April 2014

Finland's National Forest Programme (2008) More welfare from diverse forests – government resolution. Publication of the Finnish Ministry of Agriculture and Forestry, No 3b/2008

Finnish Statistical Yearbook of Forestry (2013) Official Statistics of Finland. Agriculture, forestry and fishery

Henttonen H (2000) Growth variation. In: Mälkönen E (ed) Forest condition in a changing environment – the Finnish case. Forestry sciences, vol 65. Kluwer Academic Publishers, Dordrecht, pp 25–32

Hynynen J, Ojansuu R, Hökkä H, et al (2002) Models for predicting stand development in MELA System. Metsäntutkimuslaitoksen tiedonantoja 835

Kallio AMI, Salminen O, Sievänen R (2013) Sequester or substitute – consequences of increased production of wood based energy on the carbon balance in Finland. J For Econ 19(4):402–415

Kärkkäinen L, Matala J, Härkönen K et al (2008) Potential recovery of industrial wood and energy wood raw material in different cutting and climate scenarios for Finland. Biomass Bioenergy 32(10):934–943

Kärkkäinen L, Nuutinen T, Hirvelä H, Mäkelä H (2011) Effects of administrative land-use and technical land-form constraints on timber production at the landscape level. Scand J For Res 26(2):120–127

Kärkkäinen L, Packalen T, Hamunen H (2013) Indicators of the criteria for good participation in ecotourism planning at local level – a Nordic case study. Tour Plan Dev 10(4):451–466

Korhonen KT (2016) National forest inventories: Assessment of wood availability and use: Finland. In: Vidal C, Alberdi I, Hernandez L, Redmond JJ (eds) National forest inventories: Assessment of wood availability and use. Springer, Switzerland, pp 369–384

Laasasenaho J (1982) Taper curve and volume functions for pine, spruce and birch. Seloste: Männyn, kuusen ja koivun runkokäyrä- ja tilavuusyhtälöt. Communicationes Instituti Forestalis Fenniae 108

Lappi J (1992) JLP: a linear programming package for management planning. Metsäntutkimuslaitoksen tiedonantoja 414

Mäkelä H, Hirvelä H, Nuutinen T, Kärkkäinen L (2011) Estimating forest data for analyses of forest production and utilization possibilities at local level by means of multi-source national forest inventory. For Ecol Manag 262(8):1345–1359

Matala J, Hynynen J, Miina J et al (2003) Comparison of a physiological model and a statistical model for prediction of growth and yield in boreal forests. Ecol Model 161(1-2):95–116

Matala J, Kärkkäinen L, Härkönen K et al (2009) Carbon sequestration in the growing stock of trees in Finland under different cutting and climate scenarios. Eur J For Res 128(5):493–504

Metinfo (2013) Metsäsektorin suorakäyttöinen tietojärjestelmä. Metsäntutkimuslaitos. WWW -sovellus (http://www.metla.fi/metinfo/)

Nuutinen T, Kilpeläinen A, Hirvelä H et al (2009) Future wood and fibre sources – case North Karelia in Eastern Finland. Silva Fenn 43(3):489–505

Nuutinen T, Anola-Pukkila A, Hirvelä H et al (2011) Information and communication technology connecting mathematical modelling with multipurpose forest management. J For Plan 16:91–98

Redsven V, Hirvelä H, Härkönen K et al (2013) MELA2012 reference manual, 2nd edn. The Finnish Forest Research Institute, 666 p

Repola J (2009a) Biomass equations for Birch in Finland. Silva Fenn 42(4):605–624

Repola J (2009b) Biomass equations for Scots pine and Norway spruce in Finland. Silva Fenn 43(4):625–647

Sievänen R, Salminen O, Lehtonen A et al (2014) Carbon stock changes of forest land in Finland under different levels of wood use and climate change. Ann For Sci 71(2):255–265

Siitonen M, Härkönen K, Hirvelä H et al (1996) MELA handbook – 1996 edition. Metsäntutkimuslaitoksen tiedonantoja 622

Statistics Finland (2014) Finland's National Inventory Report (NIR) under the UNFCCC (United Nations Framework Convention on Climate Change) and the Kyoto Protocol. Statistics Finland

UNECE and FAO (2011) The European Forest Sector Outlook Study II 2010–2030. United Nations, New York/Geneva

Uusivuori J, Kallio M, Salminen O (eds) (2008) Vaihtoehtolaskelmat Kansallisen metsäohjelman 2015 valmistelua varten. Metlan työraportteja/Working Papers of the Finnish Forest Research Institute 75

Chapter 13
France

Antoine Colin, Holger Wernsdörfer, Alain Thivolle-Cazat, and Jean-Daniel Bontemps

Projections of wood resources have been needed for decades by both public authorities and public and private stakeholders. Projections are requested to formulate and assess forest policy and management strategies for the coming years to decades at regional, national, and international scales. The French NFI provides information for describing the forest resources and their spatio-temporal dynamics. These data contribute to initializing two large-scale dynamic models for forest resource projection that have been developed in France and implemented since the 1980s. The decision to use diameter-class or the age-class matrix models depends on the type of forest to be simulated. Both approaches provide robust projections from short to medium time scales and at spatial scales ranging from regional to national.

Recent projections carried out with these models have highlighted the potential for French forests to remain an important carbon sink until 2030, while fellings can be increased by two-thirds. This matrix-model approach is associated with interdisciplinary

A. Colin (✉)
IGN, Direction Interrégionale Nord-Est, Champigneulles, France
e-mail: antoine.colin@ign.fr

H. Wernsdörfer
UMR LERFOB, AgroParisTech, INRA, Nancy, France
e-mail: holger.wernsdorfer@agroparistech.fr

A. Thivolle-Cazat
Institut Technologique FCBA, Grenoble, France
e-mail: alain.thivollecazat@fcba.fr

J.-D. Bontemps
UMR LERFOB, AgroParisTech, INRA, Nancy, France

IGN, Laboratoire de l'Inventaire Forestier (LIF), Nancy, France
e-mail: jdbontemps.ign@gmail.com

S. Barreiro et al. (eds.), *Forest Inventory-based Projection Systems for Wood and Biomass Availability*, Managing Forest Ecosystems 29,
DOI 10.1007/978-3-319-56201-8_13

research programs that aim at developing environment-dependent representations of forest dynamic processes such as growth, density-dependent mortality and forest regeneration, and also coupling of forest dynamics with forest sector models. Model validation is also essential, highlighting the major importance of current continuous forest inventory, and making forest inventory data comparable over time following changes in mensuration protocols.

13.1 Introduction

French forests are characterised by a high diversity in various aspects that must be considered when developing models for projecting wood resources and potential wood supply. We first provide this necessary background information and then present objectives, data and projection methods, and perspectives for further model development.

13.1.1 Key Figures on Forests and Forestry in France

Forests cover 16.8 million ha of the French metropolitan territory (30.8% of the area) and 8.2 million ha in oversea territories, mainly in French Guiana. Temperate and Mediterranean forests of the metropolitan territory thus represent two-thirds of the total national forested area. Intensive and systematic forest monitoring via the National Forest Inventory (NFI) has been carried out over the metropolitan territory since the late 1950s, enabling continuous analysis of forest features and wood resources. NFI-like monitoring is not conducted in the French tropical forests.

French metropolitan forests have developed in a variety of climates. Oceanic and sub-oceanic climates in western and north-central France favour the development of sessile (*Quercus petraea*) and pedunculate (*Quercus robur*) oaks. Maritime pine (*Pinus pinaster*) grows especially in the oceanic climate of the south-west. Semi-continental climate in the north-eastern and mountainous areas (Alps, Jura, Pyrenees, Massif Central and Vosges) are suitable for European beech (*Fagus sylvatica*), Norway spruce (*Picea abies*), silver fir (*Abies alba*), Scots pine (*Pinus sylvestris*) and European larch (*Larix decidua*). In the southern Mediterranean range, evergreen oak (*Quercus ilex*), Aleppo (*Pinus halepensis*) and Corsican (*Pinus nigra ssp laricio var Corsicana*) pines grow. As a consequence, French forests encompass all but the "boreal forests" type (13 types) according to the *European Forest Types* classification (EFT, Barbati et al. 2014), and rank first in Europe for the number of forest types covered. In addition, 18 tree species are found predominant over >200,000 ha (IFN 2010). Many other species are found in various locations, such as pubescent oak (*Quercus pubescens*), chestnut (*Castanea sativa*), hornbeam (*Carpinus betulus*), Douglas fir (*Pseudotsuga menziesii*), ashes (*Fraxinus spp.*), aspen (*Populus spp.*), etc. Mixed species forests hence form one half of the national resource, both in terms of forest area and growing stock (Morneau et al. 2008).

Stand structure also shows great variations, including pure even-aged high forests of broadleaved and coniferous species mainly in the northern plains and in the Aquitaine region, uneven-aged mixed species forests in the mountainous domain, and coppices and coppices with standards in the plains.

The heterogeneity in species and structure of French forests originates from ecological factors. Coniferous species are mainly located in the mountainous domain, broadleaved species are found in the plains, and Mediterranean species are located in southern France. It also results from historical socio-economic factors. In the plains, forests are dominated by broadleaved species, with sessile and pedunculate oak species ranking first for the growing stock (566 million m^3 representing 23% of the total national growing stock, IFN 2010). These species have been favoured since medieval and even Roman times, because of their ability to re-sprout from stumps after cutting. In the eighteenth and nineteenth century oak coppices with standards stands were managed to produce roundwood and fuelwood for domestic and industrial purposes. Later, following the introduction of coal and the need to produce timber, many of these forest types were either naturally or actively progressively converted into high forests, with high forests on stumps as a first structural step (IGN 2013).

Conifer plantations were promoted in the nineteenth century to drain the wetlands of Gascogne and Sologne and to protect fragile soils for purposes such as preventing erosion in the Alpine range or fixing dunes on the coast range. The National Forest Fund (Fonds Forestier National: FFN) policy framework was launched in 1947. It contributed to the development of a new conifer forest resource over marginal agricultural lands, and was instrumental in securing the wood supply for the country's industries (Gadant 1987). The FFN is estimated to have contributed 75% of the increase in coniferous forest plantation area, i.e. 1.5 million ha of new forests with Norway spruce (*Picea abies*), Douglas fir (*Pseudotsuga menziesii*), pines (*Pinus sylvestris*, *Pinus nigra*) and European larch (*Larix decidua*) as the main afforestation species (Pardé 1966; Dodane 2012).

The area of forests available for wood supply is 15.7 million ha in 2013, of which 11.8 million ha (74.8%) are private forests, 1.4 million ha (9%) are state-owned forests and 2.5 million ha (16.2%) are other public forests including forests owned by territorial communities and public authorities (IGN 2014a). Over the past decades, the decline of agricultural activity in low productive areas has led to further natural afforestation by pioneer species which may, however, neither correspond to demand on the wood market nor be situated in easily harvestable locations (IGN 2012). As a result of both afforestation and natural colonization, the French Forest area has more than doubled since 1788 (Brénac 1984). This reflects the typical pattern of 'forest transition' (Mather 1992) that has been experienced by most developed European countries following the industrial revolution and urbanization processes.

Concern for nature and biodiversity conservation has produced legal constraints (e.g. Natura 2000 network) that partially exclude some French forest areas from traditional forest management and harvest schemes. The impact on wood supply can be locally significant. In the Vosges Mountains and on the Lorraine plateau, 4% of the total forest area is not available for wood supply, and 8% of the forests available for wood supply are subject to management for species conservation (Colin and Lambert 2012; Thivolle-Cazat et al. 2012). Physical constraints also significantly hamper forest management in France, and cause a decrease in the profitability of harvest operations. One-third of the forest area available for wood supply is estimated to be very difficult to access, and 1 million ha of the forested area has a forwarding distance greater than 1 km, mainly in south-eastern France (IFN 2010).

Total annual fellings amount to 64 million m^3 of total aboveground volume over bark (IFN 2011c). Industrial roundwood removals (sawlogs, pulpwood, etc.) represent about 36 million m^3 over bark per year (Agreste 2011) and fuelwood about 20 million m^3 over bark per year (FAO 2011), while the remaining part corresponds to harvest residues. Total removals have increased slowly over the last 20 years and result from both an increase of removals from coniferous stands afforested between 1950 and 1980 (FFN) and a regular decrease in hardwood removals. Forest chip removals for energy have increased rapidly in recent years; the removed volumes have doubled every year since 2006, and reached 1.8 million m^3 in 2013 (round wood equivalent).

For several decades, French forests have experienced an increasing gap between total drain and increment. Hence, growing stock, assessed as bole volume up to a top-end diameter of 7 cm, has increased by about 25 million m^3 per year over a 26-year period. This increase represents an additional 650 million m^3 (IFN 2011b) for a total of 2.52 billion m^3. There is no sign of a future shift in this structural trend. Recent and severe damages caused by storms *Lothar* and *Martin* in 1999 (140 million m^3 of damage, Doll 2000 and IFN 2003) and *Klaus* in 2009 (43 million m^3 of damage, IFN 2009) represent anecdotal drains compared to the trend in growing stock on a national scale. The current expansion of the French forest area will further contribute to an increase in the growing stock in future decades. Thus, the French forest is far from a stationary state of development.

13.1.2 Motives and Objectives for Projection

The needs for projections on forest resources arise from two categories of stakeholders:

- **Public authorities,** at both national and regional levels, require projections to formulate and assess public policies, for strategic guidance and planning of the industrial roundwood and fuelwood sectors, and to report the contribution of forests to carbon sequestration as required by international policy processes on Green House Gases (GHG) emissions reduction.
- **Public or private stakeholders of the forest-wood sector**, including the French Forest Service (Office National des Forêts, ONF) that manages the public forests, saw and veneer-mills, pulp and paper, panel and fuelwood companies, and professional organisations (e.g. associations of wood producers and processers, interbranch organisations) require estimates for the wood availability and quality of a given species in a given area to formulate or support their economic strategies.

Inventory, assessment, analysis and projection of the forest resource, including wood volume, biomass or carbon, form the missions of the French NFI which is carried out by the French National Institute for Geographic and Forest Information (Institut national de l'information géographique et forestière: IGN). The French NFI consists of a repeated and spatially-systematic inventory of metropolitan forests with the objective of providing statistics at the regional and national levels (Vidal et al. 2005). It is thus well-suited for contributing to decision making and policy formulation at these scales. By contrast, it is not suitable for supporting forest management at a local scale.

Illustrations of recent wood resource projection studies based on NFI data follow:

- 2009, national level: potential wood availability for industrial and energy uses (study funded by the national renewable energy agency and the ministry of agriculture, Colin et al. 2009; Ginisty et al. 2011);
- 2010, intra-regional level: impact of the 2009 storm on the wood resource of the Landes de Gascogne maritime pine forest (study funded by the ministry of agriculture, Colin et al. 2010).
- 2012, regional level: potential wood availability in Lorraine and Aquitaine until 2025 (studies co-funded by regional and infra-regional public bodies, as well as professional organizations for pulp, paper and fuelwood, Colin and Lambert 2012; Thivolle-Cazat et al. 2012; Colin et al. 2012).
- 2014, national level: GHG emissions of the French forest until 2030 in the context of the development of renewable energies (study funded by the ministry of environment, Colin 2014).

13.2 Data Sources

13.2.1 NFI Data Used for Wood Resource Projections

Each year, the French NFI inventories the entire metropolitan forested area. The NFI is carried out in public and private forests, regardless of whether they are available for wood supply. Since 2004, the NFI design has featured a systematic sampling grid with a 10% systematic sample of the grid measured each year. Each year sampling is conducted in two phases. In the first phase, approximately 80,000 photo plots are interpreted to assess land cover and land use. In the second phase, approximately 7000 temporary ground plots are established at a subsample of first phase photo plot locations that have forest land use. All living trees on the sample plots are cored at 1.30 m height to measure radial increment over the past 5 years. Trees with diameters less than 7.5 cm are not measured. NFI statistics are calculated using a post-stratification process using the French forest map (IFN 2008) and aerial photo-interpretation of land cover/land use for the 80,000 photo plots for a given year. Typically five annual NFI samples are combined to calculate statistics for the forest resource.

NFI data used for projections include:

- Contextual data: a biophysical classification of the forest area into "sylvo-eco-regions" (IFN 2011a), administrative regions, forest land ownership, availability of forests for wood supply, topographic position, elevation, physical accessibility, soil characteristics, etc.;
- Plot level data: species composition and forest structure, stand age, number of stems, top height and site index (Seynave et al. 2005), etc.;
- Tree level data: tree species, circumference at 1.30 m height, height, wood quality classes, radial increment, bark thickness, growing stock, mortality, fellings, etc.

13.2.2 Additional Data for Spatial Contextualization of the Forest Resource

When projecting wood resources, the territory is usually divided into strata, often derived from GIS information. Both the wood resource and silvicultural scenarios are described at this level. In such studies, potential wood availability estimates and actual fellings often show a gap, due to repeatedly acknowledged physical, socio-economic, and environmental constraints (Puech 2009). Physical restrictions such as slope, presence of forwarding tracks and forwarding distance, prevail in mountain areas. Socio-economic restrictions result from the private ownership structure in France with a majority of small forest estates (Agreste 2013). Environmental restrictions arise from conservation and protection measures (soils, freshwater springs, etc.).

The initial state of the wood resource and projection scenarios can be refined using external GIS information to identify forest types with specific management features. For example, maps of large private forests with forest management plans allow identification of areas with more intensive management opportunities. Legal restrictions for environmental protection also have an impact on harvest costs or management regimes, and the areas where silviculture is legally not permitted (e.g. biological reserves with strong protective rules) are sometimes excluded from wood projections. Specific forest management practices are also implemented in areas with harvesting limitations for fauna/flora conservation. For example in the Vosges Mountains, silvicultural practices for the conservation of the Western Capercaillie (*Tetrao urogallus major*) habitat are encouraged (MEDDE 2012).

13.2.3 Newer NFI Data to Quantify Felling Regimes

Since 2010, the French NFI has provided direct measurements of fellings based on a partial re-inventory of temporary plots inventoried 5 years earlier. These estimates are much more accurate than previous estimates based on consumption surveys or stump inventories (IFN 2003). Productivity, mortality and felling statistics are now fully consistent with variations in the growing stock (IFN 2011c). These new data can contribute to the calibration of silvicultural and mortality scenarios used for projections, and especially the Business-As-Usual (BAU) scenario (Colin 2014). The silvicultural scenario derived from these measurements is relevant for BAU projections, because it takes into account implicitly all the current management and harvest opportunities and constraints (physical, economic, environmental and social).

These new data can also contribute to virtually doubling the number of NFI sample plots available for estimating the forest resource for a given year, because they can be used to update the wood resource of the forest plots inventoried 5 years ago. This practice leads to more precise NFI estimates, which finally allows more spatially accurate NFI results or more detailed results for a given spatial domain (Colin et al. 2012).

13.2.4 Data on Wood Assortments of Standing Trees and Removals

The French NFI identifies three classes of roundwood quality for each standing tree, according to potential uses such as veneer, sawn timber, pulpwood and fuelwood. In addition, market fluctuations and changes in transformation practices lead to implementation of a complementary expert-driven approach, based on analysis of data such as historical removals data.

Removal assortments are estimated from two surveys for which representativeness and statistical accuracy/bias are not assessed: (1) Since 1948, the national annual survey on commercialized wood removals (Agreste 2011) has provided harvested volumes over bark by species and assortment, at administrative-department level; and (2) the national survey on energy consumption by households (CEREN 2008) provides a raw estimate for fuelwood volume at regional level. The survey is carried out every 5 years. Finally, estimates of historical removals can be compared to wood availability estimates.

13.2.5 NFI Data on Extreme Climatic Disasters

The flexibility granted by the NFI design makes it possible to quickly update the state of forest resources for purposes including estimation of the extent of damages caused by severe windstorms such as *Klaus* in 2009 (IFN 2009). Every NFI plot was remeasured just after the storm to evaluate both volume of windthrows and remaining standing volumes. Regularly updated initial states of forest resources can hence be made available for wood projection studies as was done in Aquitaine (Colin et al. 2012).

13.2.6 *Auxiliary Data on Technical/Economic Factors of Production*

With respect to wood availability assessment and scenario simulation, independent economic data and specific expertise can also be used. Some examples are: harvesting system costs as a function of accessibility, cut types and stand characteristics, stumpage prices as a function of species and stand characteristics, and roadside wood prices as a function of species and assortments.

13.3 Tools and Simulators

13.3.1 *Main Purposes and Specification of Projections*

Five main forest policy issues are raised at national and regional levels: (1) the match between wood supply and demand by industries, (2) the opportunity for the development of new energy-wood industries, (3) the trends in the national timber sector, (4) the contribution of the French forest sector to GHG emissions reduction, and (5) as a matter of increasing concern, the impact of climate change on future forest patterns under different economic and climate scenarios.

Fundamental target variables for estimation include future quantity and quality of growing stock, the availability of wood for different uses, and projections of forest carbon stocks and fluxes.

The typical temporal horizon for forest resource projections ranges between 10 and 30 years with a usual simulation time-step of 5 years. Three spatial scales are typical: (1) a given tree species in a given regional domain, (2) all the broadleaved/coniferous species in a given regional domain, and (3) a comprehensive approach including the entire forest resource of a regional or national territory.

13.3.2 *Types of Models and Their Implementation*

In France, specific growth models are not available for many tree species (Pérot and Ginisty 2004), and when these models do exist (typically stand-scale models such as for abundant species including sessile oak, beech or maritime pine), they are usually not appropriate for large-scale projections due to French forest diversity in forest types and ecological conditions, and their non-stationary state. This context has fostered the development of large-scale forest resource models and simulators, as well as a specific expertise based on forest inventory data (Pignard 1993; Buongiorno et al. 1995; Houllier 1995; Wernsdörfer et al. 2012). The description of the current state of forest resources (area, growing stock, etc. distributed by tree species etc.)

and estimation of model parameters associated with forest dynamic processes (growth increment, natural mortality and recruitment) are based on NFI data.

Two types of large-scale forest dynamics models are currently used in France for forest resource projections: an age-class based approach for forests where stand age can be assessed, and a diameter-class approach for forests where this is not the case, or not meaningful. Both are currently developed at IGN in close association with the French NFI program, and both are used to address public and private needs (Sect. 13.1.2). The diameter-class approach is also associated with research programs (Wernsdörfer et al. 2012).

Diameter and age-class models are used in projection studies depending on the types of forests to be simulated. The age-class model applies to even-aged forests only (e.g. mono-species conifer plantations, poplar plantation, oak or beech high forests, managed coppices) while the diameter-class model can be used regardless of whether the stand is even-aged or uneven-aged, pure or mixed. Both models are implemented in specific simulation tools after being calibrated and initialised interactively on data extracted from the NFI database. They are run by an expert user.

13.4 Model Description

13.4.1 Forest Stratification

Forests are usually partitioned into homogenous subsets or 'strata' (Wernsdörfer et al. 2012), and projections are independently performed for each stratum. Stratification factors usually include geographic region or a categorization of the study area (biophysical classifications based on soil and climate such as the 'sylvo-eco-regions', IFN 2011a) with a view to encompass variations in growth conditions. Tree species composition is another mandatory stratification factor (e.g. single species, broadleaf-dominated, conifer-dominated and broadleaf-conifer mixed stands). Forest ownership category may be used as a stratification factor because management practices often differ between public and private forests and potentially lead to quite different stand characteristics for the different categories. Stand structure (high forest, coppice forest and high-forest with coppice) can also be used as a stratification factor when the available information is sufficiently accurate. Many other environmental factors can also be taken into account, depending on the context of the study. The number of strata is subject to statistical constraints including a minimum number of NFI plots per stratum to ensure an acceptable sampling error for within-stratum estimates of forest attributes.

13.4.2 *Consideration for Environmental and Silvicultural Factors*

In both the age-class and the diameter-class models, forest dynamics are not currently explicitly linked to environmental conditions such as site fertility. Spatial variations in growth conditions are implicitly accommodated in strata-specific estimates of model parameters. In addition, the models do not consider the impacts of climate and environmental changes. Last, density-dependence of forest dynamic processes is also implicit in forest stratification which may be sensitive to non-stationary contexts. This becomes very important when alternative silvicultural scenarios differing from the BAU scenario (Sect. 13.2.3) are simulated. For these reasons, the projection length should be limited to about 30 years.

13.4.3 *Diameter-Class Model*

13.4.3.1 Aim of Model Development

The tree diameter-class model was originally developed to assess future growing stock and wood availability for the medium-term (up to 30 years) within heterogeneous forested areas, i.e., uneven-aged and/or mixed-species stands. The model is also relevant for pure even-aged stands so that, unlike the age-class model, it allows for comprehensive projections for an entire territory (region, country) with various stand types. This model is particularly well-adapted to project forest resources in contexts of valuation of new afforested areas which are rarely even-aged, or where the forests are often heterogeneous such as in mountainous and Mediterranean regions. The diameter-class model is called MARGOT (MAtrix model of forest Resource Growth and dynamics On the Territory scale) and is described in Wernsdörfer et al. (2012). The next model development steps will be to incorporate competition and site as factors affecting forest dynamics.

13.4.3.2 Modelling Approach

Forest dynamics over time are described by within-stratum tree-diameter distribution and are governed by a Markov transition matrix with constant recruitment. The lower limit of the first diameter class is usually equal to the minimum diameter threshold (7.5 cm) of the French NFI. Three stratum-specific groups of parameters are defined: (1) the proportion of trees within a given diameter class that move up to the next larger diameter class (growth parameter), (2) the proportion of trees within a given diameter class that are removed through felling and natural mortality (felling and mortality parameters), and (3) the number of new trees that grow into the lowest diameter class (recruitment parameter) over a time-step. Further output variables such as merchantable tree volume, biomass or carbon content can be provided by multiplying the number of trees in each diameter class by the respective output variable value estimated for the tree of mean size (i.e. mean merchantable tree volume, biomass or carbon content, respectively).

13.4.3.3 Assumptions

The current model relies on four assumptions: (1) diameter distributions within different strata change over time independently of each other; (2) forest area, site and competition conditions are constant over time within each stratum; (3) the state of a tree at the end of one time-step depends only on its state at the beginning of that time-step (Markov hypothesis); and (4) no tree can move up more than one diameter class during a time-step, or regress towards a lower diameter class (Usher hypothesis, Vanclay 1994). Some of these assumptions such as constancy of site and competition will be relaxed in the near future.

13.4.4 Age-Class Model

13.4.4.1 Aim of Model Development

The stand age-class model applies to even-aged forests, mostly single-species conifer and poplar plantations, high forests of broadleaved species in the plains regions, and managed coppices where intensive management practices are usually implemented. In such homogeneous stands, age is an efficient proxy to estimate growth and to apply forest management practices. The model was described for the first time by Alvarez-Marty (1989).

13.4.4.2 Modelling Approach

Forest dynamics over time are described by a within-stratum age distribution of the forested area. Each stand cohort is represented by an age class and its associated area, growing stock per hectare, and volume production per hectare. The model simulates the lifespan of each cohort, from its birth (i.e., natural seedlings or plantation) to its death (i.e., clear cut of the last area). The growing stock per hectare of each cohort is recalculated at the end of each time-step of the simulation by subtracting the total volume of thinnings over the period from the initial growing stock per hectare. The mean production per hectare of the age class is finally added to this new growing stock. The area of a given cohort remains the same during the simulation, until it reaches the final cut age. From there, it is progressively reduced over time until zero, thus representing the variety in forest management practices within the stratum, and also the variety in site conditions. Silvicultural operations are defined by two parameters: for age classes up to the final cut age, a thinning rate is applied as a percentage of the growing stock per hectare; for age classes corresponding to the final cut age and beyond, a clear-cut rate is applied as a percentage of the surface area of the age-class. Importantly, reforestation and afforestation within a given stratum are represented by a net influx of new areas in the first age-class.

13.4.4.3 Assumptions

The model relies on two assumptions: (1) the site and density conditions are constant over time within each age class; and (2) the state of a cohort at the end of one time-step depends only on its state at the beginning of that time-step and not on any previous state. Contrary to the previous approach, the total forest area under study is not fixed and afforestation and land-use issues can be addressed.

13.4.5 Model Evaluation

Both the diameter-class and the age-class models provide projections that are consistent with historical trends observed in NFI data. A first sensitivity analysis based on NFI data for the years 2006–2008 was carried out by Wernsdörfer et al. (2012) for the diameter-class model. It focused on the sensitivity of forest dynamics to various stratification factors, showing that both varying growth conditions represented by nine large forest regions, tree-species composition (broadleaf-dominated, conifer-dominated and broadleaf-conifer mixed stands) and stand structure (high forest, coppice forest and high-coppice forest mixture) clearly affect forest dynamics at the larger-scale. Additional analyses on model evaluation are part of the future work envisaged in the frame of further model development.

13.5 Conclusions and Perspectives

Both large-scale age-class and diameter-class forest dynamics models are used for projecting wood resources in France, and they primarily target forest structures associated with specific silvicultural systems. While age-class models are of immediate advantage for simulation of forest areas of homogeneous stands and mapping their age-structure, diameter-class models encompass forest heterogeneity. However, the use of age as a proxy for developmental stage is questioned, because any historical factor may be confounded in this temporal proxy, including forest management practices, and environmental changes that affect soil fertility over time and create a shift in age-size relationships (Bontemps et al. 2009).

Diameter-class models also directly facilitate volume and biomass prediction using volume and biomass models. Quantification of available roundwood for different uses such as fuelwood, pulpwood and timber still requires further refinement.

Recent developments include total aboveground volume models for the main tree species in France (Vallet et al. 2006; Loustau 2010). A new generation of adaptive volume and biomass models has also been developed in the recently completed research project *Emerge*, funded by the French Research Agency (e.g. Genet et al. 2011; Dassot et al. 2012; Longuetaud et al. 2013).

Accurate projections of wood resources over time require that BAU felling regimes and initial state conditions are correctly quantified. In the past, this was difficult due to only approximate estimates of felled tree volumes resulting from the temporary-plot approach, and the occurrence of major disturbances caused by severe windstorms (e.g. *Lothar* and *Martin* in 1999). The forest inventory design implemented since 2004 (Hervé et al. 2014) is a source of major improvements: (1) plots in forest areas affected by severe storms can be quickly remeasured (e.g. in 2009), and (2) felling levels can be accurately estimated by remeasuring plots that were measured 5 years ago (IFN 2011c). Responsiveness in forest resource projections after major disturbances highlights the major role of a continuous and flexible inventory design.

Newer information sources such as aerial images and laser data should also be combined with NFI data in the near future. Multi-source inventory using maps of forest attributes as auxiliary information can improve the precision of forest resource estimates, and finally of wood availability and forest resource projections.

There is increasing demand to account for changes in the socio-economic environment in forest resource projection studies, for instance, to include a cost-approach to account for the feasibility of harvesting operations. This suggests an approach whereby forest characterisation with respect to physical and socio-economic factors is included as an input to such simulations. Additional GIS data sources may contribute to better assessments of the socio-economic context, e.g. the cadastral (ownership) data may be used to stratify projections according to ownership structure (IGN 2014b), and maps of harvest costs based on physical conditions may help to better identify the economic constraints (Clouet et al. 2010).

All models implicitly take growth conditions into account by varying growth estimates according to geographic domain or strata. In the face of ongoing environmental change impact onto forest growth as assessed from the French NFI (Charru et al. 2014), including climatic change, a future challenge will be to include environmental forcing into forest dynamics models using impact models of growth, mortality, and regeneration. Explicit approaches should rely on the environmental monitoring conducted on NFI plots (e.g. soil pH prediction using plant bioindication, Gégout et al. 2003, or mapping of soil water capacity, Piedallu et al. 2011), highlighting the interplay between forest inventory and environmental monitoring (FAO 2011).

Projections of wood resources are often conducted on a medium-term scale (about 20–30 years), either for regional strategic planning, or for national carbon accounting. In the context of global change, demand is increasing to perform simulations over the longer-term, e.g. 2050 up to 2100. Such applications should, however, not be encouraged as long as slow-developing processes such as changes in the forested area, in the growing stock, in tree species composition and site conditions are not covered at a minimum. Incorporating forest area dynamics into the diameter-class model approach is therefore needed. This also points to the importance of making forest inventory data comparable over time following changes in mensuration protocols.

References

Agreste (2011) La récolte de bois et la production de sciages en 2009. Un tiers des chablis de la tempête Klaus récoltés dès 2009. Agreste Primeur. Numéro 254. janvier 2011

Agreste (2013) Structure de la forêt privée en 2012. Des objectifs de production pour un tiers des propriétaires. Agreste Primeur. Numéro 306. Décembre 2013

Alvarez-Marty S (1989) La méthode des générations dans l'étude de la ressource d'une forêt équienne : le cas du massif landais. AFOCEL-ARMEF 3:135–146

Barbati A, Marchetti M, Chirici G, Corona P (2014) European forest types and Forest Europe SFM indicators: tools for monitoring progress on forest biodiversity conservation. For Ecol Manag 321:145–157

Bontemps J-D, Hervé J-C, Dhôte J-F (2009) Long-term changes in forest productivity: a consistent assessment in even-aged stands. For Sci 55:549–564

Brénac L (1984) Connaissance statistique des forêts françaises avant 1984. Revue Forestière Française 36:77–90

Buongiorno J, Peyron J-L, Houllier F, Bruciamacchie M (1995) Growth and management of mixed-species, uneven-aged forests in the French Jura: implications for economic returns and tree diversity. For Sci 41:397–429

CEREN (2008) Bilan national du bois de chauffage en 2006. CEREN

Charru M, Seynave I, Hervé J-C, Bontemps J-D (2014) Spatial patterns of historical growth changes in Norway spruce across western European mountains and the key effect of climate warming. Trees 28:205–221

Clouet N, Berger F, Monnet JM, Descroix L (2010) CARTUVI: un modèle sous SIG pour la cartographie des surfaces débardables en zone de montagne. Revue Forestière Française 62(2):155–170

Colin A (2014) Emissions et absorptions de gaz à effet de serre liées au secteur forestier dans le contexte d'un accroissement possible de la récolte aux horizons 2020 et 2030. Contribution de l'IGN aux projections du puits de CO2 dans la biomasse des forêts gérées de France métropolitaine en 2020 et 2030, selon différents scénarios d'offre de bois. Rapport final, mars 2014. Convention MEDDE.DGEC / IGN n°2200682886 (IGN n°10998)

Colin A, Lambert P (2012) Analyse de la ressource forestière actuelle en Lorraine. Résultats statistiques et représentations cartographiques. Rapport final de convention. Tome 2. Contrat IGN/DRAAF Lorraine n°ETU-2011-11

Colin A, Thivolle-Cazat A, Coulon F et al (2009) Biomasse forestière, populicole et bocagère disponible pour l'énergie à l'horizon 2020. Rapport final de convention. Contrat IFN/ADEME n°0601C0134, avec FCBA et SOLAGRO, 105 p. http://www.dispo-boisenergie.fr

Colin A, Meredieu C, Labbé T, Bélouard T (2010) Etude rétrospective et mise à jour de la ressource en pin maritime du massif des Landes de Gascogne après la tempête Klaus du 24 janvier 2009. Rapport final de la convention IFN/MAAP n°E18/2010 du 21 juin 2010. Décembre 2010

Colin A, Drouineau S, Cavaignac S et al (2012) Analyse prospective de la ressource forestière et des disponibilités en bois de la région Aquitaine à l'horizon 2025. Etat des lieux des forêts aquitaines à l'automne 2011. Version 1.0 du 31 juillet 2012. Rapport de conventions

Dassot M, Colin A, Santenoise P et al (2012) Terrestrial laser scanning for measuring the solid wood volume, including branches, of adult standing trees in the forest environment. Comput Electron Agr 89:86–93

Dodane C (2012) Quelle est la part du FFN dans la transformation du visage des forêts françaises au cours de la seconde moitié du XXe siècle ? Geoconfluences, 4 octobre 2010. http://geoconfluences.ens-lyon.fr/doc/territ/FranceMut/popup/Dodane3.htm

Doll D (2000) Statistiques historiques des grands chablis éoliens en Europe occidentale depuis le milieu du XIXe siècle: analyse critique. Dossiers de l'Environnement de l'INRA 20: 28–42. http://www7.inra.fr/dpenv/pdf/DollD20.pdf

FAO (2011) State of the world's forests 2011. Food and Agricultural Organisation. ISBN:978-92-5-1067550-5

Gadant J (1987) Quarante ans au service de la forêt française. Revue Forestière Française 39:10–19

Gégout JC, Hervé JC, Houllier F, Pierrat JC (2003) Prediction of forest soil nutrient status using vegetation. J Veg Sci 14:55–62

Genet A, Wernsdörfer H, Jonard M et al (2011) Ontogeny partly explains the apparent heterogeneity of published biomass equations for Fagus sylvatica in central Europe. For Ecol Manag 261(7):1188–1202

Ginisty C, Vallet P, Chevalier H, Colin A (2011) Disponibilité en biomasse ligneuse en forêt, dans les peupleraies et dans les haies pour les différents usages du bois. Evaluation à l'échelle métropolitaine à partir des données de l'Inventaire forestier national et des statistiques de consommation de bois. Revue Forestière Française LXIII-2-2011:151–162

Hervé JC, Wurpillot S, Vidal C, Roman-Amat B (2014) L'inventaire des ressources forestières en France: un nouveau regard sur de nouvelles forêts. Revue Forestière Française XLVI-3-2014:247–260

Houllier F (1995) A propos des modèles de la dynamique des peuplements hétérogènes: structures, processus démographiques et mécanismes de régulation. Revue Ecologie (Terre Vie) 50:273–282

IFN (2003) Les tempêtes de décembre 1999. Bilan national et enseignements. L'IF n°2 décembre 2003. Editeur IFN

IFN (2008) Nouvelle cartographie forestière. De la production à l'utilisation. L'IF n°20. 3ième trimestre 2008. Editeur IFN

IFN (2009) Tempête Klaus du 24 janvier 2009: 234 000 ha de forêts affectées à plus de 40%. 42,5 millions de mètres cubes de dégâts. L'IF n°21 – 1er trimestre 2009. Editeur IFN

IFN (2010) La forêt française. Résultats des campagnes 2005 à 2009. Nogent-sur-Vernisson, France

IFN (2011a) Un nouveau cadre de référence géographique pour les forestiers français : les sylvoécorégions (SER)

IFN (2011b) Volume de bois sur pied dans les forêts françaises : 650 millions de mètres cubes supplémentaires en un quart de siècle. L'IF n°27 – 2ième trimestre 2011. Editeur IFN

IFN (2011c) Prélèvements de bois en forêt et production biologique: des estimations directes et compatibles. L'IF n°28 – 3ième et 4ième trimestres 2011. Editeur IFN

IGN (2012) Quelles sont les ressources exploitables ? Analyse spatiale et temporelle. L'IF n°30 (numéro double). Décembre 2012. Editeur IGN

IGN (2013) Un siècle d'expansion des forêts françaises. De la statistique Daubrée à l'inventaire forestier de l'IGN. L'IF n°31. Mai 2013. Editeur IGN

IGN (2014a) Résultats d'inventaire forestier. Résultats standard. Les résultats des campagnes d'inventaire 2008–2012. Le palmarès

IGN (2014b) Notice méthodologique pour la détermination d'indicateurs de morcellement de la forêt privée. Version 2.0. IGN Conseil. Rapport pour le CRPF Picardie. Juin 2014. D2SI/SAI/2014.0126

Longuetaud F, Santenoise P, Mothe F et al (2013) Modeling expansion factors for European tree species in France. Forest Ecol Manag 292:111–121

Loustau D (2010) Forests, carbon cycle and climate change. Ouvrage collectif issu du projet CARBOFOR, Editions QUAE, Versailles, collection Update Sciences and Technologies. ISBN:97827592038

Mather AS (1992) The forest transition. Area 24:367–379

MEDDE (2012) Stratégie nationale d'actions en faveur du grand tétras (Tetrao urogallus major) 2012-2021. Ministère de l'Ecologie, du Développement Durable, des Transport et du Logement

Morneau F, Duprez C, Hervé J-C (2008) Les forêts mélangées en France métropolitaine. Caractérisation à partir des résultats de l'Inventaire Forestier National. Revue Forestière Française 60:107–120

Pardé J (1966) Forêts et reboisements à haute productivité en France. Revue Forestière Française 11:718–724

Pérot T, Ginisty C (2004) Bilan et perspectives sur les modèles de croissance, de dynamique forestière, et de qualité du bois. Cemagref, Rapport pour le Ministère de l'Agriculture et de la Pêche, Nogent-sur-Vernisson

Piedallu C, Gégout J-C, Bruand A, Seynave I (2011) Mapping soil water holding capacity over large areas to predict potential production of forest stands. Geoderma 160:355–366

Pignard G (1993) Eléments pour l'élaboration d'un logiciel de simulation de l'évolution des peuplements irréguliers. IFN, Cellule d'Evaluation de la Ressource (CER), Montpellier, document interne

Puech J (2009) Mise ne valeur de la forêt française et développement de la filière bois. Mission confiée à Jean Puech ancien ministre. Rapport remis à Monsieur Nicolas Sarkozy, Président de la République. 6 avril 2009. Paris

Seynave I, Gégout J-C, Hervé J-C et al (2005) Picea abies site index prediction by environmental factors and understorey vegetation: a two-scale approach based on survey databases. Can J For Res 35:1669–1678

Thivolle-Cazat A, Colin A, Lambert P (2012) Analyse de la ressource forestière et évaluation de la disponibilité en bois en Lorraine en 2025. Rapport final de convention. Tome 1. Contrats FCBA/DRAAF Lorraine n°ETU-2011-10 et IGN/DRAAF Lorraine n°ETU-2011-11

Vallet P, Dhôte J-F, Le Moguédec G et al (2006) Development of total aboveground volume equations for seven important forest tree species in France. Forest Ecol Manag 229:98–110

Vanclay JK (1994) Modelling forest growth and yield. Applications to mixed tropical forests. CAB International, Wallingford

Vidal C, Bélouard T, Hervé J-C et al (2005) A new flexible forest inventory in France. In: Proceedings of the 7th annual forest inventory and analysis symposium; 2005 October 3-6; Portland ME, McRoberts RE, Reams GA, Van Deusen PC, McWilliams WH (eds) – General Technical Report WO-77. – Washington, DC: U.S. Department of Agriculture Forest Service, 2007. – 319 p. [Online]: http://fia.fs.fed.us/symposium/proceedings/pubs/FIA2005%5Bhi%5D.pdf

Wernsdörfer H, Colin A, Bontemps J-D et al (2012) Large scale dynamics of a heterogeneous forest resource are driven jointly by geographically varying growth conditions, tree species composition and stand structure. Ann For Sci 69:829–844

Chapter 14
Germany

Gerald Kändler and Uli Riemer

14.1 Introduction

Although Germany has a long tradition of forest management based on surveys of forest conditions, a sample-based, large-scale inventory was not introduced until quite late in West Germany with the first Federal Forest Inventory ("Bundeswaldinventur") conducted in the years 1986–1990. The second inventory was conducted in the years 2001 and 2002, after the re-unification. Today, the usefulness of a sample-based inventory is generally accepted and the results obtained are in wide demand. In 2010, the Federal Forest Act was amended and now prescribes a 10-year interval between NFIs. Data collection for the third NFI was carried out in 2011 and 2012. The NFI is a joint project of the federal government and the federal states. The main task of the states is data collection; the federal government is responsible for data analysis and reporting.

The NFI provides a vast amount of data on the current state of forest conditions and on changes between consecutive surveys. These data serve as a valuable basis for controlling forest development and sustainability as well as decision-making in various sectors related to forests and forestry. NFI data, besides providing information on the current state of the forests, are also important for addressing future forest development and timber supply. For the latter purpose, a special tool was developed that facilitates assessment of forest development and timber supply for several future decades starting with the state of the forest state as described by the current NFI. WEHAM ("Waldentwicklungs- und Holzaufkommensmodell": forest development and wood supply model) is a forest growth simulator. It was developed in the 1990s by the Baden-Württemberg Forest Research Institute by order of the German Federal Ministry of Agriculture for the analysis of data for the 1st NFI

G. Kändler (✉) • U. Riemer
Forest Research Institute, Baden-Württemberg, Freiburg, Germany
e-mail: gerald.kaendler@forst.bwl.de; uli.riemer@forst.bwl.de

S. Barreiro et al. (eds.), *Forest Inventory-based Projection Systems for Wood and Biomass Availability*, Managing Forest Ecosystems 29,
DOI 10.1007/978-3-319-56201-8_14

(Bösch 1995). WEHAM was updated for the second and third NFIs. It is used as the core instrument for nationwide future forest resource assessments, sustainability checks and for various ecological, economic and political planning purposes.

14.2 The German NFI – Design and Results

14.2.1 Inventory Design

The German NFI is based on a systematic rectangular grid with clusters (tracts) as primary sampling units. The General Administrative Regulation prescribes a 4 × 4 km grid covering the complete surface of Germany with a defined starting point. In some federal states, the basic grid has been intensified to a 2 × 2 km or a 2.83 by 2.83 km grid. The purposes of the intensification were to obtain more precise estimates on a regional level in some states or acquisition of a sufficiently large sample in less densely forested regions.

The tract is a quadrangle with sides of 150 m. The sides of the tract are oriented north-south and east-west respectively. Plots are established on tract corners hitting forest land. Aerial photographs are used for a preliminary decision regarding whether the tract corners are located in a forest. When the German NFI was designed in the 1980s, use of remote sensing for a pre-stratification in a two-phase sampling framework was considered unnecessary. The main reason was that integration of remote sensing would require additional operating expense which could not be justified because sufficiently detailed information on forest area was already available and because most German forests are readily accessible.

Sample trees are selected on tract corners using multiple methods. Trees with diameters at breast-height (1.3 m, *dbh*) over bark of at least 7 cm are selected by the angle-count method (relascope-technique with basal-area factor 4 [m^2/ha]). Multiple measurements and assessments are made for these trees, and their locations are recorded by azimuth and horizontal distance from the plot centre. The attributes recorded include sample tree code (e.g. new, re-measured, removed, species, canopy class (vertical layer membership), *dbh*, tree class, age, trunk damage and pruning. Tree height and upper diameter are only recorded for a sub-sample. Smaller trees are sampled in two concentric circular plots with radius of 1 or 2 m, offset 5 m from the plot centre. Deadwood is surveyed in a plot with 5 m radius centred in the corner. In a circle of 10 m radius, trees up to 4 m in height, shrub layers, and ground vegetation are surveyed. Site characteristics and forest edges are registered in a circle of 25 m around the plot centres. In total, approximately 150 characteristics for each inventory tract are recorded. In contrast to Germany's first NFI (1986–1990 in the western states), the spectrum of data collected in the second and third NFIs has been expanded to include ecological parameters such as deadwood, closeness to nature, and ground vegetation. A detailed description of the German NFI sampling design is given in Polley et al. (2010).

14.2.2 *Forest Conditions in Germany*

Forest conditions are briefly described based on the results of the third NFI with 2012 as reference year.[1]

14.2.2.1 Forest Area and Ownership

Although Germany is one of the most densely populated countries of the European Union it also includes substantial wooded areas. Approximately 11.4 million ha, nearly one third of the surface area, is covered with forests; this area is equivalent to approximately 1 ha of forest area per seven inhabitants. Approximately 10.4 million ha are productive forests available for wood supply, and approximately 48% (5.5 million ha) of the forest area is privately owned, consisting of mostly small scale forest lots. Only 13% of the private forest land belongs to forest estates of more than 1000 ha; approximately half of the private forest area is part of properties with less than 20 ha. Municipalities own 2.2 million ha (19%) and the federal states (with a small share by the federal government) 3.7 million ha (33%).

14.2.2.2 Species Composition, Forest Stand Structure

Deciduous and conifers tree species occupy 43.4% and 54.2% of the forest land, respectively, with only 2.4% temporarily un-stocked. Norway spruce is the most common species with 26% of the total; Scots pine follows with 23%; European beech with 16%; oak species with 11%; other deciduous species with 17%; and other conifers with 7%.

The forest is almost completely high forest (99%) which originates either by natural regeneration, seeding, or planting. Approximately one-third of forest stands in Germany are single-storied, mainly dominated by Norway spruce, Scots pine, Douglas fir and birch; 57% of the forests are two-storied, and 11% are multi-storied. Two- or multi-storied stand structures occur mainly within beech, oak and silver fir stands. Young growth (trees shorter than 4 m) cover an area of 2.7 million ha, with 85% originating from natural regeneration; the remaining young growth area comes from planting (13%), while seeding and coppice play a minor role.

14.2.2.3 Growing Stock, Growth and Removals

Total growing stock consists of 3.66 billion m^3 of solid volume over bark, corresponding to 336 m^3/ha. This total represents an increase of 7% since the preceding survey of 2002. Between the second and third inventories, 2003–2012, mean annual

[1] https://bwi.info/ provides access to the results of the third German NFI.

volume increment was approximately 122 million m^3 of solid volume over bark (11.2 m^3/ha per year), and mean annual removals were approximately 106 million m^3 (9.8 m^3/ha per year), corresponding to 75.7 million m^3 of harvested volume under bark.

14.3 The WEHAM Forest Development and Wood Supply Model

WEHAM was designed to project forest conditions and future wood supply over a period of 40 years based on NFI data that define the initial conditions for each simulation period. WEHAM uses NFI sample data representing stands whose growth and silvicultural treatments are simulated over the projection period assuming common management practices and constant environmental conditions. The projection period is subdivided into 5-year time-steps with silvicultural interventions within each 5-year sub-period distributed randomly over the 5 years.

WHEAM simulator consists of three components or modules: (1) a growth module, (2) a management module, and (3) a grading module.

The growth module predicts tree dimensions (diameter and height) based on diameter growth using a trend function (Sloboda 1971) which is fitted using diameter increments obtained from re-measured sample trees in the preceding period. Mathematically, the trend function is a differential equation that takes into account age, diameter and its increment; afterwards it is integrated as an initial value problem, providing growth curves as function of age. For the projection, each sample tree is associated with its species-specific growth curve that depends on the tree's current *dbh* and age and that prescribes the tree's diameter growth trajectory for the projection period. Heights are not projected in the same way but rather are calculated from tariff functions with diameter as the predictor. The reason why no growth model is used for height is that height growth, particularly for older ages, may be overestimated by this approach with greater overestimation for longer projection periods.

The management module implements all standwise silvicultural interventions based on parameters controlling the timing and intensity of thinning and cutting activities; these parameters, in turn, depend on tree species, age, and dimensions. Thinning interventions follow different schemes which also use a control parameter. The intensity of thinnings depends on yield-tables which may be provided by users. In the upgraded version of WEHAM, new management features were implemented (Rock et al. 2013). No-management scenarios may be simulated using self-thinning models based on the concept of stand density index according to Reineke's law as modified by Sterba (1987). In addition, management control parameters can be changed over time, thus allowing modifications of silvicultural treatments in future decades. The management module includes a simplified approach to regeneration. The default procedure is the regeneration of the same species that was dominant before the harvest. An alternative is regeneration of the species that is dominant for the natural forest community associated with each stand's site conditions.

The set of control parameters (Table 14.1) that define thinning rules and intensities, target diameters for final harvests, and rotation length may be specified differently by federal state and other domains such as ownership. This feature permits typical and current regional treatment concepts to be reproduced at tree-level. Cap limits prevent unrealistic thinning volumes. By altering the control parameters different scenarios and management schedules can be simulated.

The grading module calculates merchantable volume and assortments such as logs or fibre wood for harvested trees. This module uses a software library that includes species-specific taper models. Timber volume for a given tree or stand is first estimated as the solid volume over bark, after which bark losses due to harvest and cut-to-length division are subtracted. The grading module uses tree species and the top diameter of the stem as control parameters which are modified according to different regional practices in the federal states.

Because WEHAM projects the development of single trees on a plot-by-plot basis; results can be aggregated at various levels including plot, stand, region or state levels; by species or species groups; by age classes; and for any combination of these classifications. By default, all projections are done in 5-year time-steps, and the state of the forest (plot, single tree) at the beginning and at the end of this period is given, as well as mean annual changes over the five years.

Additionally, in the new version, aboveground biomass is also provided, allowing estimation of carbon stocks and carbon sequestration effects as well as wood harvest in term of biomass or energy-equivalents. The estimation of carbon stocks is done using the same models and tools as used for Kyoto and UNFCCC reporting.

Table 14.1 Example of control parameters of the management simulator for beech and Norway spruce according to the silvicultural concept of Baden-Württemberg for the simulation based on the third NFI

Species/ownership	Beech			Norway Spruce		
Control parameters	*State forest*	*Communal forest*	*Private forest*	*State forest*	*Communal forest*	*Private forest*
Start of thinning at age	15–40	15–40	15–40	15–30	15–30	15–30
Start of thinning at average tree height [m]	15	15	15	11	11	11
Start of the regeneration cutting at age	80	80	80	80	80	80
End of regeneration cutting at age	150	150	160	150	150	150
Target diameter	65	60	60	50	50	50
Maximum percentage of cutting from target diameter	30	30	30	30	30	30

The control parameters used in the management simulator and the grading module describe the management of forests as conducted and planned at the beginning of the projection period. Actual management might differ from these assumptions due to changes in economic conditions and/or owner preferences. In new scenarios based on the third NFI, changes in management practices over time may be simulated.

14.4 Results of WEHAM Projections

We present a few results of a projection based on the third NFI (BMEL 2016). The overall time horizon of these projections is limited to 40 years because the scenario does not accommodate changes in growing conditions, silvicultural treatments or economic or socio-political framework conditions.

14.4.1 Scenarios

The potential harvestable timber was assessed for forest land with no legal harvesting restrictions. Simulations were run for 5-year periods according to the rules in the different federal states, tree species and assortments as defined in 2013. The simulation parameters related to silvicultural treatments and grading rules (assortments) were kept constant over the entire projection period. Growth dynamics were simulated as recorded in the preceding inventory period (second to the third NFI: 2002–2012).

14.4.2 Essential Trends

Over the simulation period 2013–2052, potential marketable timber supply decreases from 76.6 million to 71.2 million m^3/year (−7.6%). On average, the total consists of 62.4 million m^3 of stemwood (72.7%), 14.2 million m^3 of pulpwood (16.6%) and 9.2 million m^3 of currently non-marketable wood (10.7%). However, the volumes of the different assortment groups are shifting towards stemwood. Over the next 40 years, the share of stemwood is projected to increase approximately 4% percentage points from 72.7 to 77.5%, whereas the share of pulpwood is projected to decrease from 16.6 to 13.3% (Table 14.2). This shift in volumes is primarily attributable to the increasing average age and size of the trees.

Tree species are grouped into wood-type categories based on the four main species in Germany: Norway spruce (including Silver fir and Douglas fir), Scots pine

Table 14.2 Potential timber harvest by assortment category and projection period

	Projected merchantable wood solid volume under bark (1000 m³/year)							
Assortment category	2013–2017	2018–2022	2023–2027	2028–2032	2033–2037	2038–2042	2043–2047	2048–2052
Stemwood	62,360	62,348	55,728	57,834	57,723	58,080	58,180	60,774
Pulpwood	14,241	12,463	10,427	10,769	10,421	10,398	10,417	10,445
Marketable	76,601	74,811	66,155	68,604	68,144	68,478	68,597	71,219
Not marketable	9,154	7,995	6,893	7,043	6,884	7,044	7,039	7,215
All assortments	85,755	82,806	73,048	75,647	75,028	75,522	75,636	78,434

Table 14.3 Projected potential timber harvest by wood-type group (compared to period 2003–2012)

Merchantable wood solid volume under bark (1000 m³ per year)									
	Observed	Projected							
Species	2003–2012	2013–2017	2018–2022	2023–2027	2028–2032	2033–2037	2038–2042	2043–2047	2048–2052
Spruce	42,365	32,530	32,158	29,331	30,791	31,119	31,775	31,553	33,842
Beech	15,316	23,804	21,629	18,198	19,119	18,833	18,288	18,755	18,384
Pine	14,850	14,906	15,298	13,845	13,647	13,615	13,351	13,568	14,000
Oak	3,148	5,361	5,725	4,781	5,046	4,577	5,065	4,722	4,992
Total	75,680	76,601	74,811	66,155	68,604	68,144	68,478	68,597	71,219

(including larch), European beech (including other hardwoods), and oak (Table 14.3). The spruce wood-type group makes by far the largest contribution to the potential usable timber stock (on average 45%), followed by beech (28%), pine (20%), and oak (7%). In comparison to the period 2003–2012 spruce wood supply will drop significantly. The projected volumes will decrease from 32.5 million m³/year in the period 2013–2027 and then rise again to about 33.8 million m³/year in the period 2048–2052. On the other hand potential wood supply of beech along with other hardwoods and oak is expected to rise considerably, namely in the first decade, and then drop. Pine-wood supply will not change significantly in the first projection decade, then decrease and recover again slightly until 2052.

14.5 Discussion

The potential timber supply estimated by WEHAM is based on assumptions of constant management and environmental conditions and cannot be regarded as a forecast in a narrower sense. The model does not take into account economic factors or conditions which may select the best harvesting technique considering the costs associated. Accessibility and hauling distances are not included either. Furthermore, WEHAM does not consider other external drivers such as consumer demand which may affect wood harvest. In principle, WEHAM is management-driven; nevertheless it must be kept in mind that forest owners might not bring potential harvest amounts to the market due to individual objectives or other socio-economic reasons.

A major controversial issue that was discussed following publication of WEHAM projections was whether current wood harvests are good estimates of available timber. This issue is very important for the wood-based industry in general and for the wood-processing industry (sawmills) in particular. Moreover, the increasing demand for bio-fuel from forests makes competition for wood more intense. Obviously, the WEHAM timber supply assessments exceeded the harvesting statistics by a great deal; however, this comparison is misleading because harvesting statistics underestimate real timber production and consumption for multiple reasons including incomplete data from private forests and underestimation of fuel-wood harvest. A more realistic comparison is to contrast the projected timber supply with the amount of harvested wood as estimated from repeated NFI surveys, although these data are available only for a past period. This comparison (Table 14.3) reveals that actual Norway spruce harvests in the preceding period exceed the projections for 2013–2027. So far, WEHAM projections based on the third NFI have provided a plausible and reasonable prospect of potential timber supply and forest development; however, the results must be interpreted carefully because there are limitations in the modelling approach that may affect the validity of the results, especially concerning WEHAM's long-term projection ability.

The updated version of WEHAM includes no-management options in new projections based on the third NFI. However, hazards and resulting risks are not explicitly modelled; rather, WEHAM simulates forest development under "normal" conditions with natural mortality being implicitly subsumed under all cutting regimes or self-thinning.

Regeneration is also simulated under the simplified assumption that the same tree species will be regenerated at the same locations without regard to any changes in species composition. In principle, WEHAM is able to simulate a change of tree species in the course of regeneration; however, the rules have yet to be determined. The existing option is to switch to the tree species that is dominant for the natural forest community, but this option also requires sound rules which have yet to be formulated.

Another shortcoming results from the fact that the growth model is calibrated using increments observed in the preceding period which assumes environmental conditions that are unlikely to remain constant in future periods. Currently, an alternative growth model that is sensitive to climatic and soil variables is being developed in a research project on climate change. This model could replace the current growth model of WEHAM in future applications. Due to WEHAM's modular structure, components can be exchanged rather easily. Finally, the lack of empirical data for calibrating desired adaptations and extensions of the models is a major problem.

Generally, the uncertainty of the results should be expected to increase with the duration of the projection. Moreover, an assessment of uncertainty of the projections is not possible. Uncertainty increases due to changes in the environment, particularly due to global warming, and to socio-economic factors and silvicultural treatments. These factors must be considered when interpreting the results. Uncertainties associated with the factors that influence forest development and treatment are a major reason for limiting the projection period to 40 years with the result that estimates for only the first two decades are considered more or less trustworthy.

References

Bösch B (1995) Ein Informationssystem zur Prognose des künftigen Nutzungspotentials. Forst und Holz 50:587–593

Bundesministerium für Ernährung und Landwirtschaft (BMEL) (2016) Wald und Rohholzpotenzial der nächsten 40 Jahre Ausgewählte Ergebnisse der Waldentwicklungs- und Holzaufkommensmodellierung 2013 bis 2052, 64 p. https://www.bundeswaldinventur.de/fileadmin/SITE_MASTER/content/Dokumente/Downloads/BMEL_BWI3_WEHAM_Broschuere_RZ02_web.pdf

Polley H, Schmitz F, Hennig P, Kroiher F (2010) National forest inventories reports: Germany. In: Tomppo E, Gschwantner T, Lawrence M, McRoberts RE (eds) National forest inventories – pathways for common reporting. Springer, Berlin, pp 223–243

Rock J, Bösch B, Kändler G (2013) WEHAM 2012 – Waldentwicklungs- und Holzaufkommensmodellierung für die dritte Bundeswaldinventur. In: Klädtke J, Kohnle U (ed) Deutscher Verband Forstlicher Forschungsanstalten. Sektion Ertragskunde. Jahrestagung 13.-15. Mai 2013, Rychnov nad Kneznou/Tschechien. ISSN 1432-2609: 127–133. http://sektionertragskunde.fvabw.de/2013/Beitrag_13_17.pdf

Sloboda B (1971) Zur Darstellung von Wachstumsprozessen mit Hilfe von Differentialgleichungen erster Ordnung. Mitteilungen der Forstlichen Versuchs- und Forschungsanstalt Baden-Württemberg, 32, 1971

Sterba H (1987) Estimating potential density from thinning experiments and inventory data. For Sci 33:1022–1034

Chapter 15
Hungary

Pál Kovácsevics

15.1 Introduction

Forest land area in Hungary has been gradually increasing for the last 90 years. This is due to large-scale afforestation carried out under the supervision of professional foresters. As a result, the forest area which in 1921 was barely larger than 1 million ha exceeded 1.9 million ha by 2015 and represented 20.8% of the country area.

The share of state-owned forests is 56%, community-owned is 1% and 42% of the forests are private (1% of Hungarian forests are in mixed ownership). A long-term purpose, primarily based on afforestation, is the large-scale increase of private and community owned areas. The highlighted objective of forest policy is the structural improvement of the fragmented estate system that hinders private forest management, and the establishment of viable management organisations and partnerships.

The management of state-owned forests is primarily conducted by 22 state forest management corporations. However, other national institutions, such as water resource directorates and national parks are also managing state-owned forest areas. The share of community ownership is relatively small, mostly managed by municipalities of villages and cities (NFCSO 2014).

The majority of private forests are in undivided joint ownerships. Private forest owners often contract with private forest management companies to manage their forests, but significant areas are managed by joint management organizations such as forest cooperatives or local forest owners together with management associations.

To ensure that the interests of society and nature are satisfied, forests are managed according to district forest management plans. These plans are prepared by the forestry authority that is also responsible for verifying their correct execution for both public and private forests.

P. Kovácsevics (✉)
National Food Chain Safety Office, Forestry Directorate, Budapest, Hungary
e-mail: KovacsevicsP@nebih.gov.hu

S. Barreiro et al. (eds.), *Forest Inventory-based Projection Systems for Wood and Biomass Availability*, Managing Forest Ecosystems 29,
DOI 10.1007/978-3-319-56201-8_15

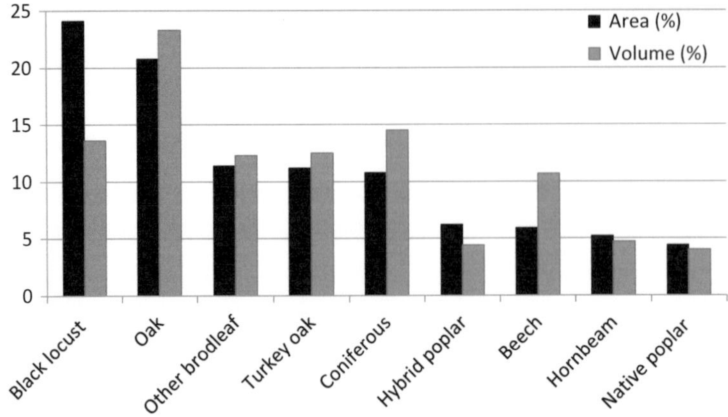

Fig. 15.1 Distribution of forest area and volume by the main tree species (Source: National Forestry Database (NFD))

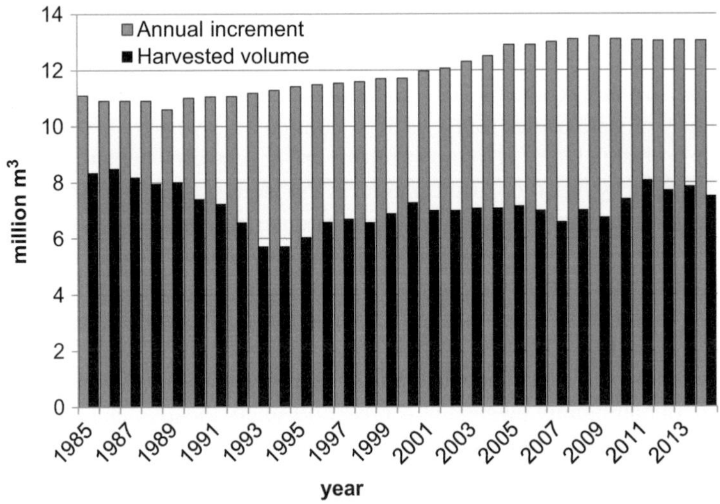

Fig. 15.2 Annual increment and harvested volume since 1985 (Source: National Forestry Database (NFD))

Hungarian forests comprise mainly Central European deciduous tree species (Fig. 15.1) of which Black locust (*Robinia pseudoacacia*) accounts for the largest area (24.1%), while noble oaks (*Quercus robur* and *Q. petraea*) have the greatest standing volume (23.3%).

The growing stock of Hungarian forests amounts to 355.8 million m^3, and has been continuously increasing for the last 30 years. This increase is due partially to afforestation, but mainly to allowable cutting levels prescribed in the approved management plans which result in considerably smaller harvested tree volume (felling) when compared to wood increment (Fig. 15.2).

15.2 Forest Inventory History

The first order to survey and map Hungarian forests was decreed by Maria Terézia and came into force in 1769. In 1935, forest owners were first required to manage their forests according to forest management plans. Forest management planning has been supported by computer databases since 1970 through the creation of the National Forestry Database (NFD). In 1978, the Forest Protection Network (FPN), a 4 × 4 km grid of sampling plots covering the entire country was established to obtain data on forest health and to serve as the basis for inventory data collection (Central Agriculture Office Centre, Forest Directorate 2009). In 1993, a new monitoring system was launched, the Growth Monitoring System (GMS), in close connection with the existing FPN. The main objective of the program was monitoring increment by creating a long time-series of repeated measurements of all trees on the same plots. However, after three remeasurement periods the GMS method was reassessed and modified to meet national and international requirements, always bearing in mind the need for and the importance of ensuring the continuity and consistency in data collection. In the framework of a European Union funded project, a Geographic Information System (GIS) that records, stores, analyses, manages, and permits visualization of data was integrated into the GMS. This new system permits integration of cartography, statistical analysis, and database technology. To distinguish among the previous and current operating systems, they will be referred henceforth as GMS and NFI, respectively. The first NFI corresponds to the 4th cycle of GMS.

15.2.1 The National Forestry Database

Creation of the National Forestry Database (NFD) had three main objectives: (1) storing data for the main attributes of Hungarian forests; (2) compiling the monitoring data necessary for forestry administration; and (3) allowing for an information technology environment. The main data sources for the NFD are Forest Planning (forest management plans) and Forestry Inspections. However, it must be emphasized that almost 90% of the data are obtained from the standwise forest planning inventories which are partially sample-based.

15.2.2 Forest Panning as a Data Source

Forest planning is regulated at the highest level by the Forest Management Plan Regulation issued by the responsible Minister. The Regulation includes the main limits of forest management activities including prescription of the maximum timber harvest. The result of the forest planning activity is the forest management plan. It is based on field surveys and prescribes tasks and timelines that must fulfilled during the next 10-year period. Each forest manager receives a forest management plan describing his rights and responsibilities. Exemptions from the forest management plan might occur but are exceptional and only upon request.

Forest management planning activities cover the entire forest area of Hungary. About 10% of the forest area of Hungary is subject to forest management planning each year, meaning that each forest sub-compartment is planned once every 10 years. The unit of forest planning is the forest sub-compartment which is the smallest management unit possible belonging to the same forest owner and is characterized by more or less homogeneous site conditions and tree composition. Consequently, the forest within a given sub-compartment can be managed according to one management schedule defined in the forest management plan. Furthermore, the area of the sub-compartment must be geographically reasonable (neither too big nor too small) so that the same management activities can be conducted in the stand. The average size of forest sub-compartments in Hungary is about 4 ha. Because the forest sub-compartment is the basic unit for forest planning, the NFD contains data by sub-compartments. Forest sub-compartments are grouped into forest compartments by geographical and administrative units (municipalities). Grouping helps spatial organization of forest management activities and forest logistics. Forest management planning is conducted in each forest district separately.

15.2.3 Forestry Inspection

The Forestry Authority inspects forestry operations for compliance with the prescriptions in the forest management plan. Forest inspection includes multiple activities: inspection of afforestation and regeneration, reviews and audits of subsidies, and imposition of sanctions for violating rules; assessment of strong natural disturbances; inspection of harvested volume; identification of land use changes such as deforestation; definition and establishment of forest sub-compartments; recording of forest manager data; and updating of the database.

15.2.4 *Growth Monitoring System*

15.2.4.1 Grid, Plot Selection, Sampling and Representativeness

Permanent GMS plots were established at the intersections of the national 4 × 4 km FPN sampling grid. To increase the sampling intensity, additional permanent GMS plots were established at the intersections of a second 4 × 4 km grid whose intersections were shifted 2.0 km to south-east of the FPN grid intersections.

The GMS sampling grid was overlaid with forest maps and any plot that fell within a forest compartment and met the requirements was selected as a GMS sampling plot. Forest management planning data were used to select sampling plots and later to evaluate growth during data processing. The land use classification of GMS sample plots was revised annually based on the maps linked to the National Forest Database. Plots that did not meet the selection criteria were left out of the network, while others that had been left out in the previous visit due to their land cover type were integrated if the criteria were met. The main purposes of the selection criterion for GMS was to ensure that the required accuracy for the of growing stock estimate of the country's forest resources was achieved, in other words, at the 95% confidence level the growing stock estimate for the country and directorate level was to be ±5% and ±10%, respectively.

The survey and assessment unit consists of four circular 'satellite' plots located at the corners of a square (tract) with side length of 200 m. The south-western corner of the tract is the centre of the sampling plot. Each grid corner and tract plot represents approximately 8 km^2 (800 ha) and 2 km^2 (200 ha) area, respectively. Each tract is assessed once every five years.

15.2.4.2 Survey Methodology

The sample plot survey is responsible for the stand description whenever stands are too young to be measured and also for individual sample tree assessment when the trees have grown enough to be measured.

When trees within a plot do not meet the selection criteria or are inaccessible, individual tree assessment is impossible. In this case, the stand description method is used to describe the forest compartment. This description includes only minimal data on tree species obtained mostly from the database but also from field assessments.

In case the GMS sample plot lies within the FPN sample plot, measurement are carried out for the four FPN circular 'satellite' plots. In GMS plots, data collected at plot level relates to plot characteristics such as the geographic coordinates, survey method and plot radius. The plot's physiography, topography, altitude, slope and soil erosion damages are described as well as the water management class, the occurrence of shrubs, stand structure and the canopy closure. During the individual tree assessment, in case a sample point overlaps a FPN point, data collection is carried out within circular plots with a radius between 4 and 25 m. The size of the circular plot depends on the number of trees within the sample plot. The number of sampled trees must be between 15 and 25 to ensure and adequate compromise between accuracy and measurement times. Therefore, in young forests circular plots have a smaller radius than in old-growth forests due to the lower density observed in these stands. At regional or county level, the lower threshold for the average number of trees is 20. As for tree measurements, tree diameters and heights are measured and a height class is assigned. Due to visibility conditions, field measurements for tree height are carried out from late autumn to spring. Additionally, tree identification number, and tree level information such as tree species, damages (level and causes), cutting year and tree age are also collected.

On the other hand, when the stand description method is applied, the stand composition, the regeneration method, the crown cover and signs of damage are assessed together with estimates for the stand's age, the number of stems per hectare, the average diameter and the average height.

The GMS field assessment was carried out for 15 years, included three periodic measurements (three GMS cycles) for each plot, and ended in the spring of 2008. The 15 years of annual data also permitted estimation using 5-year moving averages with starting dates from 1993/1994 to 2003/2004.

The huge amount of data collected required comprehensive control and consistency analysis, and therefore data evaluation took a long time.

15.2.5 The National Forest Inventory

In the course of the 4th cycle of the Growth Monitoring System (GMS), which corresponds to the 1st National Forest Inventory, some changes were implemented, namely on the type of inventory plot and on the set of sustainability indicators collected. Although the sampling grid and the selection of sampling points have not been significantly changed, NFI concentric circular plots with constant radius were assessed instead of plots with radii between 4 and 25 m depending on stand density.

In the inner circle (3 m radius) all trees with *dbh* greater than 7 cm are measured; in the intermediate circle (7 m radius) trees with *dbh* greater than 12 cm are measured, whereas for the outer circle (12.62 m radius) only trees with *dbh* greater than 20 cm are measured.

In addition to the data collected under the GMS, in the NFI survey indicators of forest naturalness were emphasised and information on standing and lying dead trees, stumps and actual and potential forest habitats were assessed. Field data collection was supported by the use of maps, and all relevant locations of points, lines, and areas are stored in GIS shapefiles.

The design of the NFI database is aimed at providing forest resource information on forests that can be distinguished according to forest region, structure, site class, tree species and management system. Therefore, NFI data can be used to characterize Hungary's forests and its resources.

15.3 Prediction of Potential Harvest

The potential final harvest volume for the next 30 years is predicted in 10-year time-steps using data from the NFD (Fig. 15.3). Predictions require the following input data: rotation age, current growing stock (V, m^3), current annual increment (I, m^3/year) obtained from yield tables and a reduction factor (r) depending on tree species and tree age also derived from yield tables. For a 30-year period, three felling classes are considered based on the difference between stand age and the final felling age (Felling Index). Class I includes stands with Felling Index less than 10; Class II is for stands to be felled in the second decade; and Class III is for stands to be felled in the last 10 years of simulation. All stands felled in Classes I, II and III are assumed to be harvested at years 5, 15 and 25, respectively, which represent the mid-points of the classes. The harvestable volume is $V_{harv\_j} = V + j \times I \times r$, where j stands for the Classes' mid-points (5, 15 and 25). The reduction factor is considered to be 1 for the next decade because stands in Class I are near the rotation age and normally not to be thinned.

The first step is to determine the stand's Felling Class using the stand age and the final felling age. After the class has been determined, the stands' standing stock must be predicted for the 5th, 15th or 25th year of simulation, depending on whether the stand belongs to Class I, II or III. The volume to be harvested for a stand in Class I is obtained by adding to the stand's standing stock and the stand's current annual increment multiplied by 5 years. As examples, suppose we have two stands A and B: stand A is of age 92 years with a felling age of 100, whereas stand B is of age 99 with a felling age of 110. Stand A has a Felling Index of 8, whereas stand B has a felling Index of 11 which makes stand A a Class I stand and Stand B a Class II stand. Assuming that both stands have a standing stock of 100 m^3 and current annual increments of 20 and 10 m^3/year respectively, their harvest volumes would be: $V_A = 100 + 5 \times 20$ and $V_B = 100 + 15 \times 10 \times r$. The potential harvest volume for the 30-year period is the sum of volumes for all stands in each Class. This prediction does not include thinning volume which will be added to the final yield.

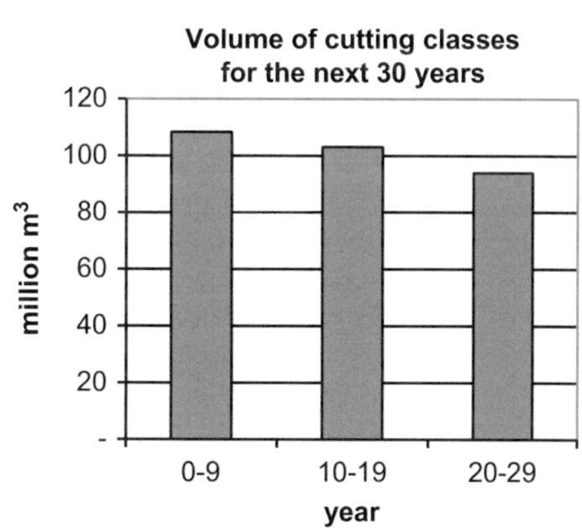

Fig. 15.3 Final harvest opportunity according to the NFD for the next 30 years

According to the felling targets defined in the district forest management plans, both the volume resulting from pre-commercial thinning and final harvest can be predicted every 10 years. Of course, this is only a harvesting potential and is not necessarily exploited by forest managers.

The increment and the age class structure allow yield control for a given forest planning district or yield regulation, if necessary.

15.4 Conclusion

In Hungary, in addition to the NFI, the Standwise system inventory (NFD) is available as well. The use of the NFD whose data are continuously updated is currently dominant. As a result, very accurate data on harvest opportunities are available.

The main advantage of the NFI is that it covers the total forest area including areas that are not recorded in the NFD and, therefore, can provide more comprehensive information about forest resources. The first 5-year NFI period ended in 2014. Subsequently, there will be opportunities to use this data for producing more comprehensive projections focusing on aspects other than just timber volume.

The importance of the NFI is expected to increase further in the future, as it enables a proper data service recently required by the forestry sector. These new survey criteria can be easily integrated into the NFI system on a year-to-year basis. The NFI is a powerful tool for meeting comprehensive and independent national or regional statistical data requirements.

References

Central Agriculture Office Centre, Forest Directorate (2009) Forest monitoring and observation system 1988–2008

NFCSO (2014) Forest resources and forest management in Hungary, 2013

Chapter 16
Iceland

Arnór Snorrason and Bjarki Kjartansson

16.1 Introduction

16.1.1 Forest History

Iceland has a long history of deforestation with extensive overexploitation of woodland resources, but in recent decades a short history of afforestation was initiated. From 1990 onwards several government subsidised projects have encouraged individual land owners to plant trees. Nowadays, despite Iceland having among the least forest cover in Europe, it shows the greatest relative increase in forest cover in the European Union. From 1990 to 2000, the annual increase in forest area was 8.0%, although the rate decreased to 4.8% in the next decade (UN-ECE/FAO 2015; FOREST EUROPE 2015).

Forests and other wooded land represent 0.4% and 1.4%, respectively, of Iceland's area, reflecting the rather small importance of forestry in the country. In 2010, the area of forest cover was estimated to be 42.7 thousand ha of which 32.0 thousand ha were cultivated forests (UN-ECE/FAO 2015; FOREST EUROPE 2015); the area of other wooded land was estimated to be 139.3 thousand ha, with natural birch woodlands covering 135.0 thousand ha.

The main tree species in Iceland are Mountain birch (*Betula pubescens)* with 43% of the forested area in 2010, Siberian larch (*Larix* sibirica) with 20%, Stika spruce (*Picea sitchensis*) with 11%, Lodgepole pine (*Pinus contorta*) with 11% and Black cottonwood (*Populus trichocarpa*) with 7% (FOREST EUROPE 2015).

According to the 2015 Forest Resource Assessment report (UNECE/FAO 2015), Icelandic forests include 5.86 thousand ha designated for wood production, 15.21 thousand ha assigned to soil and water protection, 25.88 thousand ha for multiple

A. Snorrason (✉) • B. Kjartansson
Icelandic Forest Research, Mógilsá, Iceland
e-mail: arnor@skogur.is; bjarki@skogur.is

S. Barreiro et al. (eds.), *Forest Inventory-based Projection Systems for Wood and Biomass Availability*, Managing Forest Ecosystems 29,
DOI 10.1007/978-3-319-56201-8_16

uses, 0.60 thousand ha for nature conservation and 6.19 thousand ha for social services (e.g. recreation, summerhouses) (UN-ECE/FAO 2015). About 66% of the forested area in Iceland is under private ownership either by individuals, private business or non-governmental organisations. The remaining 34% is under public ownership and is managed by the Icelandic Forest Service, the Icelandic Soil Conservation Service or by municipalities.

The growing stock of trees with diameters of at least 10 cm is increasing from 170,000 m^3 over bark in 2005, to 290,000 m^3 over bark in 2010. Because most of the trees in planted forests are still young and small, the growing stock reflects only part of the woody biomass, which was estimated as 570,000 metric tonnes in year 2005 and 790,000 metric tonnes in 2010.

After the Kyoto protocol was signed and ratified, reducing Greenhouse Gas (GHG) emissions and increasing CO_2 sequestration through afforestation became one of the main national targets (Ministry for the Environment 2010) and led the Icelandic State to request forest growth projections. In the period 1999–2001, a country-wide measurement campaign was conducted. The aim was to describe the growth potential for the 11 most common tree species in Icelandic forestry in five different regions (Snorrason and Einarsson 2001; Snorrason et al. 2001a, b; Snorrason and Einarsson 2002; Snorrason et al. 2002). These data were also used to develop general growth models for stem volume and aboveground biomass for these species which were later integrated into a simple forest simulator. The first projection aimed at estimating the annual GHG fluxes related to forest and forestry activities, with special focus on CO_2 emissions/removals and how different afforestation rates could affect these findings (Sigurðsson et al. 2005; Snorrason 2006). An improved version of the same simulator was later used to make a long-term projection of growing stock and possible wood removals in Iceland under a business as usual scenario (Snorrason 2011).

The objective for the National Forest Inventory (NFI) is to obtain reliable information about the current status of forest and other wooded land and to use it as a benchmark in Icelandic forestry. Similarly, the main reason for making projections is to predict expected long term GHG emissions, forest area, growing stock and wood removals. Such information is valuable for governmental decision making at regional and national levels and supports the small, but increasing, forestry sector.

16.1.2 NFI History and Data

The aim of the Icelandic NFI is to assess the national forest cover and to estimate carbon stock changes (Snorrason 2010). The data from the first survey, which was carried out between 2005 and 2009, provided the first official information on the status, growth and total resources of Icelandic forests at the national level. A systematic sample plot design was used with two strata: (1) Natural Birch Woodlands (NBW); and (2) Cultivated Forests (CF), which were mostly planted and included both native and exotic tree species. The two strata have different sampling densities: for NBW, a grid of 1500 × 3000 m is used, resulting in 203 plots; while for CF forest a grid of 500 × 1000 m is used resulting in 663 plots (Snorrason 2010). Concentric, circular, permanent plots are used for both strata but with different diameter and height thresholds. In cultivated forests, plot size depends on the stem density and is chosen to guarantee a minimum number of 20 trees per plot. In the second NFI cycle (2010–2015), only CF plots were re-measured, because a re-mapping of the natural birch forests was needed and therefore carried out in the same period.

16.2 NFI Data Used for Resource Projections

Diameter at breast height is measured on all trees that reach heights of 2 m, although heights of all the trees are not measured. Data for trees with both diameter and height measurements were used to developed height-diameter models so that the heights of trees without height measurements could be predicted. Single tree models for predicting volume over bark above stump volume were developed for the 11 most common tree species grown in Iceland and used to predict single tree volume (Snorrason and Einarsson 2006). Plot-level growing stock is predicted as the sum of individual tree volume predictions. Afterwards, the total volume (m^3) and the mean volume per unit area (m^3/ha) for each stratum are calculated as area-weighted sums and area-weighted averages, respectively.

Although the forest was mapped before the first NFI started, no good estimates of historical annual afforestation rates were available. However, accurate records for seedlings of different tree species produced annually and used in plantations have been kept since 1899 (Pétursson 1999). The first attempt to reconstruct afforestation history used the annual number of seedlings to estimate the area planted annually.

In the first afforestation programmes, the practice was to establish dense plantations, but later the planting densities were reduced. Based on a further analysis of the average planting density for year 2003 (Snorrason and Kjartansson 2004), the average planting density before and after 1990 was estimated in 4000 and 2350 seedlings per ha, respectively.

16.3 Methods

16.3.1 The Simulator

A simple forest simulator has been constructed to make projections for as long as 120 years into the future in one year time-steps. This tool was specifically developed for simulating cultivated forests and is not applicable to the natural birch forest. The main outputs of the simulator consist of forest area, C-stock and C-stock changes, growing stock and possible wood removals. Because the focus here is on the wood stock change and wood utilization, the C-stock module will not be further described. The simulator consists of two modules: a growth module and a management module that carries out thinnings and clear-cuts. Future annual afforestation area is estimated from annual seedling production, assuming a fixed number of seedlings per unit area.

16.3.1.1 The Growth Module

For each time step, the new growing stock per simulated stand is estimated using its age and standard volume models for each species. These models were used in the second version of the simulator (Snorrason 2006) and updated with new models for stem volume and aboveground biomass (Snorrason and Einarsson 2006; Bjarnadottir et al. 2007; Jónsson 2007).

16.3.1.2 The Management Module

With the current version of the tool, potential wood removals are projected using a fixed regime of thinnings and clear-cuts for each species, depending on their age. This regime represents standard thinning interventions based on recommendations, observations and clear-felling at age 35–105, depending on the species.

16.3.2 The Scenarios and Assumptions

Usually business as usual scenarios are used for which the annual number of seedlings planted as well as the species compositions considered are based on the average values observed for the last 5–10 years. Other scenarios consider increased afforestation levels which assume increase in the number of seedlings, but have only been tested for biomass and carbon stock projections. Different rates of wood availability for projected forest area have been set, but comparisons of reported and projected removals always show differences with the reported removals less than the potential.

For the current version of the simulator, the basic inputs are based on the forest area estimated in the first NFI (2005–2009). Calibrations have been done for the total forest area as well as for the forest area by species groups and by time periods.

The current approach relies on several assumptions and correction factors. Annual afforestation area is estimated based on the number of seedlings produced and assumed planting density. However, this resulted in an overestimation of plantation areas, because it didn't take into account failures of plantations with high mortality. After correction, the area generated by 100,000 seedlings is now fixed at 18.8 and 31.9 ha of successful plantations before and after 1990, respectively.

16.4 Results

The projections for growing stock and possible wood removals obtained with the current version of the simulator are presented in Table 16.1.

For this projection, the planted area, which was estimated based on the number of seedlings used either in new plantations or in reforestation after clear-cutting, was set based on the average observed area of plantations by species for the period 2005–2009. Equilibrium is reached when plantation areas and clear-cutting areas are of equal size. In this projection, the five most common species and one species group of slow growing conifers planted in Icelandic forests were considered. In total, these species' groups cover about 89% of the area and 93% of the growing stock of the CF.

The potential wood removals were also simulated (Table 16.2). Harvesting residues as well as wood not meeting the quality requirements were discounted (10%) from the potential wood available.

The current version of the simulator is more fine-tuned and better calibrated, but so far it has only been used for biomass and carbon stock projections.

16.5 Discussion and Conclusions

Comparisons of simulation results and NFI estimates showed that the simulator overestimates standing volume for the six species classes presented in Table 16.1. Further work must be done to improve the species models that integrate the growth module of the simulator and NFI data; this will be a crucial task. The comparison of NFI estimates and projection results for 2010 indicate that the simulator is capable of reproducing the relative importance of each species/species' groups for the

Table 16.1 Projected standing volume (thousand m^3 over bark) for the years 2010, 2020, 2060 and 2110; for the most common species/species groups planted in Iceland

	Projected standing volume (thousand m^3 over bark)			
Species/species groups	2010	2020	2060	2110
Sitka spruce (*Picea sitchensis*)	70	174	1446	2731
Lodgepole pine (*Pinus contorta*)	82	197	1471	2641
Siberian larch (*Larix sibirica*)	110	272	1657	2627
Slow growing conifers[a]	49	76	185	270
Black cottonwood (*Populus trichocarpa*)	40	92	347	339
Mountain birch (*Betula pubescens*)	33	76	467	1031
Total	385	886	5573	9639

[a]Norway spruce (*Picea abies*), white spruce (*Picea glauca*) and Engelmann spruce (*Picea engelmanni*)

Table 16.2 The potential wood removal (thousand m^3 over bark) for the years 2010, 2020, 2060 and 2110; for the most common species/species groups planted in Iceland

Species/species groups	Projected potential volume removals (thousand m^3 over bark)			
	2010	2020	2060	2110
Sitka spruce (*Picea sitchensis*)	1.9	1.9	21.3	70.7
Lodgepole pine (*Pinus contorta*)	1.2	2.7	13.5	53.0
Siberian larch (*Larix sibirica*)	1.0	5.4	22.8	71.3
Slow growing conifers[a]	0.4	1.0	7.6	7.3
Black cottonwood (*Populus trichocarpa*)	1.4	0.8	12.0	13.2
Mountain birch (*Betula pubescens*)	0.2	0.7	4.5	20.6
Total	6.0	12.5	81.8	236.1

[a]Norway spruce (*Picea abies*), white spruce (*Picea glauca*) and Engelmann spruce (*Picea engelmanni*)

short-term. Again, future NFI data will be essential for validating and improving the simulator accuracy.

Regarding the projection of the potential wood removals, no attempt was made to estimate the portion of wood actually available for wood supply other than a simple exclusion of stem wood left in the forest. Actual commercial wood removals were reported to be 4.2 thousand m^3 over bark. in the 2010 annual report of forest activities (Gunnarsson 2011), which is 70% of the projected potential wood removal for that year. In the SoEF2015 report (FOREST EUROPE 2015), available growing stock (2010) was estimated to be 66% of the total growing stock of the CF forest in that year. The first five species/species' groups in Table 16.1 provide round wood of sizes and straightness well-suited for industrial use, although the sawn wood output will be relatively small for some of these species/species groups. Native mountain birch has a low yield and produces small and rather crooked wood not suitable for industrial use. However, the natural birch forest has an important role as protection forests in Iceland.

As mentioned before, the Icelandic forest cover is small, but is growing each year. With political interest in increasing the forest cover as well as obtaining more detailed information on forest wood and non-wood resources, the Icelandic NFI has been conducting measurements since 2005. Methods and practices on data collection and processing have been slowly evolving since the inventory surveys were started, and it is clear that they can still be improved to a great extent. All the same, the method for projections and the simulator can also be improved by using more detailed growth models. Possible improvements could include developing regional models for each age class or including site index as a dependent variable in the growth models. Currently, the same growth model is applied to each species-specific age-class regardless of the location of the stand or the site index. Despite the present limited use of forest products, removals will be more relevant and important in the future, and information on removals should be improved. Another desirable improvement to the current version of the simulator would be adding the effects of a changing climate.

References

Bjarnadottir B, Inghammar AC, Brinker M-M, Sigurdsson BD (2007) Single tree biomass and volume functions for young Siberian larch trees (Larix sibirica) in eastern Iceland. Icel Agric Sci 20:125–135. http://www.landbunadur.is/landbunadur/wgsamvef.nsf/key2/index.html. Accessed 16 Apr 2016

FOREST EUROPE (2015) State of Europe's Forests 2015

Gunnarsson E (2011) Skógræktarárið 2010. Skógræktarritið 2011(2):96–101

Jónsson JÁ (2007) Áhrif skógræktaraðgerða á viðarvöxt og flæði kolefnis í asparskógi. Háskóli Íslands, Reykjavík 84

Ministry for the Environment (2010) Iceland's Fifth National Communication on Climate Change Under the United Nations Framework Convention on Climate Change. Ministry for the Environment Iceland. http://unfccc.int/resource/docs/natc/isl_nc5_resubmit.pdf. Accessed 16 Apr 2016

Pétursson JG (1999) Skógræktaröldin. Ársrit Skógræktarfélags Íslands:49–53

Sigurðsson BD, Snorrason A, Kjartansson BÞ, Bjarnadóttir B (2005) Kolefnisbinding með nýskógrækt. Hvar stöndum við og hverjir eru möguleikarnir? In: Fræðaþing landbúnaðarins, Hótel Saga, pp 20–24. http://landbunadur.lbhi.is/landbunadur/wgsamvef.nsf/6d3d18e301de1f5e0025768c00561c33/d27c62073222efd900256f96004ef2c2/$FILE/04.pdf. Accessed 16 Apr 2016

Snorrason A (2006) Langtímaspá um kolefnisbindingu nýskógræktar. Skógræktarritið:58–64

Snorrason A (2010) National forest inventories reports: Iceland. In: Tomppo E, Gschwantner T, Lawrence M, McRoberts RE (eds) National forest inventories – pathways for common reporting. Springer, Berlin, pp 277–289

Snorrason A (2011) Prediction of reference level for the period 2013–2020 for Forest Management in Iceland. Icelandic Forest Research, Mógilsá. https://unfccc.int/files/meetings/ad_hoc_working_groups/kp/application/pdf/awgkp_iceland_2011.pdf. Accessed 16 Apr 2016

Snorrason A, Einarsson SF (2001) Landsúttekt á skógræktarskilyrðum. Áfangaskýrsla 1997-2001 fyrir Vestfirði. Rit Mógilsár Rannsóknastöðvar Skógræktar 7:64

Snorrason A, Einarsson SF (2002) Landsúttekt á skógræktarskilyrðum. Áfangaskýrsla 1997-2002 fyrir Suðurland og Suðvesturland. Rit Mógilsár Rannsóknastöðvar Skógræktar 14:68. http://www.landbunadur.is/landbunadur/wgsamvef.nsf/key2/index.html. Accessed 16 Apr 2016

Snorrason A, Einarsson SF (2006) Single-tree biomass and stem volume functions for eleven tree species used in Icelandic forestry. Icel Agric Sci 19:15–24

Snorrason A, Kjartansson BÞ (2004) Íslensk skógarúttekt. Verkefni um landsúttekt á skóglendum á Íslandi Kynning og fyrstu niðurstöður Skógræktarritið:101–108

Snorrason A, Einarsson SF, Traustason T, Fanney DB (2001a) Landsúttekt á skógræktarskilyrðum. Áfangaskýrsla 1997–2001 fyrir Norðurland. Rit Mógilsár Rannsóknastöðvar Skógræktar 6:71. http://www.landbunadur.is/landbunadur/wgsamvef.nsf/key2/index.html. Accessed 16 Apr 2016

Snorrason A, Traustason T, Einarsson SF, Fanney DB (2001b) Landsúttekt á skógræktarskilyrðum. Áfangaskýrsla 1997–2001 fyrir Vesturland. Rit Mógilsár Rannsóknastöðvar Skógræktar 5:70

Snorrason A, Heiðarsson L, Einarsson SF (2002) Landsúttekt á skógræktarskilyrðum. Áfangaskýrsla 1997–2002 fyrir Austurland. Rit Mógilsár Rannsóknastöðvar Skógræktar 13:68

UN-ECE/FAO 2015 Global Forest Resource Assessment (2015) Food and agriculture organization of the United Nations (FAO)

Chapter 17
Ireland

Henry Phillips, Mark Twomey, and John Redmond

17.1 Introduction

17.1.1 Forest Inventory

The Irish national forest estate has increased from a modest 89,000 ha or 1.3% of the land area in 1928 (Minister for Lands and Agriculture 1928) to 732,000 ha in 2012, which represents 10.5% of the land area. In the 1920s grave concerns were expressed regarding the deforestation of private forests, and the 1928 Forestry Act sought to address these concerns by regulating tree felling and establishing a legal replanting obligation.

The first inventory of growing stock in State forests was carried out in 1958/1959 and formed the basis for forecasting of future volume yields (O'Muirgheasa 1967). Up to 1980, 80% of stocked forests were in State ownership and hence stand level assessments, as well as long term analyses of woody biomass projections, were concentrated in State forests. The first statistical sample survey estimating standing volume in the national forest estate was conducted in 2007 by the National Forest Inventory (NFI). In 2011, the first national forecast of the national forest estate was conducted which also included Northern Ireland.

H. Phillips (✉)
Forestry Consultant, Cloot na Bare, Rathonoragh, Sligo, Ireland
e-mail: hprphillips@gmail.com

M. Twomey
Forest Service, DAFM, Clogheen, Clonakilty, Co, Cork, Ireland
e-mail: mark.twomey@agriculture.gov.ie

J. Redmond
Forest Service, DAFM, Johnstown Castle Estate, Co, Wexford, Ireland
e-mail: johnj.redmond@agriculture.gov.ie

S. Barreiro et al. (eds.), *Forest Inventory-based Projection Systems for Wood and Biomass Availability*, Managing Forest Ecosystems 29,
DOI 10.1007/978-3-319-56201-8_17

The Irish NFI is based on a randomised systematic grid sample design and has the task of describing the state and changes in Ireland's forests. At present the periodic sample consists of 1827 permanent sample plots, each representing approximately 400 ha. Within each plot a variety of primary attributes are assessed from the tree top to the soil underneath. All recent woody biomass projections have primarily relied on geospatial data which have been available at local forest level since 1998. However, research is currently underway to use the NFI data for national woody biomass projections.

17.1.2 Descriptive Statistics

Up to the eighties almost all afforestation was undertaken by the State, but with the introduction of financial incentives for afforestation from the State/EU private landowners, mainly farmers, began to afforest significant amounts of land, doubling the area of forests. Ireland has 732,000 ha of stocked forest land of which approximately 342,000 ha (46.8%) are privately owned while the remaining 53.2% is publicly-owned (Forest Service 2013a).

Non-native species represent 76.2% of the forest area and native species 23.8% (Forest Service 2013a). The most common forest type is Sitka spruce, *P. sitchensis,* which covers 52.5% of Ireland's stocked forest area. Other important forest types are Lodgepole pine, *P. contorta,* (9.7%) and willow, *Salix* sp., (7.3%). The total growing stock is estimated as 97 million m^3, with 63% in publicly-owned forests. In 2012, 56% of the total stocked forest estate was less than 20 years old. Young and thinning stage forest are the dominant maturity classes with 61.8% and 38.2%, respectively of the stocked forest area. The average growing stock is 148 m^3/ha, which is small in comparison with many European countries (Ministerial Conference on the Protection of Forests in Europe 2007) and is a reflection of the young age structure of the forest estate. Productivity is high in Ireland with an average annual increment of 13 m^3/ha in public forests and 11 m^3/ha in private forests (Forest Service 2013a).

Standing volume in Ireland's forests has increased consistently since the surveys of 1968 and 1973 from 12.7 million m^3 to 97 million m^3 in 2012. The proportion of deadwood is about 6% of the total standing volume (Forest Service 2013a).

17.1.3 Forecasting in Ireland

Forest roundwood supply forecasting for both private (Purcell 1979) and State owned forests (Forest Service 1987) was undertaken periodically by the Forest Service up to 1988. The forecast period was typically 20 years and provided annual volumes overbark by assortment classes and by species group. Due to the relatively small proportion of broadleaved species within the forest estate and their use primarily for recreation and landscape, they were not included in the forecast data.

Every 5 years, Coillte publishes detailed five year volume forecasts, based on a continuous Geographic Information System (GIS) stand level inventory for the forests under its stewardship,[1] with the most recent being Forecast 2011 (Coillte 2011). In 2001, a desk study (Gallagher and O'Carroll 2001) on the production of private forests was made, and in 2009 a more detailed geospatial production forecast for the private sector was published (Phillips et al. 2009). Unlike previous forecasts, it included broadleaved species.

The first NFI cycle, which was completed in 2007, did not provide a roundwood production forecast. The second cycle, which was completed in 2012 included the estimation of increment and harvesting for the first time at a national level.

In 2011, an All-Ireland forecast was produced for the period 2011–2028 (Phillips 2011) using production forecast data from (a) Coillte, (b) Northern Ireland Forest Service (NIFS) and (c) the private sector in the Republic of Ireland (ROI). This provided woody biomass forecast data for varying size assortments and species groups and an estimate of wood fibre potentially available for wood energy. The Republic of Ireland component of this forecast is discussed in more detail in the following sections.

17.1.4 Data Sources

Two main sources of data were used in woody biomass projections for the Republic of Ireland: the Coillte stand inventory and Forest Service geospatial data which varies in the level of detail regarding species and age category. A 10 m Digital Elevation Model (DEM) was also used along with a forest soils productivity dataset. In addition, a database of Forest Service planting records was used in combination with the above geospatial data.

17.2 Forecast Methods

17.2.1 General

Roundwood volume forecasts are a function of (a) productivity (yield class[2]), (b) management regime (rotation length and thinning prescription), (c) forest area and stocking, (d) volume reduction factors and, (e) growth model.

Two types of growth models are used when forecasting roundwood volumes in Ireland, i.e. static (tabular) and dynamic. Until recently, forecasting in Ireland relied

[1] In addition to State-owned forests, Coillte manages areas where it has entered into a partnership arrangement with the landowner and areas where the harvesting rights have been sold to pension vehicles such as the Irish Forestry Unit Trust (IForUT).

[2] Yield class is a measure of site productivity and is expressed in terms of m^3 growth per ha per annum. The higher the yield class the more productive the site and the greater the overall volume.

on UK static models (Forestry Commission 1981) for all species except for Lodgepole pine for which an Irish model was used (Forest and Wildlife Service 1975).

Dynamic models are more flexible and can accommodate changes in the timing, intensity and frequency of thinning. Such models were initially developed for Sitka spruce (Broad and Lynch 2006) and based on Irish data subsequently for the other main species and more recently for ash, *F.excelsior*. The suite of stand-level dynamic models is known as *GrowFor*.[3] To estimate the volume in each of the three standard assortment sizes, the approach was to use UK assortment tables which apportioned the volume based on mean diameter at breast height (*dbh*) for all conifer species up until the late nineties. Research data around that time indicated a difference in the growth pattern for Sitka spruce between Ireland and the UK (Jordan 1992), and an Irish assortment table for Sitka spruce was incorporated into the forecasting model used by Coillte.

17.2.2 Forecasting Methodology 2011

The objective of the 2011 National Forecast was to enable woody biomass projections to provide public authorities, organizations and industry a broad base for strategic decision making.

17.2.2.1 Coillte

The starting point for production forecasting is the stand-based inventory and the series of Forest Management Plans (FMPs). Together they are the inputs to the forecast model including species, age, area, productivity, stocking, accessibility and management regime. Forecast volumes are generated at a sub-compartment level.

Coillte uses a combination of Forestry Commission static yield models (Forestry Commission 1981), Irish static models and Irish dynamic models (*GrowFor*) to forecast stand volume. The dependent variables are tree species, mean *dbh*, top height and stems per hectare (Fig. 17.1).

The starting point for forecasting volume is the estimate of the gross over-bark standing volume to a top diameter of 7 cm provided by the stand-level volume model which is then subject to a series of reductions to approximate the volume that will eventually be invoiced arising from the sale of the timber crop (thinning or clearfell). *GrowFor* has four components (Broad and Lynch 2006): (1) volume model, (2) thinning model, (3) growth model, and (4) assortment model.

The thinning model proposed by Garcia (1984) allows for estimation of post-thinning basal area *(resp.* stocking) when the top height, pre-thin basal area, pre-thin stocking and post-thin stocking *(resp.* basal area) are known. *GrowFor* allows the user to specify the thinning in terms of stems removed, volume removed, or basal

[3] Available for download at http://www.coford.ie/toolsservices/growfor/

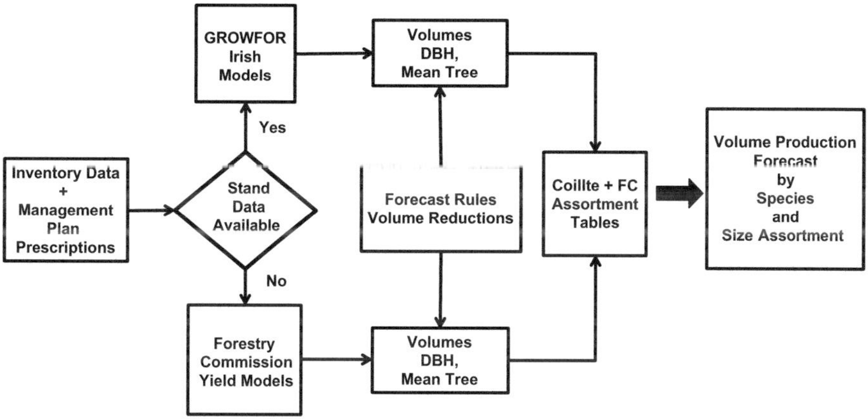

Fig. 17.1 Coillte forecasting schema (Coillte 2011)

area removed during thinning. Alternatively, the user can specify the thinning in terms of the post-thin desired state for volume, basal area or stocking.

The growth projection mechanism consists of a system of differential equations formulated by Garcia (1979). The system is sufficiently flexible to permit both empirical growth modelling, where no assumptions are made as to relations between stand variables, and modelling in limited data situations, where known biological principles are used as a basis for fanning models (Garcia and Ruiz 2003).

The volume is initially reduced by 15% to allow for roads, ridelines (unplanted areas between forest compartments), firebreaks and other non-productive areas. The volume is then reduced to compensate for losses arising from additional factors such as wind blow and disease.

Finally, the volume is reduced to take into account harvest losses, including: stump wood above ground level; bark removed during harvesting; waste wood arising from cutting into product assortments; logs left in forest and not extracted; and logs left at roadside following sale closure.

Coillte applies a volume reduction factor based on a combination of (a) species, (b) stage of harvesting (thinning/clearfell) and (c) tree size. The factor is based on the detailed field analysis of a series of volume sales. The factor will be updated over time in line with further analyses.

17.2.2.2 Private Sector

There was no national private sector database with species, area, productivity and other parameters to serve as input to the 2009 forecast for the private sector, so the datasets used to run the GIS-based roundwood production forecast reflected what was available from the Forest Service and consisted of:

(a) The **Premiums** dataset which covers private plantations in receipt of premiums between the years 1989 and 2007.
(b) The **WP08** dataset which covers private plantations that were in receipt of grant aid prior to 1989–1990.
(c) **FIPS98** which covers the remainder of the private estate, i.e. Private Grant Aided area (PGA) and Private Non-Grant Aided area (PNGA).

The **Premiums** and **WP08** datasets contain species and age data suitable for forecasting. The **FIPS98** species groups are too general for forecasting and do not include age. Estimates of discrete species and age classes were, therefore, derived for each of the FIPS98 species categories using the NFI (Forest Service 2007) and the 1973 Inventory of Private Woodlands (Purcell 1979).

For each polygon in the forecast dataset, a centroid was established using the easting (m) and northing (m) Irish National Grid coordinates which were imported as point data. Centroids were intersected with a Digital Elevation Model (DEM) to obtain elevation (m). The Irish Forest Soil (IFS) type data (Fealy et al. 2009) was appended using the same approach.

To determine yield class, to identify the most appropriate range and type of management regimes, and to determine a relationship between site factors and management regime (rotation length and thinning prescription), Coillte made available a sample of their Thinning and Rotation Classification (TRC) inventory. In total, 21,260 records of sub-compartment details were provided. The TRC assigns a thinning regime and a rotation type. Any deviation from the standard thinning treatment or standard rotation is accompanied by an explanation.

The Coillte TRC inventory data were made available, providing information on yield class by species by soil type and by elevation class. A multiple regression model was developed to predict yield class for Sitka spruce using elevation class and soil type as independent variables. The model was used to assign a yield class to areas of Sitka spruce in the private estate.

For other species, the TRC data were analysed and a relationship determined between the yield class for Sitka spruce and that for other species within the same sub-compartment. This was converted to a simple look-up table, validated through consultation with forestry professionals and then used to assign a yield class for all other species including broadleaved species.

A default thinning prescription was assigned to each polygon based on an analysis of the area and frequency of thinning types by main species (Sitka spruce, Norway spruce (*P.abies*), Douglas fir (*P.menzeisii*), lodgepole pine (south coastal), Lodgepole pine (north coastal), Japanese larch (*L. kaempferi*) and Scots pine (*P. sylvestris*)), by Coillte soil type classification by elevation class (<100 m, 100–200 m, 200–300 m and >300 m) within the TRC dataset. With this approach approximately 30% of the private forest estate was classified as no thin, 50% as standard thin and the remaining 20% as receiving two thinnings.

The current convention on rotation lengths for conifers is, site permitting, to grow crops to a rotation of Maximum Mean Annual Increment (MMAI) with the exception of Sitka spruce (20% below age of MMAI), Norway Spruce (30% below

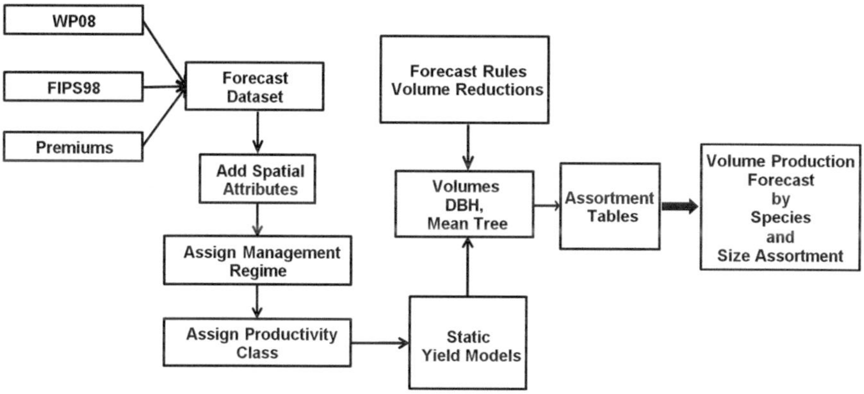

Fig. 17.2 Private forecasting schema (Phillips 2011)

age of MMAI) and lodgepole pine (coastal) (30% below age of MMAI) (Forest and Wildlife Service 1976). The underlying assumption used in the private sector forecast is that owners will, on average, wish to manage their plantations to maximize returns and use a financial rotation. An economic analysis of rotation length and thinning regime was undertaken and confirmed with minor modifications to the current convention (Phillips 2008). To simplify the underlying assumptions and spatial modelling, areas having standard thinning were assigned a standard rotation, and areas with either no thinning or two thinnings were assigned a local or reduced rotation (Fig. 17.2). Typically reduced rotations are practised where crop stability is a concern.

17.2.2.3 Net Realisable Volume

The All-Ireland forecast (Phillips 2011) combined data from the following sources into a standard reporting format, expressing forecast production in terms of Net Realisable Volume. The Net Realisable Volume is estimated roundwood volume that will potentially be available to the end-user.

Private Sector ROI The private sector forecast (Phillips et al. 2009) highlighted the lack of information on the accessibility of private forests. The forecast volumes were adjusted to exclude thinnings from small forest areas and plantations with a potentially uneconomic forest roading requirement. These areas were assigned a no-thinning regime and assumed to be harvestable at time of clear-fell. The overall net impact was a reduction of 4.3% in total volume.

Coillte Coillte revised its forecast estimates to: (1) include only volumes from Coillte-owned forests, (2) take account of harvest losses, (3) exclude those areas[4]

[4] Coillte estimates based on historical analysis that these areas can account for between 100,000 and 150,000 m^3 per annum.

which, for a variety of reasons, principally accessibility, are unlikely to be harvestable based on a continuation of current conditions, and (4) provide a further volume assortment category (tip–7 cm). In addition, Coillte adjusted the forecast volumes to provide an estimate for the last 2 years, i.e., 2027 and 2028, of the forecast tables. Coillte does not currently estimate the volume for broadleaved species in their forecasts but has plans to do so in the near future.

Harvest losses may decrease over time due to improvements in technology and or harvesting practices. No reduction in harvest losses was assumed and the factors were applied equally to all years within the forecast period.

17.3 Forecast Results

17.3.1 Roundwood

The forecast produced statistics on the net realisable volume production for the Republic of Ireland over the forecast period, 2011–2028. The forecast also included the future sustainable harvest levels between 2014 and 2028 by assortment and ownership type, as well as the evolution of volume available from clear-fells.

The total forecast of net realisable volume production for the Republic of Ireland over the forecast period, 2011–2028, is estimated as being 87.58 million m^3 over bark with an additional 2.51 million m^3 potentially available in the tip–7 cm category. Despite the almost doubling of the total forecast total net realisable volume by 2028, there is only a modest increase in volume in the 7–13 cm assortment of the order of 20%, with the volume within this category. Forecast volume in the 14–19 cm assortment shows a significant increase from 2014 to 2028 while the volume in the 20 cm + assortment more than doubles from within the same period.

There is an increasing need to inform policy makers, academia and potential investors regarding the potential availability of wood fibre volumes for energy use. To address this requirement, an addendum providing an estimate of potential wood fibre availability for energy was included in the All-Ireland forecast.

17.3.2 Wood Energy

Wood energy forecasts are based on the assumption that there are three main sources of raw material for wood energy – small roundwood from thinnings (7–13 cm top diameter), wood residues from the processing sector and post-consumer recycled wood; additional raw material is potentially available through the harvesting of tree tops (tip–7 cm) and through the harvesting of lop and top (including branches and some harvest loss material) on clear-fell suitable sites.

The net realisable volumes in the Republic of Ireland forecasts the potential wood fibre available for energy totals 23.749 million m^3 over the forecast period (2010–2028). It is important to note that the total available for energy is not an estimate of new or additional volume available for wood energy over and above current usage. Wood energy will have to compete with other end uses for the volumes indicated. Market price will ultimately determine whether the material goes to energy.

17.4 Discussion

17.4.1 Assumptions and Limitations of the Present Methodology

According to the latest forecast, the volume from private sector forests will increase almost eightfold within the next 15 years and be mainly responsible for increasing supply volumes for all parts of the forest sector. However, the information available for the private sector forecasting is limited. In particular, the datasets do not include information on accessibility, the quality of broadleaved species, and age, and limited species information is available for **FIPS98** areas. There is limited information on stocking and little information on the management intentions of private forest owners. In addition, the productivity estimate for each plantation is based on a sample analysis of the Coillte forest estate which may not be fully representative of private forests.

17.4.2 Error Estimates and Uncertainties

While the *GrowFor* volume estimates based on Sitka input parameters, have been validated, the model predictions for future volumes based on varying management (thinning) regimes have not as yet been validated.

There are no error estimates for either the *GrowFor* or Forestry Commission model forecasts of volume. Research undertaken as part of the Forecast project has attempted to provide error estimates for yield class models derived from Irish data (Lekwadi et al. 2011), but the models are as yet unsuited to form part of a robust forecast process.

The ability to provide error estimates will increase confidence in the forecast by all users including industry, Government agencies, academia and grower organisations.

17.4.3 Future Improvements

Future forecasts should be able to address most but not all the shortcomings.

The Forest Service is in the process of updating the ***Forest07*** database which contained the three datasets used to forecast private sector volumes. This process will append recent afforestation and update both the spatial and attribute data on the remaining forest areas.

The Forest Service supports the drafting of management plans at age 10 years for grant-aided plantations of more than 10 ha (5 ha for broadleaved species). Enhancements to the management plan format have been introduced in line with the requirements of certification to support the principles of Sustainable Forest Management (SFM). The management plans also request information on the appropriate management regime. The use of this information to support future forecasting process will provide for more reliable roundwood production information at national, regional, county and catchment levels, thus facilitating improved investment decisions by the processing and wood energy sectors.

The *GrowFor* models based on Irish data are more appropriate for Irish conditions than the UK Forestry Commission models. Over time, Coillte intends to reduce its reliance on the UK models and move exclusively to *GrowFor* as stand data and models become available. The growth model has never been validated. Doing so would require measurement of a series of representative plots now with re-measurement in 4–5 years. With such data, the growth model could be recalibrated to reflect the actual growth measurements.

Accessibility could be interpreted based on orthophotomaps, Ordnance Survey Ireland's spatial datasets on road networks and proximity of adjoining plantations. Work in this area is being undertaken as part of the SupplyChip[5] and Forecast projects (Whelan 2011). Site productivity could be based on the Teagasc model for Sitka spruce (Farrelly et al. 2009) which has been shown to be robust in the field and should help improve the reliability of production forecasting. The draft forest policy completed in 2012 seeks to address the limited information on owner intentions through the introduction of a standardised and integrated forest management planning process.

17.4.4 Research into Using NFI Data for Forecasting

Forecast, a research project on forecasting methodologies and their application in Ireland, is currently underway. One objective of the project is to assess the possibility of generating a reliable forecast of production from privately owned forests using the existing NFI plot data. The NFI data contain no information on top height

[5] SupplyChip – Facilitating the supply of wood chip from forest plantations for a major heat user. Project funded by COFORD and being undertaken by Teagasc.

or yield class, which are traditionally used for forecasting in Ireland. The data were analysed to determine whether it would be possible to apply a surrogate top height based on a mean height derived from the NFI data. Robust relationships between mean height and top height have been developed for the main coniferous species based on permanent sample plot data provided by Coillte. Models were also developed to predict the upper and lower confidence intervals. These mean height top height models allow the NFI plot data to be populated with top height, yield class with the precision of the estimates quantified. Code for the *GrowFor* growth model has been modified to allow forecast estimates to be generated for any number of input vectors for both thinned and unthinned stands. This capacity will be extended for all coniferous species for which *GrowFor* models exist.

Within the past decade there have been multiple major advances and initiatives to support the development of forecasting roundwood volume production from Irish forests. These include: (1) the availability of dynamic growth models based on Irish growth data which will over time decrease reliance on UK models, (2) the development of methods and datasets to increase the reliance of forecast volumes from the private sector, and (3) research into the most appropriate forecasting methodology and use of NFI data for forecasting purposes.

Further development and capacity is required to enhance the reliability of volume forecasts including incorporation of error estimates and integration of forest management plan data from the private sector into the overall national forecast process.

While the sawmilling sector relies on volume forecasts to aid their planning and investment decisions, they ultimately are interested in log volume as opposed to conventional assortment volumes. With further research into the relationship between assortments and product outturn, it should be possible to provide volume forecasts which include product volumes for a standard range of product specifications.

References

Broad LR, Lynch T (2006) Growth models for Sitka spruce in Ireland. Irish For 63(1):53–79

Coillte (2011) In: Roundwood Supply Forecast 2011–2015.Coillte. http://www.coillte.ie/fileadmin/user_upload/pdfs/Timber_Sales_Forecast/Forecast_2011-2015_document__2_.pdf. Accessed 21 Sept 2013

Farrelly N, Ní Dhubháin A, Nieuwenhuis M, Grant J (2009) The distribution and productivity of Sitka spruce (Picea sitchensis) in Ireland in relation to site, soil and climatic factors. Irish For 66(1–2):51–73

Fealy RM, Green S, Loftus M, et al (2009) Teagasc EPA soil and subsoils mapping project-final report. Volume I. Teagasc Dublin

Forest and Wildlife Service (1975) Research Communication No. 16. Revised yield tables for Coastal Lodgepole Pine. Research Branch, Crop Structure, Bray Ireland

Forest and Wildlife Service (1976) Rotation lengths and thinning regimes for conifers. Operational Directive (1/77) rotation lengths and thinning regimes for conifers. Department of Fisheries, Dublin

Forest Service (1987) A forecast of volume production. Forest Service, Dublin

Forest Service (2007) National Forest Inventory Republic of Ireland Results. Forest Service, Dublin

Forest Service (2013a) National Forest Inventory – Republic of Ireland – field procedures and methodology. Department of Agriculture, Fisheries and Food, Dublin

Forest Service (2013b) National Forest Inventory – Republic of Ireland – results. Department of Agriculture, Fisheries and Food, Dublin

Forestry Commission (1981) Yield models for forest management. Forestry Commission Booklet 39. HM Stationery Office, London

Gallagher G, O'Carroll J (2001) Forecast of roundwood production from the forests of Ireland 2001–2015. COFORD, Dublin

Garcia O (1979) Modelling stand development with stochastic differential equations. In Elliott DA (Camp.) Mensuration for management planning of exotic forest plantations. FRI Symposium No. 20. New Zealand Forest Service, pp 315–333

Garcia O (1984) New class of growth models for even-aged stands: *Pinus radiata* in Golden Down's forest. NZ J For Sci 14:65–88

Garcia O, Ruiz F (2003) A growth model for eucalypt in Galicia, Spain. For Ecol Manag 173:49–62

Jordan P (1992) Volume assortment tables for Sitka spruce (*Picea sitchensis* Bong. Carr.) in Ireland. M AgrSc (Forestry) thesis. National University of Ireland, Dublin

Lekwadi S, Nemesova A, Hunter A, Mac Siurtain M (2011) Quantile top height-age site class growth models for state-owned Sitka spruce forest plantations in Ireland. In: International Biometric Society, 3rd Channel Network Conference, France, April 12 2011

Ministerial Conference on the Protection of Forests in Europe (2007) The state of Europe's forests 2007. MCPFE Warsaw

Minister for Lands and Agriculture (1928) Dáil Éireann, vol 23, 3 May 1928

O'Muirgheasa N (1967) Forest research review 1957–1964. Forestry Division, Department of Lands, Ireland. Government Publications, Dublin

Phillips H (2008) Review of rotation lengths and thinning regimes for conifer species. Internal report, FORECAST project. COFORD, Dublin

Phillips H, Redmond J, Mac Siurtain M, Nemesova A (2009) Roundwood production from private sector forests 2009–2028: a geospatial forecast. COFORD, Dublin

Phillips H (2011) All-Ireland Roundwood Production Forecast 2011–2028. COFORD, Dublin

Purcell T (1979) Inventory of private woodlands, 1973. Department of Fisheries and Forestry, Forest and Wildlife Service, Dublin

Whelan, A (2011) Spatial constraints on geospatial forecasts of Private Sector Timber Supply. ESRI Forestry GIS Solution Conference Redlands, California May 24–26, 2011

Chapter 18
Italy

Alessandro Paletto, Sandro Sacchelli, and Patrizia Gasparini

18.1 Introduction

The first Italian National Forest Inventory (NFI) was carried out between 1983 and 1986, and the results which were published in 1988 refer to the year 1985 (IFNI85). IFNI85 adopted a systematic sampling scheme based on a regular 3 × 3 km grid, with a single field survey (MAF-ISAFA 1988).

The second Italian NFI, called National Inventory of Forests and Forest Carbon Pools (INFC), was designed in early the 2000s to meet national and international needs for updated statistics on Italian forests (Tabacchi et al. 2005). INFC adopted a three-phase sampling for stratification design: the first two phases provided the data to estimate forest area and its partition into inventory categories (e.g. plantations, high forests and coppices, sparse forests, scrublands, shrubs) and forest types, and the third phase was devoted to tree-level measurements (Gasparini and Tabacchi 2011; Gasparini et al. 2010). The first phase sampling units are approximately 300,000 photo-points and are randomly located at the intersections of a 1 × 1 km grid (tessellation sampling) covering the whole Italian territory. They were photo-interpreted on digital orthophotos to assign the land cover and land use class. For the second phase, a sub-sample of 30,000 points was randomly selected from the inventory categories "forest" and "Other Wooded Land" (OWL) and stratified by

---

A. Paletto (✉) • P. Gasparini
Council for Agricultural Research and Economics, Research Centre for Forestry and Wood (CREA-FL), Trento, Italy
e-mail: alessandro.paletto@crea.gov.it; patrizia.gasparini@crea.gov.it

S. Sacchelli
Department of Agricultural, Food and Forest Systems Management (GESAAF), University of Florence, Florence, Italy
e-mail: sandro.sacchelli@unifi.it

S. Barreiro et al. (eds.), *Forest Inventory-based Projection Systems for Wood and Biomass Availability*, Managing Forest Ecosystems 29,
DOI 10.1007/978-3-319-56201-8_18

the 21 administrative regions of Italy. The sampling units were visited in the field in year 2005 to collect information on 40 qualitative plot-level attributes (e.g. forest category, crown cover, exposure, slope, stand origin, stand structure, ownership). In particular, 23 forest categories and 91 sub-categories were used according to a national classification scheme (Gasparini and Tabacchi 2011). In the third phase, quantitative information (e.g. tree diameter at breast height, tree and crown height, size of lying deadwood, standing dead trees and stumps) was collected together with additional qualitative information (i.e. tree and shrub species, tree vitality and integrity, silvicultural treatments, harvesting and extraction system). The measurements were taken for approximately 6600 sampling plots to derive estimates on quantitative stand attributes (e.g. growing stock volume and biomass, standing and lying deadwood volume, regeneration and shrubs abundance). The third phase sampling units were randomly selected from the second phase sample and stratified by administrative region and forest category (Gasparini and Tabacchi 2011). The results of INFC are important for national forest policy and provide essential data to comply with international commitments (e.g. Kyoto Protocol and United Nations Framework Convention on Climate Change).

In Italy, the forest sector contributes just 0.7% of the Italian Gross Domestic Product (Lasserre et al. 2011). The main reasons for this small value are linked to the declining timber market value, the highly fragmented private ownership, and the high costs of harvesting (Cesaro et al. 2010).

According to national legislation (DPR n.616/77), the forest departments and services of the 21 administrative regions in Italy are responsible for the forest sector. At the state level, the Ministry of Agricultural, Food and Forestry Policies (MiPAAF) is responsible for the national forestry policy, the coordination of regional policies and representation of the country internationally. Moreover, other activities concerning forest resources such as forest condition monitoring, forest fire protection and the NFI are carried out by National Forest Service that is part of MiPAAF. The Ministry for Environment and Protection of Land and Sea (MATTM) is responsible for environmental issues and land protection. An agency of MATTM, the Italian National Institute for Environmental Protection and Research (ISPRA) is in charge of reporting to the United Nations Framework Convention on Climate Change (UNFCCC) and the Kyoto Protocol.

NFI estimates are part of the Italian official statistics collected by Italian National Institute of Statistics (ISTAT) and are used for the main reporting activities at the international level (UNFCCC and Kyoto protocol, FAO-Global Forest Resources Assessments, and State of Europe‘s Forests).

Although tools for predicting the woody biomass resource from inventory data under different management conditions and for different time scales have been developed for some case studies, a resource projection system applicable at national scale has not been developed yet.

18.2 Data

18.2.1 Descriptive Statistics

According to the second Italian NFI (INFC-2005), the total forest area was estimated as 10,467,533 ha (8,759,200 ha of forest land and 1,708,333 ha of other wooded land, FRA 2000 definitions; FAO 2001).

Forest ownership is an important feature in the Italian context because small private property is prevalent, and this aspect must necessarily be taken into account in the development of a woody biomass resource projection system. The results of INFC show that 66.2% of forest land is private property and the remaining 33.8% is public forest (Gasparini and Tabacchi 2011).

The most common forest categories are oak dominated forests (sessile oak, downy oak and pedunculate oak) covering 12.4% of the forest land, followed by beech forests (11.8%) and Turkey oak forests (11.5%). The most widespread coniferous forests are Norway spruce dominated forests representing 6.7% of the forest land, and European larch and stone pine forests (4.4%) mainly located in the Alps (Gasparini and Tabacchi 2011).

The average basal area of Italian forests is 20.4 m^2/ha. The growing stock volume, defined as the volume over bark and above stump of stem and main branches with diameters of at least 5 cm (Gasparini et al. 2010), is 1269 million m^3 corresponding to 144.9 m^3/ha, while the total above-ground biomass is 874.4 million Mg (99.8 Mg/ha). The annual volume increment is 35.9 million m^3 (4.1 m^3/ha per year), and its value differs significantly among forest types (Gasparini and Tabacchi 2011).

The volume of deadwood amounts to 8.7 m^3/ha, divided into the three components: 5.3 m^3/ha for standing dead trees, 1.9 m^3/ha for lying deadwood, and 1.5 m^3/ha for stumps. Regarding forest categories, the greatest values of deadwood volume are found for the Alpine coniferous forests (e.g. silver fir) with 21.0 m^3/ha and for chestnut forests with 26.9 m^3/ha (Gasparini and Tabacchi 2011).

Regarding the volume of removals, INFC estimated total fellings' volume during the 12 months before the inventory survey of 13.8 million m^3 over bark (1.6 m^3/ha per year). The regions with the greatest values are Trentino (3.6 m^3/ha per year) and Umbria (3.5 m^3/ha per year), while other three regions (Lazio, Alto Adige and Campania) show values in the range of 2–3 m^3/ha per year. On the other hand, the smallest values of removals, close to zero, are found in the regions Valle d'Aosta, Basilicata, Sicilia and Sardegna.

18.2.2 NFI Data

The second Italian NFI collected tree, stand and site data using different fixed area plots. For tree and deadwood measurements two concentric plots of 13 and 4 m radius, centered at the sampling point, were used, while regeneration and shrubs were measured on two smaller plots with 2 m radius and located 10 m distant from the sampling point (Gasparini et al. 2010). Trees with diameter at breast height ≥9.5 cm were measured on the 13 m radius plot, while trees with a diameter at breast height between 4.5 and 9.5 cm were measured on the 4 m radius plot. The 13 m radius plot was also used for deadwood measurements (lying deadwood with minimum diameter ≥9.5 cm, standing dead trees with a diameter at breast height ≥4.5 cm and stumps with a diameter ≥9.5 cm). Saplings and shrubs from 50 cm height up to 5 cm of diameter at breast height were counted by three diameter-height classes in the two 2 m radius plots. Stand and site data (e.g. damage, microhabitat, forest category) were collected on a larger 25 m radius plot centered at the sampling point as were the plots for tree and deadwood measurements (Gasparini et al. 2010).

The field data collected in the second and third inventory phases were used to calculate NFI estimates of growing stock volume, annual increment, standing and lying deadwood volume, above-ground biomass and volume of removals at the national and regional level, by forest category.

Regarding information usable for woody biomass resource projections, INFC provides two useful categories of data:

1. data on growth and harvest (quantitative: annual volume increment and removals volume; qualitative: forest management regimes and extraction systems);
2. data on accessibility and availability for wood supply (mainly qualitative: degree of accessibility, slope class, roughness, distance to roads or forest tracks' class, and availability for wood supply).

Forest Growth and Harvest

Data on growth and harvest were collected in the third phase of INFC. The information useful for woody biomass resource projections are annual volume increment and volume of removals.

Annual volume increment was assessed using sample cores extracted from 5–10 sample trees per plot for which the diameter increment was measured for the last 5 years before the inventory survey (Gasparini et al. 2010; Gasparini and Di Cosmo 2016). As a consequence of the procedure adopted, the estimated volume increment is the periodic annual volume increment of the living trees at the time of the survey (reference year 2005).

The volume of cut trees was estimated by the INFC by measuring the diameter of the stumps left after harvest (due to thinning, fellings or cuttings of dead/broken trees after wind storms, avalanches, etc.) for the last year before the survey (<12 months). Stumps with a diameter at the cutting section ≥9.5 cm , located within the 13 m radius plot, were counted and measured (two diameters at the cut section and average stump height from the ground); also the tree species was identified. A model with the diameter at the cut section as independent variable was used to predict the diameter at breast height of each felled tree, which was then used to estimate the stem plus large branches volume of the individual tree. Consequently, the sum of the volumes of felled trees in the plot does not refer exactly to the removals volume because no information is available on which part has been actually removed and which one has been left in the forest.

The forest management regime and the extraction system were classified considering the forest stand characteristics and the local silvicultural practices (traditions and customs). The information sources were: direct observation in the field, interviews with local experts and other official documents (e.g. forest plans, GIS layers). In particular, INFC also recorded the forest management regime using six classes: high forest clearcutting or coppicing, high forest clearcutting with reserves or coppicing with standards, patch cutting, shelterwood cutting, selective or uneven-aged cutting and other forest management regimes (Gasparini and Tabacchi 2011). Lastly, INFC classified the extraction system using five classes (downhill, animal skidding or tractor without skidder, direct or indirect skidding, cable systems and aerial logging).

Accessibility and Availability for Wood Supply

Data on accessibility and availability for wood supply include two main types of information: the first concerns slope and roughness, the second concerns some accessibility parameters (i.e. distance to roads and tracks). Additionally, the sampling units were also classified as accessible or inaccessible due to physical obstacles (orographic causes) or legal restrictions as defined by national and international legislation (e.g. forests in strictly protected areas or military areas). The information on the accessibility was considered when defining the availability for wood supply. The classification of availability for wood supply was based on a synthetic evaluation about limitations on forestry activities due to regulations (e.g. presence of integral reserves in protected areas) or to physical features implying high costs for logging.

The distance to roads or forest tracks was measured as the distance between the sampling point and the nearest road, and was then aggregated into four classes (<500 m, 501–1000 m, 1001–2000 m, >2000 m). The distance to roads or forest tracks was calculated as the difference between the plot coordinates and the coordinates of the closest road/track directly measured in the field, without considering the roughness and asperity of the terrain. The roads were divided into four categories: roads passable for trucks, forest roads passable by tractors with trailers, forest tracks passable by tractors without trailers, and paths or trails.

The slope was measured on the plot as a continuous variable (degrees of the angle of inclination) and classified using five classes (0–20%, 21–40%, 41–60%, 61–80%, >80%).

The asperity considers the micro-morphology of the terrain and the presence of obstacles such as rocks, ditches and gullies. The roughness is classified into three classes considering the presence of small and large obstacles: (1) absence of obstacles or small obstacles on less than 25% of the plot, (2) small obstacles on 26-75% of the plot or large obstacles on less than 25% of the plot, and (3) small obstacles on more than 75% of the plot or large obstacles on more than 25% of the plot. The presence of obstacles was expressed as a percentage of the plot area.

18.2.3 ISTAT Data

The Italian National Institute of Statistics (ISTAT) reported data on forest areas and industrial roundwood or firewood removals for the reference period 2001–2006. The figures on removed woody assortments are provided for forest and no-forest areas and are derived from annual questionnaires completed by the Provincial Offices of the National Forest Service.

Data are presented at regional level by wood use typology:

- Industrial roundwood (sawlogs, round and split pulpwood, other industrial roundwood)
- Wood fuel (included wood for charcoal)

Data on volume and areas of removals authorized by the National Forest Service are also reported at regional level and classified by forest ownership: State and Region, Municipality, Other public ownership and Private ownership.

The ISTAT data useful for woody biomass resource projections are limited to the percentage of felled woody assortments and their market prices. The wood market prices (timber and fuelwood) are annually collected by the ISTAT through the use of an *ad hoc* questionnaire compiled by Local Offices of the National Forest Service. The prices refer to the timber and fuelwood prices at the roadside distinguishing between conifers and broadleaves. The timber prices are also subdivided by species, assortment and geographical location (administrative regions).

18.3 Woody Biomass Resource Projections Based on NFI: Potential Fields of Research

In Italy, woody biomass resource projection procedures based on detailed input variables have not been sufficiently developed at the national scale until now. The only application was implemented by Pilli et al. (2013). The authors proposed the use of the Canadian Carbon Budget Model (CBM-CFS3) to predict the carbon sink of Italian forests at the national scale under different scenarios of natural disturbance and fires using aggregated NFI data. Even though the prediction of woody biomass availability is not the main aim of this study, the dynamic of aboveground biomass in the years 1995–2020 is provided as one of the outputs of the model application. In addition, some local and regional models have been developed to assess ecologically and technically-logistically exploitable forest resources (see e.g. Corona et al. 2002; Bernetti et al. 2004; Scrinzi et al. 2007; Lasserre et al. 2011). In two of these studies the BIOMASFOR model (Zambelli et al. 2012; Sacchelli et al. 2013), a Geographic Information System (GIS) based tool, was used. BIOMASFOR is a population growth-based tool for decision making at different administrative levels. It is regarded as suitable for implementation with data available at the country level such as INFC data (Sect. 18.2.2), integrated with additional geo-referenced and non-geo-referenced information.

The last version of the BIOMASFOR model, currently named *r.green.biomassfor* (Garegnani et al. 2015), was implemented using GRASS GIS 7.0 open-source software. The model applies economic and forest multi-function criteria to predict annual forest energy-biomass and timber availability for a defined territory. The forest multi-function criteria currently considered in the model are: soil fertility maintenance; soil and water protection; biodiversity maintenance; fire risk reduction; touristic-recreational valorization and CO_2 emission reduction. The *r.green.biomassfor* model performs a multi-step analysis that can lead to different estimates of bioenergy and timber availability defined as ecological, technical, economic, and sustainable production. The outputs can be exported as text files and geo-referenced maps (see Sacchelli et al. 2013, for more details).

The input variables needed for running the model can be classified as mandatory or optional. If one variable is optional, a default value can be used (if not available). This approach permits the model to process the data also in cases when information on some input variables is lacking. The mandatory input variables are slope and elevation, distance from main roads and forest roads, total yield, yield of *n-th* forest category (e.g. Norway spruce forest, Turkey oak forest etc.). The current version of the model permits to consider a maximum of five forest categories, forest management, and woodchip collection point. Examples of optional variables are: forest treatment, roughness, mean tree diameter, mean tree volume, soil productivity as well as fire risk index.

18.4 Discussion

The main potential users of woody biomass resource projection at the larger scales are the policy makers, the forest planners and managers of the regional departments and services (forestry, energy and environmental services), and the Ministries (MiPAAF, MATTM). Woody biomass resource projections could be potentially used by political decision makers for implementation of national policies and programs for the forestry and energy sectors. In addition, the results of biomass resource projections could provide useful support to forest planners and land managers in developing forest management plans at the landscape level.

The *r.green.biomassfor* model is regarded as suitable for development into such a projection system using INFC plot data as input. For example, the ecological annual (or periodic) availability could be predicted from annual volume increment data and information on forest management regime. Technical availability could be predicted from data on the extraction system and accessibility, slope, roughness and distance to roads or forest tracks (also available as a continuous variable). The addition of information on the efficiency of forest processes (related to forest typology, accessibility, forest management regime etc.), unitary worker and machine costs and assortments and woodchips prices, would enable assessment of economical availability. Furthermore, with the help of national and international literature and datasets, the sustainability of the biomass production could be assessed. Forest data used in *r.green.biomassfor* tool have to be represented with continuous map surface. Given that INFC data refer to NFI sampling units and are spatialized as point pattern, application of *r.green.biomassfor* model should be tested firstly with non-spatialized data to project the predicted variables at plot level and produce estimates of the same variables at the national or regional level, using INFC algorithms. Additional work should be necessary to test the *r.green.biomassfor* model on previously spatialized INFC data.

The structure of *r.green.biomassfor* (implemented in sub-models) and the open-source based approach provide the opportunity to further develop the integration of sub-models, criteria and variables as well as to perform its evaluation. For example, a future improvement could be the inclusion of a dynamic forest growth sub-model based on short-medium term forest planning. The available user friendly graphical interface facilitates the application of the model for users.

18.5 Preliminary Conclusions for Woody Biomass Resource Projections in Italy

In Italy, woody biomass resource projection is severely hampered by the lack of information concerning logistical variables (e.g. a national database regarding the spatial distribution of forest roads and landing sites) and the high variability of economic parameters. In fact, Italian territory is characterized by high differentiation of geomorphology, forest features, level of mechanization of harvesting operations and socio-economic aspects. This variability does not allow generalizing the information necessary for woody biomass resource projection. Although accurate evaluations were carried out for local studies, up-scaling of the models used in these cases for the national level presents the same above mentioned difficulties in input generalization.

A possibility for producing national-scale projections could be depicted in the availability of recently updated data from the second Italian NFI. In particular, the Italian NFI provides useful information concerning accessibility and availability of wood (degree of accessibility, slope, roughness, distance to roads or forest tracks, and timber availability), tree growth and harvesting. On the other hand, accurate information on harvesting costs and assortments and in general on logistical and economical aspects at the plot level are not available in the NFI which should be integrated with other data sources and models. The development of a model to predict the ecologically and technical-logistically exploitable forest resources at national scale should be further investigated. In addition, an allocation model could permit division of the predicted timber production into traditional assortments and wood residues for energetic uses.

Finally, different GIS-based analyses could be used to model economic restrictions related to, for instance, transportation of wood from forests to roads and along roads to heating plants, sawmills, etc. as well as restrictions related to harvesting and extraction operations.

References

Bernetti I, Fagarazzi C, Fratini R (2004) A methodology to anaylse the potential development of biomass-energy sector: an application in Tuscany. For Policy Econ 6:415–432

Cesaro L, Florian D, Marongiu S, Tarasconi N (2010) Forest profitability measurement: a pilot project to extend FADN to forestry sector in Italy. IUFRO World Congress, Paris, 26–30 May 2010

Corona P, Marziliano PA, Scotti R (2002) Top-down growth modelling: a prototype for poplar plantations in Italy. For Ecol Manag 161:65–73

FAO (2001) Global forest resources assessment 2000: main report. FAO forestry paper 140, Food and Agriculture Organization of the United Nations, Rome

Garegnani G, Geri F, Zambelli P et al (2015) A new open source DSS for assessment and planning of renewable energy: r.green. In: Proceedings of FOSS4G Europe, Como 2015, 14–17 June 2015. Geomatics Workbooks ISSN 1591-092X, pp 39–50

Gasparini P, Di Cosmo L (2016) Chapter 26 Italy: national resource availability reports. In: Vidal C, Alberdi I, Hernández L, Redmond J (eds) National forest inventories – assessment of wood availability and use. Springer. (in press)

Gasparini P, Tabacchi G (eds) (2011) L'Inventario Nazionale delle Foreste e dei serbatoi forestali di Carbonio INFC 2005. Secondo inventario forestale nazionale italiano. Metodi e risultati. Ministero delle Politiche Agricole, Alimentari e Forestali, Corpo Forestale dello Stato, Consiglio per la Ricerca e la sperimentazione in Agricoltura, Unità di ricerca per il Monitoraggio e la Pianificazione Forestale. Edagricole, Milano

Gasparini P, Tosi V, Di Cosmo L (2010) National forest inventories reports: Italy. In: Tomppo E, Gschwantner T, Lawrence M, McRoberts RE (eds) National forest inventories – pathways for common reporting. Springer, Cham, pp 311–331

Lasserre B, Chirici G, Chiavetta U et al (2011) Assessment of potential bioenergy from coppice forests trough the integration of remote sensing and field surveys. Biomass Bioenergy 35:716–724

MAF-ISAFA (1988) Inventario Forestale Nazionale Italiano 1985 (IFNI 85). Istituto sperimentale per l'Assestamento forestale e l'Alpicoltura, Trento

Pilli R, Grassi G, Kurz WA et al (2013) Application of the CBM-CFS3 model to estimate Italy's forest carbon budget, 1995–2020. Ecol Model 266:144–171

Sacchelli S, Zambelli P, Zatelli P, Ciolli M (2013) Biomasfor – an open source holistic model for the assessment of sustainable forest bioenergy. iForest Biogeosci For 6:285–293

Scrinzi G, Marzullo L, Galvagni D (2007) Development of a neural network model to update forest distribution data for managed alpine stands. Ecol Model 206:331–346

Tabacchi G, De Natale F, Floris A et al (2005) Italian national forest inventory: methods, state of the project, and future developments. In: Proceedings of the seventh annual forest inventory and analysis symposium, 55–66

Zambelli P, Lora C, Spinelli R et al (2012) A GIS decision support system for regional forest management to assess biomass availability for renewable energy production. Environ Model Softw 38:203–213

Chapter 19
Lithuania

Andrius Kuliešis, Albertas Kasperavičius, Gintaras Kulbokas, Vilis Brukas, Edmundas Petrauskas, and Gintautas Mozgeris

19.1 Introduction

The total forest land area in Lithuania increased by 135,000 ha (2.1% of the territory) since 2003. In 2015, according to the Lithuanian forest resource assessment, the total forest land area was 2,179,900 ha covering 33.4% of the country (Lietuvos miškų ūkio statistika 2015). The area of forest stands increased from 1,951,000 ha to 2,056,000 ha during the same period. The remaining 123,900 ha include forest land not covered by stands such as clear cut areas, dead stands, forest blanks, seed orchards, and forest roads. Coniferous stands cover 56.0% of the forest area followed by 40.4% of soft deciduous tree species. From 2003 until 2015, the area of soft deciduous tree species increased 132,600 ha, while the areas of hard deciduous and coniferous tree species decreased 18,000 and 9600 ha, respectively (Table 19.1). The total growing stock volume increased from 453.4 million m^3 to 528.9 million

A. Kuliešis (✉) • A. Kasperavičius
State Forest Service, Kaunas, Lithuania
e-mail: andrius.kuliesis@amvmt.lt; albertas.kasperavicius@amvmt.lt

G. Kulbokas
State Forest Service, Kaunas, Lithuania

Aleksandras Stulginskis University, Kaunas, Lithuania
e-mail: gintaras.kulbokas@amvmt.lt

V. Brukas
Southern Swedish Forest Research Centre, Swedish University of Agricultural Sciences, Alnarp, Sweden
e-mail: vilis.brukas@slu.se

E. Petrauskas • G. Mozgeris
Aleksandras Stulginskis University, Kaunas, Lithuania
e-mail: edmundas.petrauskas@asu.lt; gintautas.mozgeris@asu.lt

S. Barreiro et al. (eds.), *Forest Inventory-based Projection Systems for Wood and Biomass Availability*, Managing Forest Ecosystems 29,
DOI 10.1007/978-3-319-56201-8_19

Table 19.1 Species representativeness in Lithuanian forests (Lietuvos miškų ūkio statistika 2015)

Tree species	Year 2015		Increase/decrease (+/−) (since 2003)	
	Forest Area (hectares)	Growing Stock Volume (million m^3)	Forest Area (hectares)	Growing Stock Volume (million m^3)
Scots pine	718,600	218.5	+7100	+38.5
Norway spruce	429,300	88.6	−16,000	+12.8
Birch species	458,400	89.0	+66,200	+10.2
Aspen	85,300	36.6	+28,800	+2.6
Black alder	149,800	50.7	+30,300	+13.0
Grey alder	125,200	22.1	+3200	+0.5
Oak	43,400	11.5	+7700	+0.2
Ash	24,100	3.5	−27,300	−6.1
Other tree species	21,900	8.4	+5800	+3.8

m^3 during the same period (Lietuvos miškų ūkio statistika 2015). Table 19.1 shows the distributions of area and growing stock by tree species.

During the same period growing stock volume of mature stands available for commercial use increased from 109.9 to 142.7 million m^3. The gross annual increment rose from 16.0 to 18.8 million m^3 corresponding to 8.9 m^3/ha. State forests, i.e. those managed by the state and where the state has exclusive rights, cover 1,084,500 ha. Private forests cover 866,200 ha, and 229,200 ha of forest are reserved for restoration of ownership rights to former owners (Lietuvos miškų ūkio statistika 2015). Forests reserved for restoration of ownership rights, but which will not be restored, are planned to be sold at auctions. Lithuania has about 248,000 private forest owners with an average private holding size of only 3.49 ha.

All Lithuanian forests are divided into four functional groups according to multifunctional objectives. Group I includes forests for which no management is applied and all cuttings are prohibited (26,300 ha or 1.2%). Group II represents ecosystem protection and recreational forests (266,500 ha or 12.2%) for which cuttings are allowed at the age of natural maturity. Forests with water, soil, landscape or valuable natural features such as genetic and culture reserves as well as forest seed stands are under Group III (331,000 ha or 15.2%). Finally, Group IV represents commercial forests (1,556,100 ha or 71.4%). The principles of forest grouping are regulated by Lithuanian forest law and the distribution of forest area by functional group which is quite stable. The maps assigning each forest area to a specific group at the compartment level currently must be approved at a Ministry of Environment level.

Lithuania has developed a wood availability and use projection system based on long-term (more than 30 years) planning combined with management planning for 10–20 years. Forest resources' assessment, management planning and long-term wood availability, and wood use scenarios are usually based on two types of inventory data: stand-level or Standwise Forest Inventory (SFI) and the National Forest Inventory (NFI) which uses sampling methods.

SFI and NFI data are combined for wood-use planning in the country. SFI data are used for 10–20 year's wood-use planning of forest management units (enterprises, private holdings, etc.), whereas NFI data are used for forest resources assessment in the country and also for forest management control using gross increment balances. The proportion of dead trees decomposing in the forest is used to estimate forest growth efficiency, while the ratio of removed and felled trees allows estimation of harvest efficiency. The level of sustainable wood-use is estimated as the proportion of stem volume accumulated from the gross increment. The combination of data from both inventories is used to develop long-term wood availability and wood-use scenarios.

The current text aims to report the status of woody biomass resource projection methods for Lithuania based on forest inventory data. First, the inventories providing inputs to the wood resource projection models are introduced. Next, the methodological principles behind the projection methods are discussed. Finally, the simulator *Kupolis* is explained in detail.

19.2 Data Sources

19.2.1 National Forest Inventory

The Lithuanian NFI is based on sampling methods and integrated Geographic Information System (GIS) technology and was launched in 1998 to conduct a thorough monitoring of Lithuanian forests for an efficient assessment of the main forest variables in the country and its regions (Kasperavičius and Kuliešis 2002; Kuliešis et al. 2010, 2016). High priority was especially given to estimation of volume increment, structure and wood-use balance. The allocation and measurement of permanent plots was finished in 2002 (Kuliešis et al. 2003). Re-measurement of these plots and establishment of new temporary plots started in 2003 (Kuliešis et al. 2009), and the third cycle of Lithuanian NFI was finished in 2012 (Anonymous 2011). The fourth cycle of NFI will be finished in 2017.

The NFI provides data necessary for the preparation of various forest statistical reports as well as for forest management planning and long-term wood availability and wood-use projections (Kuliešis et al. 2010). The estimation of current forest statistics is based on a combination of data resulting from the NFI sample plots which are either remeasured permanent plots, temporary plots, or special temporary plots for inventory of fellings. Measurement of permanent sample plots facilitates estimation of the efficiency of forest management and estimation of total drain, removals, changes in biodiversity and land use changes. Estimates of a large number of variables as required for various NFI users and especially for wood-use projections, are based on a combination of tree measurements from plots and aerial plot assessments (Table 19.2).

Table 19.2 Structure of NFI data measured and evaluated on the plots during field measurements

Dimension of plants	State of plants	Areal variables
Trees	Species	Ownership
dbh	Storey	Land use category
Height	Living	Functional forest group
Lenght of crown	Dead, windthrown, windbroken	Administrative region, county
Age	Cut	Site type
Increment	Kind of cuttings	Forest type
Volume	Quality class	Age class
Understorey, natural regeneration	**Damages**	Site index H_{AB}[a]
	Type	Site index D_{AB}[b]
Height	Cause	Stocking level
Age	Location	Species composition
Underbrush	Intensity	
Height	Defoliation	

[a]H_{AB} is estimated according to current mean height and age of the prevailing tree species in main storey and expresses mean height at reference age
[b]D_{AB} is estimated according to current mean diameter at breast height (*dbh*) and age of the prevailing tree species in main storey and expresses mean *dbh* at reference age

Each tree measured on a sample plot represents 8000 trees in the field. Measurements of 6000 permanent plots produce estimates with 1% accuracy ($p = 0.683$) for the main forest resource attributes (growing stock volume and gross increment) for the entire country.

To conduct an IPCC National Greenhouse Gas inventory in the Land Use, Land-Use Change and Forestry sector, Lithuania has chosen the stock change method which is based on information from the NFI. Above ground biomass estimation is based on the volume of living trees stems with bark and wood density and Biomass Expansion Factor (BEF) values. Root-to-shoot ratios are used to estimate below ground biomass. The default value of 0.5 ton C per ton dry biomass is used for estimation of the carbon fraction.

NFI data are the major information input for forest resources assessments, which are carried out every year (Valstybinė miškų apskaita 2012) to produce an annual forest statistics report (Lietuvos miškų ūkio statistika 2003; Lietuvos miškų ūkio statistika 2015) which is available at www.amvmt.lt.

19.2.2 Standwise Forest Inventory

SFIs have been carried out for the entire country on a regular basis since 1922. The seventh SFI was started in 2012. According to the Lithuanian forest law (1994), SFIs and mapping should be executed every 10 years throughout the country independently of forest ownership. Nowadays, one-tenth of the country's territory is inventoried every year. Forest mapping using GIS techniques was started in 1995 (Brukas et al. 2000; Mozgeris et al. 2008; Anonymous 2012). The main SFI operations are delineation of forests into compartments or stands, stand inventories, and the inventory of other attributes such as soil properties and non-timber forest resources including game, mushrooms, herbs and berries. Delineation of forests into compartments consists of preparing a draft forest map based on information from orthophotomaps, stand maps, forest soil maps, maps of cuttings and reforestation from the past inventory, and also maps of protective territories and objects. Subsequently, this map is validated in the field using GPS equipment. Forest stand borders are usually delineated after assessing forest soil, site and forest type conditions.

SFIs are the basis for tactical forest management planning. SFI inventories are based on visual methods to evaluate and describe forest stand variables including land cover type, forest group/subgroup, tree species composition, stand age, understorey and bushes, forest ownership, soil type, silvicultural treatments, wild animals' presence and the damages they cause, aesthetics, recreational and protective value of stands and an estimate of the value of the forest land. Although visual methods are used, some variables such as tree height and tree diameters are measured and basal area and volume are estimated. Growth models and information on silvicultural treatments are used for annual updating.

Data, collected within the frames of SFI, are stored on an SQL server database management system since 2001 (Brukas et al. 2002). Since 2003, these data have formed the basis for Lithuania's State Forest Cadastre and are managed by the State Forest Service.

19.3 Methods

19.3.1 Forest Management Planning

Forest management planning in Lithuania is based on rigid routines and plays a major role in defining forestry practices (Brukas et al. 2011; Anonymous 2012). Planning relies on data from both SFIs and the NFI. Forest inventory experts visit each forest stand and subsequently elaborate a forest management plan for 10 years (lately up to 20 years for private forest estates). For a State Forest Enterprise (SFE) the plan is to be followed by SFE managers and includes detailed management provisions such as sequencing of stands for thinning and final fellings.

Planning experts estimate the allowable annual cut using the method of area control. Several variations are available, but basically the area control method entails dividing the available area of commercial stands dominated by a given species by the number of years within the rotation, where the rotation is traditionally estimated by adding 10 years to the Minimum Allowable Rotation Age (MARA). The resulting annual cut area for the country is distributed among SFEs. Annual final cut area by prevailing tree species is estimated for each SFE every 5 years. The proposed norm of annual cut area should guarantee a stable harvesting level for the analyzed tree species for two decades. For every SFE, the area of annual final cut is specified

Table 19.3 Minimum allowable rotation ages for Lithuanian forest stands by functional forest group

	Forest group		
Prevailing tree species	II	III	IV
Scots pine, Ash	171	111	101
Norway Spruce	121	81	71
Birch	91	61	61
Aspen	81	41	41
Black alder	91	61	61
Grey alder	51	31	31
Oak	201	141	121

assuming the area of mature stands. The mature stands of the country occupy more than 20% of the area of all stands. This means that mature stands in a particular SFE are planned to be felled in a sustainable, long-term manner (13–16 years). The underlying aim is to continuously obtain as much timber in the long-run as possible.

MARA is a crucial variable because it determines the minimum rotation length. No stands can be final felled earlier than MARA, except in cases of extraordinary calamities. MARA is defined by prevailing tree species and forest group (Table 19.3).

In commercial forests the rotation age is set according to the criterion of technical maturity which refers to the annual increment of specific valuable assortments (saw logs with top diameter of 19 cm) for average site conditions.

Forest management plans for SFEs must be checked by the regional subdivisions of forest management control department of the State Forest Service and finally approved by the Minister of Environment. As of 2006, regional forest management schemes should also be elaborated, encompassing environmental impact assessments with the possibility of public hearings. Annual cuts for each of 42 SFEs and the total national annual cut are examined by the Scientific-Technical Board of Forest Management Planning. The Board includes representatives from the Ministry of Environment, Directorate General of State Forests, SFEs, universities and colleges, the Forest Inventory and Management Institute and State Forest Service. After approval by the Board, the annual cutting norm is approved by the Minister of Environment. In addition, the 5-year annual norm for cutting area and maximum amount of merchantable timber must be approved by the Lithuanian Government for all state forests.

The procedure is somewhat less complex in private forestry. Final felling can only be carried out by a private forest owner if the owner has a forest management plan prepared by an officially registered inventory expert and a cutting permit that is issued at the regional agencies.

Forest management plans are not mandatory for forest holdings of less than 3 ha of forest area, where the national average is 3.49 ha, and other cases described in Regulation of private forest management and use. The compulsory parts of the forest management plan are the 10-year final cutting norm, forest regeneration, and environmental requirements. For estates of less than 150 ha, all mature stands can be typically included in the 10-year cutting norm. However, various spatial restrictions apply. For example, the area and number of clear cutting occasions during 5-year period is limited for individual estate. The forest management plan must be approved by regional subdivisions of the forest management control department and registered by the State Forest Service. When planning a final felling, the forest owner must obtain a cutting permit that is issued by the regional subdivisions of the forest management control department of the State Forest Service.

Energy wood resources are estimated for every compartment for which felling, either final or intermediate, has been projected. Energy wood in Lithuania includes full stems of small trees in young forests (available from thinning of young stands), round fuel wood and dead trees (not suitable for saw logs), felling residues (tops, branches, stumps), all available from other types of fellings. Energy wood is accounted including the bark. The volume of energy wood depends on the felled stem volume and is differentiated by tree species, felling types, fertility, and hydrology regime of sites.

19.3.2 Growth Model

Stand growth predictions are based on regression models developed for eight dominant tree species (pine (*Pinus silvestris L.*), spruce (*Picea abies (L) Karst.*), birch (*Betula pubescens Ehrh. + Betula verrucosa Ehrh.*), aspen (*Populus tremula L.*), black alder (*Alnus glutinosa*), grey alder (*Alnus incana*), oak (*Quercus robur L.*), ash (*Fraxinus excelsior*) by Kuliešis (1993). For each species and each stand, the same models are used to estimate mean gross annual increment and its components including wood left in the stand as the result of management thinning and self-thinning mortality. The growth of other species is predicted using a model for the most similar species.

The growth model for Lithuanian forests is based on general regularities of the change and growth of mean parameters of trees and stands. It is based on materials acquired within the frames of sampling based inventory of state forests in 1969 (Kuliešis 1993). The growth model consists of the following sub-models which aim to predict the following parameters (Kuliešis et al. 2014):

- changes of mean tree height, depending on stand age, site index H_{AB} and peculiarities of site (mineral, organic),
- site index H_{AB} (mean height at reference age, depending on current mean height and age of the main storey),
- changes of mean tree diameter at breast height (*dbh)*, depending on stand age and site index D_{AB}
- site index D_{AB} (mean *dbh* at reference age, depending on current mean diameter at breast height and age of the main storey),
- changes of growing stock volume, depending on mean height of trees and stocking level,

- increment of mean diameter of trees, depending on site index D_{AB} and mean *dbh* of trees,
- general interrelations between tree height and tree *dbh* and mean *dbh* of trees per stand,
- general interrelations between growth (increment) and changes of mean stand parameters (height, *dbh*, volume) as well as the interrelations between survived, dead and all trees of the same variables,
- estimation of stem volume over bark including above ground part of stump and top, depending on tree *dbh* and height.

The growth model is used to predict: (1) the change in mean diameter, mean height, number of trees, and basal area; (2) growing stock volume including surviving trees after n years, gross volume increment accumulated in stand and any parts well as removed from stand during n years; (3) mean diameter and mean height increment; (4) mean diameter, mean height and volume of dead trees; and (5) mean diameter and mean height of trees surviving during n years. The growth model is also used to predict stand-level stocking based on the portion of the gross volume increment removed or accumulated in stand. The growth model is applied in SFI for forecasting stand parameters and for the estimation of gross volume increment in every stand (corresponding to its age, mean height, stocking level or growing stock volume) as well as for the simulation of forest stand development and thinning regimes.

19.3.3 Wood-Use Control Based on Increment Balance

Wood-use control based on increment balance is applied in all stages of forest management planning – during the implementation of accepted Forest Management Plans (FMP) or for making new 10–20 year FMP as well as for forecasting of long term wood-use changes. Wood-use results are controlled in various ways, but the most efficient way uses directly estimated gross increment balance from measurements of permanent plots of NFI. Using data from repeated measurements of permanent plots, increment balance equals:

$$ZM = \Delta_M + M_K + M_0; \tag{19.1}$$

where

Δ_M – volume change, during the period between two consecutive inventories. Volume change includes ingrown trees or trees that changed storey;
M_K – volume of trees felled by all types of felling, usually is distributed into final and intermediate;
M_0 – volume of dead trees between consecutive measurements (natural losses):

$$M_0 = M_{0K} + M_{00} \tag{19.2}$$

M_{0K} – volume of removed dead trees (used natural losses)
M_{00} – volume of dead, left for natural wood decomposition, trees (unused natural losses).

Gross annual increment and its balance have been estimated every year since 2007 using data from two successive measurements of permanent NFI plots. Mean gross annual increment and its components are estimated for all forests – state forests, private forests and forests reserved for the restoration of ownership rights. Gross annual increment and its components are estimated separately for commercial and protective forests as well as for strict reserves, protected and recreational forests. The drain of gross annual increment represents felling of living trees and natural losses, both used (removed) and unused (left in forest). The volume of felled living trees can result from final harvest or thinning, whereas the volume of felled (removed) dead trees represents natural losses that are used. It is estimated as the volume of felled dead trees by various types of cuttings that were recorded as dead during the previous inventory.

19.4 Long-Term Wood-Use Scenarios

19.4.1 Tools to Simulate Forest Resource Development and Wood-Use

A large scale forestry scenario simulator *Kupolis* is structured following the recommendations by Pretzsch et al. (2002). It was developed in the last decade of the twentieth century with the primary aim of predicting forest resource development under different economic and environmental conditions for SFEs and to produce summary statistics at the country level. Currently it can be used for all levels of forest estates, ranging from one stand up to all forests in the country. The basic unit of simulation in *Kupolis* is a forest stand or compartment, but any aggregated or sample plot data can be used if the structure of the data records is compatible with stand data. The model may also work on a random sample of data. The model is designed to work using the data structure of the database system "L", which was originally developed in Lithuania and used as a standard solution to process SFI and forest management planning data throughout the entire former Soviet Union (Brukas et al. 2002). Some post-Soviet countries still successfully use this database system. Growth models (Kuliešis 1993) are adapted for Lithuanian conditions, but could be used in neighboring countries as well.

Kupolis was planned to consist of three sub-systems aimed to simulate the development and use of: (1) wood resources and their dynamics, (2) non-wood forest resources as well as (3) forest environmental and recreational functions. At the

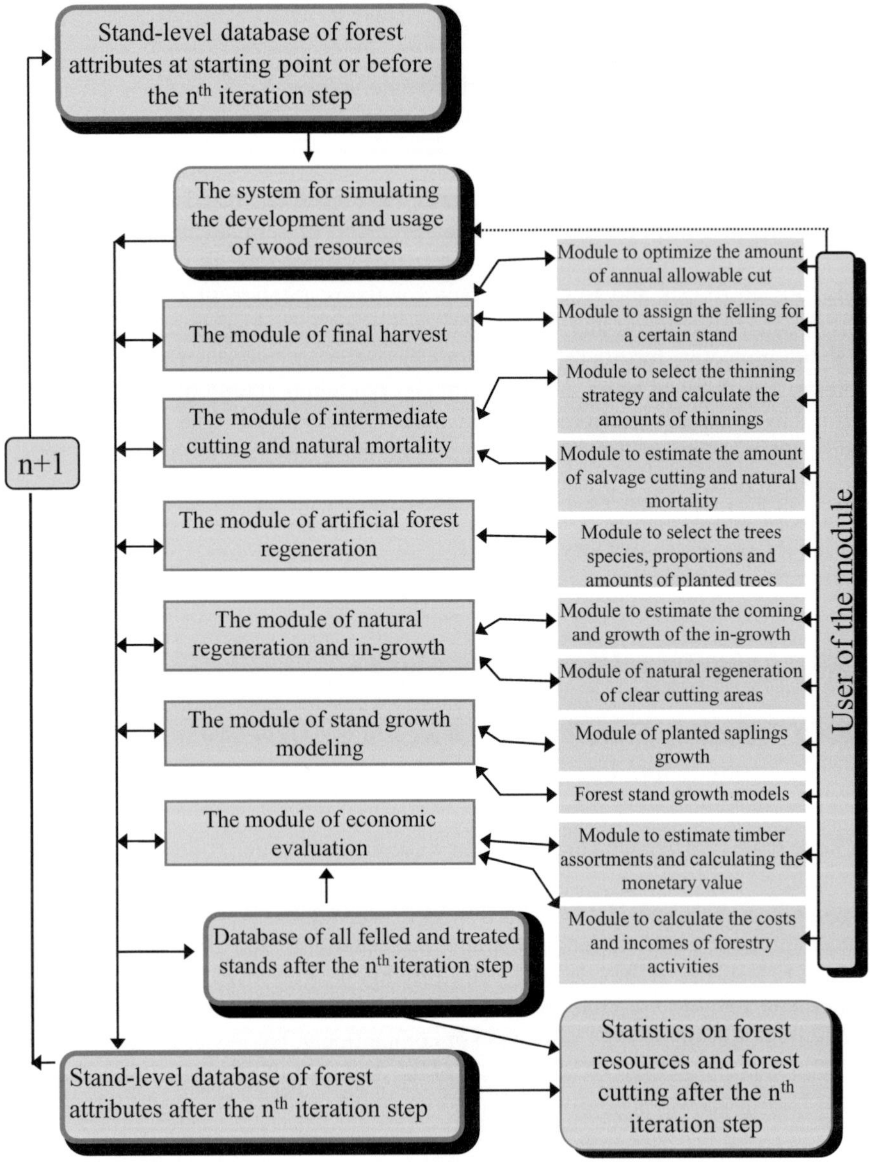

Fig. 19.1 The structure of *Kupolis* (Adapted from Petrauskas and Kuliešis 2004)

moment, just the first sub-system is operational; in particular, non-timber ecosystem services cannot be accommodated currently. The simulator has six modules to model the development and use of wood resources (Fig. 19.1): (1) final cutting, (2) intermediate cutting, (3) artificial and (4) natural forest regeneration, (5) stand growth, and (6) economic evaluation.

All the modules are combined in such a way that changing the parameters for one changes the simulation results of the next steps of the iteration. The simulation period may range from 1 to 20 years; the number of periods is practically unlimited, i.e. it is limited only by the computer data storage capacities. As a result of each step, two new virtual databases are generated: (1) a database of stands that are left to grow, that describe the condition of forest resources after a chosen time period in the same format as used for input, and that are the initial data for a new step, and (2) a database of wood removed during all kinds of cuttings and natural mortality. Special programs to extract various statistics from the simulation are used, as well as whole data manipulating functionality of the system "L" is available, too.

Data input formats are compatible with the formats of the old database management system "L" that was used in Lithuanian forest inventory and management planning system a decade ago, that is still operational at the Lithuanian State Forest Service, and that does not require any preliminary processing before using in *Kupolis*. The main stand characteristics used as input for the growth models, are species composition, age, mean height, mean diameter, stocking level, and growing stock volume, each of which is estimated for each species in every storey of the stand. However, each compartment has a detailed description with more than 100 variables used in *Kupolis*. Input stand attributes belong to the following groups:

- General data (manager, owner, estate address, forest block number, forest group and subgroup, defining the management regime, administration unit);
- Forest compartment characteristics (id, area, land category, tree species, site index, soil type, forest subgroup, inventory peculiarities, land drainage status);
- Forest stand description (storey, species composition, age, height, diameter, relative stocking index, origin, volume per 1 ha, basal area);
- Description of forestry operations (type of operation, percentage of volume cut, volume of dead and wind fallen trees);
- Forest plantations and natural regeneration (type, tree and brush species, amount of tree plantings, condition);
- Forest damages (tree species damaged, type, percentage of trees damaged, damage degree);
- Non-wood products (herbs and berry plants, species, availability in percent, commercial importance);
- Assessment of performed cuttings (cutting year, quality, validity);
- Agricultural lands (type of land, condition, use);
- Understorey (species, density);
- Protected objects, such as protected trees, bird nests, objects cultural heritage, recreational facilities, etc. (type, sub-group, title, recommended treatment, number of objects for recreation).

The simulator consists of several parts that are handled interactively. The basic module is used to select the input database, the simulation step (years, ranging from 1 to 20, usually 5 or 10. The experience is that longer than 10 years simulation steps need to be avoided as they may produce biased results), and the number of steps (technically unlimited, practically depending on the objectives of simulation exercise). Next, the query is constructed to select specific records from the database such as forest group, forest owner, and administrative unit, etc., i.e. all parameters described in their own attribute fields can be used to query. Database management, printing of the results, and export of data are carried out using the system "L" functions of special programs.

Parameters for scenarios are stored in a special normative database as ASCII files and may be adjusted using simple text editors. The normative files are arranged into following thematic groups:

- Administrative classifiers (forest enterprise, district, year of Standwise inventory, administrative districts, protected areas, functional zones, forest groups and sub-groups);
- Environmental and dendrometric classifiers (soil productivity groups, forest land categories, tree and brush species, stand stories, soil types, condition of planted trees, stand damage, stand productivity, effectiveness of cuttings, herbs and berry plants, characteristics of meadows, roads, Red Book species, wetlands, target tree species, forest types, ecological evaluation, recreational objects, characteristics of forestry activities, etc.);
- Growth model (parameters of growth models, (see the previous section on stand growth models).
- Scenario definitions (describing the stand growth: target tree species and tree species compositions, age of plantings, maximum time period to reforest clear cutting areas, lowest amount of undergrowth, succession of target tree species, target planting types, average age and age structure of undergrowth, user's proposed tree species to be planted, average tree species compositions in the country, densities of planted trees by species, surviving probabilities of planted trees; describing the timber cuttings: maturity ages, minimum allowable cutting ages, model validity limits, target diameters for cutting, selection of final cutting types, optimized reserve of final cuttings, parameters to optimize the final and intermediate cuttings, parameters to define the thinnings).
- Economic variables (timber prices on the stump, log prices, forest land prices, prices of exploitable stand volume, prices for herbs and berry plants, costs of planting, thinning of saplings, intermediate and final cutting and the incomes from intermediate and final cutting).

Because *Kupolis* is an open system consisting of elementary modules, each subsystem can simulate a different type of forest management program at the stand level. An exception is the final cutting budget optimization module which generates optimal solutions at forest management unit level, and uses aggregated data based on the age class principle. Within a forest management unit, an entire forest estate or a group of stands within the same estate is assumed to share the same tree species and functional group (reserves, recreational forests, etc.).

The forest regeneration module can simulate regeneration in four ways: (1) planting tree species following valid silvicultural rules and goals, (2) business as usual which uses average country-wise inputs, (3) user-defined regeneration characteristics, and (4) maintaining the same tree species composition as in the former stand. The natural regeneration module foresees natural regeneration on all areas, including those with artificial planting. Planting is not assumed in three cases: (1) when enough undergrowth is already available, (2) on histosol soil type, and (3) when experts allow natural regeneration due to economic reasons. Random selection of tree species composition and density from within the limits of variation in real stands in the specified soil type is used in the natural regeneration and in-growth modules to sustain the current tree species composition of simulated stands. The number of in-growing trees is corrected using a coefficient that depends on the relative stocking level (based on basal area) of the present stand.

The thinning module can operate following two methodological approaches: (1) thinning is planned by analyzing the stand characteristics and choosing the treatments that best suit the present silvicultural recommendations at each simulation step; and (2) defining the species composition of the target trees and the stocking level of the stand at a rotation age, and planning the treatment when the stand stocking level starts exceeding the targeted trends.

The main requirements for the objective function in the final harvest module are continuous and sustainable use, smoothing of age class structure, and balance between cutting and increment. The best long term forecasts meeting these requirements are provided using the model *Optina* developed by Lithuanian forest researchers in the 1970s and 1980s (Deltuvas and Miseikis 1975; Vitunskas 1988). The economic evaluation module calculates three parameters: forest land value (it does not apply discounting, i.e. should not be regarded as a variant of net present value), stumpage price, and revenues and costs of all forest operations starting from growing seedlings until timber logging on a roadside at stand level.

Kupolis is based on two methodological principles, dynamic programming and iterative simulation. This enables the user to minimize accumulation of deviations of simulated results for a long time horizon (100 years). The annual budget of final cuttings that fulfills the requirements of sustainability for a full rotation period is re-optimized at each step using the principles of dynamic programming, while other forest management activities are modeled using the iterative simulation.

Several types of output data may be generated:

- The simulation results are output using system "L" formats; thus the default way to study the results is to use the analysis functionality of the system "L". More than 50 different summary tables may be constructed, as well as interactive, user-defined report structures and contents. Results are usually output as ASCII text files.
- The "L" databases may be exported into MS Excel or dbf formats and further analyzed or merged with a GIS using the functionality of external software. For

example, exported data on simulated future characteristics of stands are used to estimate the amounts of biomass and carbon, evaluate forest environmental and recreational functions using external software tools.
- Key summary statistics may be generated as MS Excel or dbf files using specially developed *Kupolis* tools.

Attributes of forest compartments exported into dbf files may be joined with the borders of compartments in any standard GIS package such as ArcGIS. However, as yet there are no spatial considerations implemented in *Kupolis*.

19.4.2 Kupolis *Simulator and the NFI*

Kupolis basically works with data originating from SFIs. Currently it is being improved to incorporate sample plot data from the Lithuanian NFI (Petrauskas and Rupšys 2013). The key issue for using NFI data in *Kupolis* is to define the primary simulation unit which could be any of a single tree, a sector of a sample plot, a sample plot, or a stratum of sample plots. Currently, the smallest unit describing the forest stand and fitting the concepts on which *Kupolis* is based is the sector of an NFI sample plot. However, the areas of some sectors delineated in the NFI are small and the variances of stand parameters are too large to simulate the growth and use of forest resources at the level of sample plot sector. The sector-area problem disappears if the sample plot is used as a primary simulation unit. However, some sample plots are segmented (divided into sectors), leading to the issue discussed above. Thus, the recommended option is to use the NFI data in *Kupolis* to simulate the growth and timber usage by strata. The stratum or management unit is constructed by aggregating data from NFI sample plots or sectors of the plots based on the type of forest ownership, forest group, soil type, age, prevailing tree species, stand structure and tree species composition, density and the silvicultural treatments during the last 5 years. The area of each sector or sample plot is used as a weight to describe the stratum. Thus, this simplified version of the NFI database matches the structure of the SFI database used for simulation in *Kupolis*. Single-tree data may be used to generate some characteristics of strata, too.

19.5 Discussion

The simulator *Kupolis* has found application at both national and SFE levels (Kuliešis and Petrauskas 2000; Petrauskas and Kuliešis 2004). At the forest enterprise level, the model has been successfully used as a tool for forest management planners to determine the minimal level of cuttings that is needed to ensure enterprise profitability and present the boundaries of sustainable and maximal wood use. However, there are many open questions and issues regarding its usability. A spatial location component for setting rules and scenarios is completely missing. The simulator was developed more than a decade ago, is far from user-friendly, and is open to new scenario and rule definitions.

New applications of *Kupolis* have been discussed in several recent research projects. First, the simulator was tested for the usability of NFI data as an input (Petrauskas and Rupšys 2013). This study indicated that the simulator can accept input from the NFI, although the NFI data must be post-processed to satisfy the *Kupolis* concepts and input formats. The potential of *Kupolis* to integrate the behavior of forest owners or managers for modeling the development of forested landscapes under conditions of various scenarios was demonstrated by the European Union's Seventh Programme for research, technological development and demonstration project INTEGRAL – Future-oriented integrated management of European forest landscape (Mozgeris et al. 2016). This project estimated future flows of timber products, carbon sequestration, recreation, and environment protection at the landscape or even forest estate levels, and aimed to judge the potential combinations of ecosystem services as an outcome of contextual scenarios as well as of alternative mixes of forest policy instruments. Further, the functionality and user-friendliness of *Kupolis* was significantly improved, especially with respect to defining new forest management programs.

The simulator accuracy depends on the accuracy of the input data, the data representativeness, and the accuracy of the growth models. Thus, the simulation results are heavily influenced by the characteristics of the data provided both by the SFI and the sample-based NFI. The error of a volume estimate for a forest with area more than 25,000 ha should be less than 6% ($p = 0.683$) using SFI data. However, recent sample-based inventories of mature forests have demonstrated that SFIs tend to underestimate the volume of compartments (Kasperavičius 2009; Anonymous 2010). Considerable effort has been made in recent years to improve the accuracy of SFIs including introduction of a double quality control system that includes: (1) thorough accuracy assessment of all of dendrometric characteristics in inventoried compartments, and (2) sampling method based on volume estimation in mature stands.

The accuracy of the growth models is rather difficult to estimate. The current growth models were developed more than 20 years ago. Although the influence of climate change on stand growth is widely discussed, no validated models to express this potential influence have been incorporated in *Kupolis*. *Kupolis* simulations are based on the condition that the longest time period to predict stand growth corresponds to the length of the iteration step, i.e., 1–20 years. Thus, new stand growth prediction at the beginning of each iteration step potentially reduces the bias in the results. Nevertheless, the accuracy of the growth models should be inspected every decade. The NFI based on permanent sample plots could offer a solution. Trees in these plots are re-measured at least three times, thus providing good material to validate and calibrate existing growth models as well as to develop new models.

19.6 Conclusions

The Lithuanian forest inventory system confidently supplies enough reliable (NFI) and detailed (SFI) data on growing stock volume, gross mean annual increment and their structure, to facilitate wood use projections.

The forestry scenario simulator *Kupolis* is used to model forest resource and timber use development at levels ranging from forest compartment to the entire country. The simulator uses input data satisfying the formats of the Lithuanian SFI.

NFI data can be used in *Kupolis* if they are aggregated to strata that match the structure and contents of the SFI database. The stratum is constructed by aggregating data for NFI sample plots or sectors based on the type of forest ownership, forest group, soil type, age, prevailing tree species, stand structure and tree species composition, density and the silvicultural treatments during the last five years.

References

Anonymous (2010) Mokslinės studijos „Nuotolinių metodų taikymo miškų inventorizacijoje bei monitoringe galimybėms ištirti ir rekomendacijoms dėl šių metodų naudojimo parengti", ataskaita (Potential of remote sensing application for forest inventories and monitoring and development of recommendations for operational use of remote sensing, final report). Lithuanian Forest Inventory and Management Planning Institute, p 184

Anonymous (2011) Nacionalinė miškų inventorizacija atrankiniu metodu, 2007–2011. Ataskaita (National forest inventory based on sampling method, 2007–2011. Report). Aplinkos ministerija. Valstybinė miškų tarnyba. Kaunas, p 207

Anonymous (2012) Miškotvarkos darbų vykdymo instrukcija (Manual of forest management planning), p 21

Brukas A, Galaunė A, Rutkaukas A et al (2000) Remote sensing and GIS in Lithuanian forestry. In: Zawila-Niedzwinski T, Brach M (eds) Remote sensing and forest monitoring. Proceedings of IUFRO conference, 1–3 June, 1999, Rogow, Poland. Office for Official Publications of the European Communities, Luxembourg, pp 124–132

Brukas A, Jakubonis S, Kuliešis A, Rutkauskas A (2002) Lietuvos miškotvarka ir jos raida (Forest management planning and its development in Lithuania). Kaunas, "Naujasis lankas", p 188

Brukas V, Kuliesis A, Sallnas O, Linkevicius E (2011) Resource availability, planning rigidity and realpolitik in Lithuanian forest utilization. Nat Resour Forum 35(2):77–88

Deltuvas R, Miseikis F (1975) Model normalizaciji vozrastnoj struktury lesov (Model to normalize the age structure of forests). In Proceedings of scientific conference on the improvement of methods to estimate and plan the extent of final and intermediate forest felling. Kaunas, pp 22–25

Kasperavičius A (2009) Brandžių medynų inventorizacijos ateitis – pasirinkimo kryžkelėje (Reliable inventory of mature stands – a topic for discussions among foresters). Mūsų girios 4:20–22

Kasperavičius A, Kuliešis A (2002) Atrankinės miškų inventorizacijos atrankos schemos efektyvumo įvertinimas pagal lauko darbų laiko sąnaudų analizę (Estimation of sampling inventory design efficiency on the base of time consumption for field work analysis results). Miškininkystė 1(51):48–61

Kuliešis A (1993) Lietuvos medynų prieaugio panaudojimo normatyvai (Forest yield models and tables in Lithuania). Kaunas, Girios Aidas, p 384

Kuliešis A, Petrauskas E (2000) Lietuvos miškų naudojimo XXI amž. prognozė (Lithuanian forest resources in the XXI century). Kaunas, Naujasis lankas, p 146

Kuliešis A, Kasperavičius A, Kulbokas G, Kvalkauskienė M (2003) Lietuvos nacionalinė miškų inventorizacija 1998–2002. Atrankos schema, metodai, rezultatai (Lithuanian national forest inventory 1998–2002. Sampling design, methods, results). Aplinkos ministerija, Valstybinė miškotvarkos tarnyba, Naujasis laukas, Kaunas, p 255

Kuliešis A, Kasperavičius A, Kulbokas G, Kvalkauskienė M (2009) Lietuvos nacionalinė miškų inventorizacija 2003–2007. Miškų ištekliai ir jų kaita (Lithuanian national forest inventory 2003–2007. Forest resources and their dynamic). Aplinkos ministerija, Valstybinė miškotvarkos tarnyba, Lututė, Kaunas, p 284

Kuliešis A, Kasperavičius A, Kulbokas G (2010) National forest inventory reports: Lithuania. In: Tomppo E, Gschwantner T, Lawrence M, McRoberts RE (eds) National forest inventories – pathways for common reporting. Springer, New York, pp 351–374

Kuliešis A, Kulbokas G, Kuliešis AA (2014) Validation of generalized height – diameter model based on Lithuanian NFI data. Baltic Forestry 20(2):287–300

Kuliešis A., Kasperavičius A., Kulbokas G. (2016) Chapter 28. Lithuania// National Forest Inventories. In: Vidal C, Alberdi I, Hernandez L, Redmond J (eds) Assessment of wood availability and use. Springer, pp 521–547. doi:10.1007/978-3-319-44015-6

Lietuvos miškų ūkio statistika (2003) (Lithuanian statistical yearbook of forestry). Aplinkos ministerija, Valstybinė miškų tarnyba. Kaunas, Lututė, p 112

Lietuvos miškų ūkio statistika (2015) (Lithuanian statistical yearbook of forestry). Aplinkos ministerija, Valstybinė miškų tarnyba. Kaunas, Lututė, p 184

Mozgeris G, Galaunė A, Palicinas M (2008) Systemy informacji geograficznej w urządzaniu lasu na Litwie – dekada praktycznego stosowania (Geographic information systems in Lithuanian forest management – a decade of operational application). Sylwan 152(1):58–63

Mozgeris G, Brukas V, Stanislovaitis A et al (2016) Owner mapping for forest scenario modelling – a Lithuanian case study. For Policy Econ. http://dx.doi.org/10.1016/j.forpol.2016.02.002

Petrauskas E, Kuliešis A (2004) Scenario-based analysis of possible management alternatives for Lithuanian forests in the 21st century. Baltic For 10(2):72–82

Petrauskas E, Rupšys P (2013) Development of Lithuanian forest growth scenario models based on the data of National forest inventory. Report, Aleksandras Stulginskis University, p 117

Pretzsch H, Biber P, Dursky J et al (2002) Recommendations for standardized documentation and further development of forest growth simulators. For Cbl 121:138–151

Valstybinė miškų apskaita (2012) (State forest resource assessment). Aplinkos ministerija, Valstybinė miškų tarnyba, Kaunas, p 130

Vitunskas D (1988) Naucno-metodiceskoje obosnovanije i razrabotka eskizov drevesinopolzovanija (Methodological framework and improvement of timber use planning). In: Forest inventory and management planning. Inter-University proceedings, Kaunas, pp 40–51

Chapter 20
The Netherlands

Mart-Jan Schelhaas and Sandra A.P.P.M. Clerkx

20.1 Introduction

Forest coverage in the Netherlands reached its minimum at the beginning of the nineteenth century when only 2% of the land was covered with forest. In the first half of the nineteenth century, afforestation was started, first by private landowners and later, in the second half of the century, by the state. These afforestations aimed at improving the economic value of unproductive areas, mainly drift sand and heathlands. Afforestation continued in the twentieth century, but the main purpose then was to create employment in times of economic crisis (1930s). Later, when the large polders were created, plantations were established to shape the new landscape. At first, new forests in the polders were located only on soils that were not good enough for agriculture, but after 1975 they were also located on better/richer soils. In the last decades, some agricultural land was afforested as part of set-aside arrangements. As a result, forest area increased from 270,000 ha in 1900 to 340,000 ha in 2000, with the largest expansion registered between 1960 and 1980; currently, forest area is 373,480 ha (11% of the land area). In the same period unproductive areas decreased from 620,000 to 140,000 ha. Some forested areas have again been cleared to provide corridors between remaining fragments of heathlands, or to increase windiness in the last patches of drift sand. Additions to forest elsewhere compensate for these losses.

Almost half of the Dutch forest is public property, one third is private, whereas 19% is owned by private nature conservation organisations (Schelhaas et al. 2014). During the second half of the twentieth century, many private owners sold their forest. Because most forests are located on poor soils, the financial returns have been very small and in many cases even negative. Before the 1970s these forests were mostly sold to the state, whereas in later decades the buyers were mostly nature conservation organisations.

M.-J. Schelhaas (✉) • S.A.P.P.M. Clerkx
Wageningen Environmental Research (Alterra), Wageningen, The Netherlands
e-mail: martjan.schelhaas@wur.nl; Sandra.Clerkx@wur.nl

S. Barreiro et al. (eds.), *Forest Inventory-based Projection Systems for Wood and Biomass Availability*, Managing Forest Ecosystems 29,
DOI 10.1007/978-3-319-56201-8_20

As in many European countries, the standing stock in Dutch forests increased enormously over the last 30 years, from 40 to 81 million m^3 overbark. This increase is attributed to a combination of aging forest and a decrease in harvesting activities. The area of conifers, especially Scots pine and Norway spruce, is decreasing, while broadleaved forests are increasing fast, from 29% area in 1984 to 45% in 2013. As the consequence of a more nature-oriented management, the percentage of mixed forests has increased, as well as the percentage of uneven-aged forest (16% of the area in 2012–2013). Standing stock in trees outside the forest such as road plantings is estimated at an additional 7–10 million m^3. In the nineties, the annual harvest was approximately 1.25 million m^3, but decreased to approximately 1 million m^3 in the early 2000s. Currently it is again at 1.3 million m^3. Private owners in particular tend to harvest less. Many of them own only small patches, do not live close to their forest, and do not depend on the forest for their income; therefore, they are not very active.

Forest industry demands mainly coniferous and poplar wood and has an estimated capacity of approximately 1 million m^3. The majority of the wood harvested is thus from conifers (64%). Part of the wood used in industry is imported due to limited year-round availability of Dutch wood. The main reasons for this increase in imported wood are limitations on domestic harvest due to laws for protecting flora and fauna (no summer fellings allowed), increased harvesting costs due to more nature-oriented forest management and greatly fragmented forest ownership. The industry also reported that increased domestic demand for biomass for bioenergy affects the wood market (Oosterbaan et al. 2007).

In 2005, the Dutch Ministry for Agriculture, Nature and Food safety requested a projection for demand and supply of wood for the period 2005–2025, aiming at mapping risks and opportunities for forest owners as well as the woodworking industries (Oosterbaan et al. 2007).

20.2 Data Sources

Basic data for the supply projections were derived from the fifth National Forest Inventory (NFI5), carried out in the period 2001-2005. NFI5 includes 3622 sample plots based on a 1 × 1 km unaligned systematic sampling design. Trees are measured and recorded on a circular plot. The plot radius is established so that each plot includes at least 20 trees, but with a minimum of 5 m and a maximum of 20 m. All trees with a diameter of at least 5 cm at a height of 1.3 m above the stump are measured, including dead and lying trees. Half of the plots are permanent and are the only plots for which tree coordinates are mapped. In addition to the tree measurements, characteristics of the plot and/or stand are assessed including ownership, stand size, forest type, soil type, and age. Forest type assessment is rather subjective and includes a mixture of appearance of the forest and management aim such as even-aged, uneven-aged, natural afforestation, lanes, nature-oriented management, and parks.

20.3 Methods

Two forest models were applied in the study by Oosterbaan et al. (2007), the individual tree-based model ForGEM (Kramer et al. 2010; Kramer and Van der Werf 2010) and the large scale scenario model EFISCEN (Schelhaas et al. 2007). The advantage of EFISCEN is that it was designed for this type of study and can be easily parameterised using NFI data. However, EFISCEN is especially suitable for even-aged forests, but Dutch forests have tended to become more mixed and uneven-aged. Therefore, parallel to EFISCEN, the more suitable ForGEM model was used because it accommodates any mixture of species, age and tree size. However, ForGEM cannot readily use NFI data.

Both models were applied to the forest area where timber production could play a role ('production forest'), based on the forest type assessment. Production forest excludes forests where other goals are likely to be more important such as nature-oriented forests and parks. The area of production forest totals 240,000 ha. The remaining 120,000 ha of forest, classified as other forest, can in principle at least partly contribute to the market. Estimates for the latter forests were obtained under the assumption of an average increment of 6 m^3/ha per year based on the simulated increment for oak forest. Otherwise, the estimated potential supply from trees outside forest was estimated using known information on area distribution over landscape types, and assumed densities and increment of trees in different landscape types, totalling 443,000 m^3 per year. Two scenarios were studied, a low-supply and a high-supply scenario.

20.3.1 ForGEM

ForGEM is a process-based model that tracks the characteristics of individual trees over time. Trees produce biomass from intercepted light, which is subsequently allocated to different tree compartments. Trees produce seeds, which are then distributed over the area. When seeds germinate, they are treated as one cohort until they reach a user-defined height threshold (usually 1 m), after which they are treated as individuals. Trees can die due to old age, competition and late season frost (seedlings only). A range of management interventions are available including thinning from above or below, clearcut, shelterwood and selection system. For some processes the user has options regarding the level of simulation detail, depending on the question of interest. These options include, among others, the choice between the ray tracing method or the gap-type approach for light interception, and processed-based photosynthesis or the simpler light-use efficiency approach. Moreover, it is possible to make specific parameters dependent on the genetic structure of trees. This facilitates studies of the effects of natural or human selection on those parameters, and on the forest in general (Kramer and Van der Werf 2010). The version employed uses the gap-type approach for light interception with 20 × 20 m grid cells and the light use efficiency approach to convert radiation into biomass. The genetic module was not used. The time-step depends on the detail of the processes simulated; for our application a monthly time-step was used. The main drivers are temperature and radiation, which were derived from meteorological data from measurement station de Bilt, located centrally in the Netherlands, for the years 1975–2005 (KNMI 2007).

ForGEM is programmed in the language NSM (Nested Simulation Model), a language developed at Alterra and is written in C++. It can be run from a typical computer, but requires considerable memory capacity; a minimum of 2 Gb of RAM is recommended. Simulation times largely depend on simulation length, and level of detail required for the different processes. Typically the time ranges from several hours to several days, but can range up to weeks for long and detailed simulations. ForGEM has no specific user interface, but uses the general NSM interface.

The light use efficiency version of the model must be calibrated against local growth and yield tables, with different productivity classes represented by different light use efficiencies. In this case medium-productivity Dutch yield tables (Jansen et al. 1996) were used. After calibration on total productivity, results were visually compared to diameter and height development in the growth and yield tables. No further validation was done. ForGEM has been used mainly as a research tool for studies with emphases such as the effect of forest management system on wind damage (Schelhaas 2008), interacting effects of forest management and climate change (Kramer et al. 2008), and expected development of forest reserves (Schelhaas et al. 2005).

ForGEM can simulate an area up to a few hectares, but usually patches of 1 ha are simulated. Exact tree position, species, and stem and crown dimensions are needed for initialisation. Dutch NFI plots have radius between 5 and 20 m, but tree positions are recorded only for the permanent sample plots which comprise 50% of the total sample. Therefore, because ForGEM could not run for each sample plot individually, NFI plots belonging to production forests were grouped to produce a manageable number of forest types to be simulated by ForGEM. Grouping was done by dominant species, species mixture, and type of management (even-aged, uneven-aged, transition). In total, 19 representative groups were distinguished, with a minimum area per group of 3000 ha. Each forest type was simulated as a single stand for one single rotation, starting with an idealised, young, even-aged forest or a typical, uneven-aged situation for the group under consideration.

For each group, the low and a high supply scenario were implemented at harvest levels of 40% and 80%, respectively, of the basal area increment. These simulations represented the typical development of a 1 ha forest per group for the two scenarios for a full rotation. These idealised development curves as functions of age were then combined with the actual area distribution over age classes from the NFI to estimate the total potential of removals over the next 20 years. For example, if 20 plots were present in the forest type “Scots pine mixed with oak” in the age class 40–50 years, the removals for the simulation between ages 40 and 60 year of the corresponding forest type were used as an estimate of the removals on the corresponding 2000 ha for the next 20 years. ForGEM can provide output both at the tree level and the stand level for intervals of months to several years. For this application, the focus was harvest volume by diameter class per tree species. The main stochastic processes included in the model are seed dispersal, age-related mortality and selection of the simulation year to which the year of observed weather is applied. No replicates were used, because seed dispersal and age-related mortality are not important in these simulations, and weather had little influence on the results.

20.3.2 EFISCEN

The European Forest Information Scenario model (EFISCEN V3.1) simulates the development of forest resources at scales from provincial to European level (Sallnäs 1990; Nabuurs et al. 2007; Schelhaas et al. 2010; Verkerk et al. 2011). Forest resource analyses have been successfully conducted at the pan-European scale with the EFISCEN model for a range of applications.

Input data are usually obtained from NFIs in aggregated form. They can be stratified by province, tree species, site class and owner class, depending on the level of detail required and the size of the resulting groups, hereafter referred to as stand types. The input data are used to construct the initial age class distribution and growth as a function of age for each stand type. Each stand type is assigned a management regime defined in terms of the probability that a thinning or final harvest can be carried out as a function of stand age. For each 5-year time-step, the timber demand from the simulated area must be defined separately for thinnings and final fellings. This total demand is then obtained for the different stand types, according to the felling possibilities as defined by modelled age class distributions and the management regime.

Principal outputs of EFISCEN are age class distributions, growing stock volumes, harvesting levels and increment. Biomass Expansion Factors (BEFs) for converting growing stock volume to biomass for different tree compartments and turnover rates are used in EFISCEN to estimate carbon stocks in living tree biomass and litterfall from those trees. The litterfall rates are used in the build-in YASSO model (Liski et al. 2005) to estimate soil carbon stocks.

EFISCEN has been validated on historical inventory data for Finland (Nabuurs et al. 2000) and Switzerland (Thürig and Schelhaas 2006). Accurate predictions were obtained at the national scale, but deviations occurred at the provincial and tree species levels due to differences in the distribution of harvest over the stand types. The model can reasonably be used for 50–60 year projections, but the projection horizon is commonly limited to 20–30 years.

The EFISCEN model was applied to the same NFI data again but only production forests were considered. For the EFISCEN simulations, the data were aggregated into eight groups based on dominant species, ignoring species mixtures and uneven-aged structures. Although a fraction of the plots is classified as uneven-aged, they still have a single age which is estimated from the time of regeneration of the original stand. The inputs for EFISCEN are the area, average volume and average increment per age class for each of the groups distinguished. Because no increment data were available in NFI5, we used the increment functions derived from the HOSP study and as implemented for the first European Forest Sector Outlook Study (Schelhaas et al. 2006). Simulations were done over the period 2005–2025.

Table 20.1 Total wood supply according to the projections by ForGEM and EFISCEN (average over the period 2005–2025), plus estimated supply from forest not managed for wood supply and other sources (1000 m^3)

Source of wood supply	Supply low scenario ForGEM	Supply high scenario ForGEM	Supply low scenario EFISCEN	Supply high scenario EFISCEN
Production forest	681.0	1240.1	1117.4	1352.3
Other forest	129.1	258.3	129.1	258.3
Trees outside the forest	177.3	354.6	177.3	354.6
Total	987.4	1853.0	1423.8	1965.2

For forests not specifically managed for wood supply (other forest in Table 20.1), potential wood supply is estimated using estimated increment and the specific scenario percentages, 40% and 80% of increment removed for low and high scenarios, respectively. Because no model projections were involved, removals from this area are assumed to be the same for both the ForGEM and EFISCEN projections. Further, the same 40% and 80% increment removal levels were applied to the potential supply from trees outside the forest.

20.4 Results

The low supply as projected by EFISCEN just overlaps with the high supply as estimated by ForGEM (Fig. 20.1; Table 20.1). The range between the low and high supply scenarios is much greater for the ForGEM model than for EFISCEN. Under a high demand scenario (+2% increase per year), EFISCEN would be able to supply enough wood under the low supply scenario until 2010, and until 2020 under the high supply scenario. The high supply scenario of ForGEM provides sufficient supply until 2017, while the low supply scenario stays well below the demand throughout the whole period.

The potential supply from other forest and trees outside the forest is estimated at 300,000–600,000 m^3 per year. This extra demand would help to compensate for the discrepancy between the supply from the production forest and the high demand scenario.

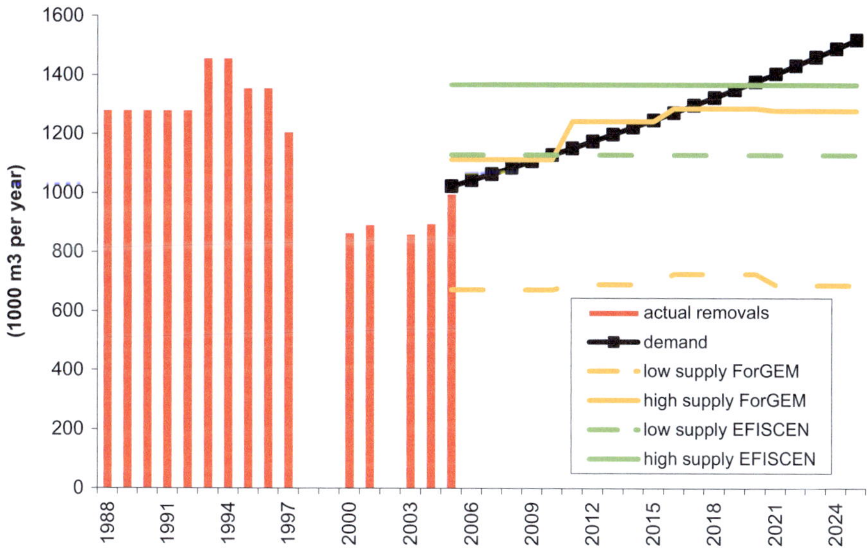

Fig. 20.1 Historical harvest (actual removals), forest supply from production forest according to low and high supply scenarios for the two models and projected demand assuming an annual 2% increase (Reproduced from Oosterbaan et al. 2007)

20.5 Discussion and Conclusions

20.5.1 Conclusions From the Report by Oosterbaan et al. (2007)

The projected developments of demand and supply were interpreted as opportunities for forest owners. Shortage in supply would mean increasing prices, which would make operations more profitable. Consequently, a larger amount of roundwood could be brought to market, especially from forests that are currently not harvested for economic reasons. According to industry, the quality of Dutch timber was adequate for their purposes.

Potentially, more timber can be harvested from the Dutch forests. For the 240,000 ha of production forest for which supply was estimated, possibilities for increased harvest are limited, because a high percentage of the increment is already harvested. Therefore, any harvest increase must be realised from the 120,000 ha of other forests where the management goals allow harvest of trees and from trees outside forest such as landscape forest and roadside plantations. The realisable harvest increase from these sources is estimated at 300,000–600,000 m^3.

The timber market will be good for all conifer species. Also the supply of timber for broadleaved species will just exceed demand. Broadleaved species can be used to fulfil the anticipated greater demand for energy purposes, but might also be used for replacing species such as spruce and poplar which are likely to experience considerable future supply shortage.

20.5.2 Reflection

The differences in outcomes between the two models are quite large. For a large part, these differences can be attributed to the uncertainty surrounding the increment. ForGEM is calibrated with yield tables that are known to underestimate increment, whereas EFISCEN was calibrated with more recent data. Moreover, the increment level in ForGEM is influenced by the management during the simulation where low removal rates lead to lower increment. A preliminary comparison with the NFI6 increment values show that projections for conifers, except Scots pine, match best with those projected by EFISCEN, while broadleaves are generally closer to the projections by ForGEM. The overall increment (7.3 m^3/ha per year) was closer to the projected value by ForGEM (6.8 m^3/ha per year) than EFISCEN (8 m^3/ha per year).

There was also considerable uncertainty in area estimates from the simulations due to missing information for a considerable number of plots. Furthermore, the division of the forest into "production forest" and "other forest" was based on the NFI field assessments. However, these assessments are highly subjective and add more uncertainty. The NFI6 estimate of the area of production forest is 287,000 ha, 18% greater than in NFI5, although it is unlikely that much has changed in the meanwhile.

Altogether the outcomes of the study are very uncertain and can serve only as an indication for the possible supply. Meanwhile, NFI6 was finished, including a re-assessment of the permanent sample plots. This allows a much better assessment of the increment and how much of the area is really harvested. From the 1235 plots that were re-assessed, 38% showed no signs of harvest (Clerkx et al. 2015), indicating that harvest took place only on about 230,000 ha during the last decade. Reasons not to harvest on the other plots varied greatly, including priority to other functions (mostly nature) and the silvicultural state of the forest (Clerkx et al. 2015). A major problem remains in identifying the management goals of the owners, the reasons for acting or not acting, and how owners and managers could be stimulated to increase their harvest level. A future study should take owner behaviour into account, and should involve additional GIS analysis to better delineate areas where harvest is allowed and possible. Also, incorporating cost estimates would help to produce a more realistic picture. Furthermore, it would be beneficial to develop a model that allows each NFI plot to be simulated separately, rather than simulating an average development of a group of plots over an entire rotation. However, the most uncertain factor will remain the forest owners' behaviour, especially how they will react to changes in prices and policies.

References

Clerkx APPM, Schelhaas MJ, Zwart J (2015) Oogst in het Nederlandse bos: analyse van niet-geoogste plots uit de Zesde Nederlandse Bosinventarisatie. Alterra rapport 2610, Wageningen

Jansen JJ, Sevenster J, Faber PJ (1996) Opbrengsttabellen voor belangrijke boomsoorten in Nederland. IBN rapport 221, IBN-DLO, Wageningen

KNMI (2007) Koninklijk Nederlands Meteorologisch Instituut. www.knmi.nl. Accessed 23 July 2007

Kramer K, van der Werf DC (2010) Equilibrium and non-equilibrium concepts in forest genetic modelling: population- and individually-based approaches. For Syst 19:100–112

Kramer K, Buiteveld J, Forstreuter M et al (2008) Bridging the gap between ecophysiological and genetic knowledge to assess the adaptive potential of European beech. Ecol Model 216:333–353

Kramer K, Degen B, Buschbom J et al (2010) Modelling exploration of the future of European beech (Fagus sylvatica L.) under climate change-Range, abundance, genetic diversity and adaptive response. For Ecol Manag 259:2213–2222

Liski J, Palosuo T, Peltoniemi M, Sievänen R (2005) Carbon and decomposition model Yasso for forest soils. Ecol Model 189:168–182

Nabuurs GJ, Schelhaas MJ, Pussinen A (2000) Validation of the European forest information scenario model (EFISCEN) and a projection of Finnish forests. Silva Fenn 34(2):167–179

Nabuurs GJ, Pussinen A, van Brusselen J, Schelhaas MJ (2007) Future harvesting pressure on European forests. Eur J For Res 126:391–400

Oosterbaan A, van den Berg CA, Schelhaas MJ (2007) Ontwikkelingen in vraag en aanbod van rondhout in Nederland en aangrenzend gebied en mogelijke knelpunten en kansen voor de bos- en houtsector in de periode 2005–2025. Alterra rapport 1510, Wageningen

Sallnäs O (1990) A matrix growth model of the Swedish forest. Studia Forestalia Suecica 183

Schelhaas MJ (2008) The wind stability of different silvicultural systems for Douglas-fir in The Netherlands: a model-based approach. Forestry 81(3):399–414

Schelhaas MJ, Wijdeven SMJ, van der Werf DC (2005) Zelfregulerende bossen. Een modelstudie naar effecten van 'niets doen' en actief beheer op ontwikkelingen in bosstructuur. Wageningen, Alterra, Alterra-rapport 1270

Schelhaas MJ, van Brusselen J, Pussinen A et al (2006) Outlook for the development of European forest Resources. A study prepared for the European Forest Sector Outlook Study (EFSOS). Geneva Timber and Forest Discussion Paper, ECE/TIM/DP/41. UN-ECE, Geneva

Schelhaas MJ, Eggers J, Lindner M et al (2007) Model documentation for the European Forest Information Scenario model (EFISCEN 3.1). Alterra report 1559, Wageningen, EFI Technical Report 26, Joensuu, Finland

Schelhaas MJ, Hengeveld G, Moriondo M et al (2010) Assessing risk and adaptation options to fires and windstorms in European forestry. Mitig Adapt Strat 15:681–701. doi:10.1007/s11027-010-9243-0

Schelhaas MJ, Clerkx APPM, Daamen WP et al (2014) Zesde Nederlandse Bosinventarisatie; Methoden en basisresultaten. Alterra rapport 2545, Wageningen

Thürig E, Schelhaas MJ (2006) Evaluation of a large-scale forest scenario model in heterogeneous forests: a case study for Switzerland. Can J For Res 36(3):671–683

Verkerk PJ, Antilla P, Eggers J et al (2011) The realisable potential supply of woody biomass from forests in the European Union. For Ecol Manag 261:2007–2015

Chapter 21
Norway

Clara Antón-Fernández and Stein Tomter

21.1 Introduction

21.1.1 Forest Resources

Forest covers approximately 37% of the Norwegian land area. The most important tree species with respect to volume and economic value are Norway spruce (*Picea abies*), Scots pine (*Pinus silvestris*) and birch (*Betula* spp.). For the 2007–2011 period, the Norwegian National Forest Inventory (NFI) estimated forest area to be 12.2 million ha, with annual volume increment under bark of 25.6 million m^3.

Annual fellings have been relatively stable for the last 80 years (around 10 million m^3/year), while mean annual increment has doubled during this period. This means that annual harvested volume is considerably less than the potential level, and potential harvest in Norway continues to increase each year. Thus, it is not surprising that current growth is now more than twice as what it was 80 years ago. Growing stock has increased significantly since the first Norwegian NFI, when the growing stock of the Norwegian forest was estimated at slightly more than 300 million m^3 (reference year 1925), while the most recent inventory (2010–2014) estimate of volume under bark of Norwegian forests is approximately 920 million m^3. This increase is due to better forest management and harvest levels that, over time, have been less than growth.

Approximately 80% of the forest area is privately owned, mainly by farmers who combine small-scale forestry with agriculture. Forest industry and private enterprises own only approximately 4% of the forest area, while the state and the municipalities own approximately 12%.

C. Antón-Fernández (✉) • S. Tomter
Norwegian Institute of Bioeconomy Research, Ås, Norway
e-mail: caf@nibio.no; stt@nibio.no

S. Barreiro et al. (eds.), *Forest Inventory-based Projection Systems for Wood and Biomass Availability*, Managing Forest Ecosystems 29,
DOI 10.1007/978-3-319-56201-8_21

Multiple reports have aimed at characterizing the local or national biomass supply situation in Norway (e.g. OED 1997; NOU 1998; Berg et al. 2003; Bernard and Bugge 2006; Vennesland et al. 2006; Hobbelstad 2007a, b; Langerud et al. 2007; Gjølsjø and Hobbelstad 2009; Tomter 2016a, b), but no countrywide projections have been reported on a regular basis. Most of the existing woody biomass resource projections are long-term projections based on the Norwegian NFI and conducted using forest scenario analysis tools such as GAYA [NorFor, SGIS] (Hoen and Eid 1990) and AVVIRK2000 (Eid and Hobbelstad 2000), with a focus on potential harvest (e.g. Eriksen et al. 2006). The last official countrywide projections for Norway were published in 2006 and were based on the eighth NFI (2000–2004). Although countrywide projections for Norway are not conducted on a regular basis, county-wise projections are published on a 15-year cycle, which means that some of the projections are very recent and some are 15 years old. These projections use data from the temporary and permanent Norwegian NFI plots, and are long-term maximum sustained yield estimates.

21.1.2 NFI History and Data

The Norwegian NFI dates back to 1919. The first inventories (NFI1 and NFI2) used strip sampling; from 1957 until 1986 the NFI used cluster sampling with temporary plots; and from 1986 until 1993 circular, fixed-area, permanent sample plots were installed in all counties except Finnmark. Starting with the seventh NFI (1994[95]–1999), the NFI became a continuous forest inventory, with 20% of the plots remeasured each year (Tomter et al. 2010).

The first NFI was motivated by concerns related to forest resource development and thus, the main focus of the first inventories was economic exploitation of the forest. In recent inventories, the assessment of environmental values has been increasingly emphasized and now represents a central task for the NFI. Recently also a number of adjustments have been made to the NFI to facilitate carbon reporting for the LULUCF sector.

The NFI sampling design is based on a 3 × 3 km grid that covers all of Norway and includes about 12,000 plots lying within forest. Certain areas have smaller or larger sampling intensities than produced by the basic 3 × 3 km grid. Plots above the coniferous forest line are distributed using a 3 × 9 km grid. The northernmost county of Norway, Finnmark, has traditionally been excluded from the NFI. However, after the first inventory in Finnmark was completed in 2011, Finnmark has been included in the regular schedule of 5-year remeasurements. Most of Finnmark county uses a 9 × 9-km grid, except for the coniferous forest area which uses a 3 × 3-km grid. Starting in 2012, the protected areas have a greater sampling intensity than the rest of the country.

Detailed tree measurements are obtained for the 250 m^2 circular plots, while land use and stand characteristics are assessed for a 1000 m^2 area surrounding each plot.

21.2 Data

21.2.1 NFI Data Usable for Resource Projections

The main source of data for woody biomass resource projection is the NFI permanent and temporary plots. Both permanent and temporary plots are 250 m^2 circular plots. On each permanent plot all trees with at breast height (*dbh*) of at least 5 cm are measured and their species recorded. On plots with 10 or fewer trees, all tree heights are measured, while for plots with more than 10 trees a subsample is selected proportionally to basal area with a target sample size of 10 trees per plot. For the selected trees, information on damage, discoloration signs, crown color, and crown density is also collected. Stand density and species composition are assessed, among other variables, for a 1000 m^2 area surrounding the plot. For all trees on the 250 m^2 plot, *dbh*, species, status, position, and damage are measured. For each plot, other measures are also taken including distance to road, skidding distance, high accuracy GPS coordinates, and slope. Current site index is estimated for the dominant species at each plot in productive forest (annual yield capacity of at least 1 m^3/ha of wood including bark), where site index is defined as the average height of the 100 largest trees per ha at age 40. At each re-measurement, treatments carried out in the past 5-year period are recorded. Possible treatments include final fellings, thinnings, selective cuttings, regeneration treatments, early stand tending treatments, drainage and pruning.

Total volume per plot is estimated as the sum of individual volume estimates for all trees with *dbh* of at least 5 cm. Individual tree volumes are estimated with species-specific, individual tree volume models with tree height and *dbh* as independent variables. For trees without measured heights, species-specific height models are used to estimate their basic height (Hhc). Hhc is then used to estimate the individual tariff tree volumes (V_{Hhc}). To account for differences in the diameter-height relationship between sites, V_{Hhc} is multiplied by a plot- and species-specific correction factor. The correction factor (tariff) is calculated using only trees with measured height as the ratio between the sum of the V_{Hhc} and the sum of the volumes estimated using the measured heights. The individual volumes in both the numerator and denominator of the correction factor are weighted to account for the unequal sampling probability originating from the selection of trees proportional to the basal area for height measurement.

Total biomass per plot is estimated as the sum of the individual biomass components (stump, roots, stem, bark, dead and living branches, and foliage) for all individual trees with *dbh* of at least 5 cm. Individual tree biomass is estimated using species-specific allometric models with tree *dbh* and tree height as independent variables.

21.2.2 Other Statistical Information Related to Wood Resources and Use Wood

The Statistics Norway (Norwegian statistics bureau) website (www.ssb.no) publishes information on growing stock (based on NFI), annual increment (based on NFI), roundwood removals, average price per m^3 of industrial roundwood for sale, and commercial roundwood removals by assortment group, among other forestry statistics. Some of these data go back to the 1920s.

21.3 Methods

AVVIRK2000 is a deterministic simulation model for large-scale, long-term (100 years) forestry scenario analyses. AVVIRK2000 and its predecessor AVVIRK3 have been the main tools used for long-term management planning and forestry scenario analysis in Norway in recent decades. The time scope of the system focuses on strategic planning with projections made for a fixed period of 100 years, divided into 10 periods of 10 years. All treatments are considered as having occurred at the midpoint of each 10-year period.

Growth projections are based on the development of the "average tree", i.e., the development of a tree with diameter at breast height equal to the basal area mean diameter (D_{ba}), and with height equal to the mean height weighted by basal area, also known as Lorey mean height, (H_L). The basal area mean diameter, also known as quadratic mean diameter, is calculated for each plot as the diameter of the tree of average basal area, that is $D_{ba} = \sqrt{\frac{4BA}{N}}$, where BA is the plot basal area (e.g. 2 m^2), and N is the number of trees on the plot. The empirical models used to describe the state of the forest and stand dynamics in AVVIRK2000 are fitted with data from experimental permanent sample plots. The core of the growth model consists of three sets of models: diameter increment models (Blingsmo 1984), height development models (Strand 1967; Tveite 1967, 1976, 1977; Braastad 1977) and a mortality model (Braastad 1982). The volume of the "average tree" is estimated using models developed by Braastad (1966, 1974, 1980), Brantseg (1967) and Vestjordet (1967), and the volume per hectare of the stand is estimated as the product of the volume of the "average tree" and the number of trees per ha, N. In young stands, site index, age and number of future trees and tree composition are the basis for yield estimation. No volume or volume increment is estimated until the dominant height of the stand has reached 9 m. At this time, mean basal area is estimated according to Braastad (1975, 1977, 1982), and then "average tree" diameter and height are estimated.

Timber value and the proportions of pulpwood and saw timber are estimated using the Blingsmo and Veidahl (1992) models which use tree species, D_{ba} and H_L as independent variables.

Four harvest strategies are available: (1) a non-declining harvest path or net income path for the period of 100 years, (2) a user-defined harvest level or net income level for any number of 10-year periods up to 10, (3) a harvest path according to user-defined harvest ages for all stands, or (4) a harvest path according to removal of stands with relative annual value increment lower than a user-defined percentage. Several environmental restrictions such as an "ecologically orientated regime", where no treatments are allowed until the forest reaches a user-defined maturity, can also be selected.

Because AVVIRK2000 estimates are based on the development of the "average tree" and the number of trees per ha, all results are at stand- or forest-level:

Forest-Level

- volume of fellings distributed by thinning and final cutting,
- volume of fellings distributed by tree species and assortments (sawlogs, pulpwood),
- income,
- regeneration cost,
- area distribution by site classes and development classes,
- volume distribution by site classes and development classes, and
- increment distribution by site classes and development classes.

Stand-Level

- volume, income, cost for cutting (thinning, clear cutting, seed tree cutting etc), regeneration, tending etc. for all treatments estimated in the stand for each 10-year period,
- state of the stand, for example, age, number of trees, tree species composition, mean diameter, mean height, volume, increment, etc.

In addition, for each forest/stand, the net present value is calculated as well as the soil expectation value for a newly established stand.

The most common biomass models for Norway are the Swedish models, Marklund (1987) and Marklund (1988) for the aboveground biomass and Petersson and Ståhl (2006) for belowground biomass. Although AVVIRK2000 does not estimate biomass, GAYA, a similar forest scenario analysis tool that uses the "average tree" approach, does estimate biomass using the Swedish individual tree models.

21.3.1 Value of the Timber

In AVVIRK2000 all prices and costs are assumed constant over time, i.e., the discount rate is a real rate, and it is defined by the user. Timber value is estimated using models (Blingsmo and Veidahl 1992) with tree species, D_{ba} and H_L of each estimation unit as independent variables. The models are standardized, i.e., the initial values are adjusted according to existing or assumed timber prices and assortment distribution. The model also provides an option that permits the user to specify the timber value per m^3 for the stand, reflecting the timber value of all assortments in the stand jointly.

21.3.2 Costs

Variable costs can be determined according to manual or mechanized logging systems. The model handles clear cutting, harvests when seed trees or shelterwood are established and phased out, and thinnings. For both systems, the costs rely on variables such as tree species, D_{ba} and H_L, as well as variables such as transport distance and cost factors. The net present value of a silvicultural regime is determined by the cash flow originating from the estimated timber values and harvest costs, and from the costs for silviculture. Fixed costs (administration, planning, etc.) are included in the net present value.

21.3.3 Uncertainty

Up to now uncertainty analysis have not been part of any of the woody biomass projections for Norway.

21.3.4 Dissemination

The Norwegian NFI regularly reports on growing stock and biomass resources for the country as a whole, for regions (i.e. several counties), and for individual counties. Results for smaller geographical units than a county are usually not reported because the requirements for maximum standard errors may be violated due to the low numbers of sample plots.

21.4 Discussion

AVVIRK2000 growth projections are based on the development of the "average tree". This means that the variability within and between stands with similar D_{ba}, H_L and number of trees per ha, is ignored. Furthermore, mixed-species stands are assumed to grow as if they were mono-species stands. The reality of the Norwegian forest is that uneven-aged stands with multiple species are not uncommon. Also, diversified tree structures including multi-layered canopies are not uncommon. Neither of these types of stands is correctly represented by the "average tree".

As a result of Norwegian forestry history, the current area of over-mature forest is relatively small. Hence, over-mature forests are barely represented in the NFI data and in the datasets used to fit AVVIRK2000 models. Because the current woody biomass projections are made for a period of 100 years, and some areas are restricted from harvesting, estimates for some are extrapolations beyond the limits of the AVVIRK2000 growth models. Most over-mature forests in Norway are currently on low productivity sites, so gathering data on growth of very old stands that cover the wide range of forest conditions across Norway is difficult. An alternative to solve this issue would be establishment of international networks with the aim of sharing data on over-mature forest growth and yield.

Norway reports potential harvest levels. These potential levels do not reflect what might be realistically available in the market. For example, for the 2007–2012 period, the annual potential harvest was estimated to be more than 19 million m^3, while the annual felling volume in Norway has been approximately 10 million m^3 for the last 80 years.

Current published results lack any measure of uncertainty. Potential sources of uncertainty that could be considered include sampling variability, model prediction uncertainty, the effects of climate change on growth, mortality, and regeneration, and uncertainties related to owner behavior, economic parameters, and catastrophic events such as wind or fires.

A new framework for forest development scenarios, called "sitree", is being developed in Norway. The new framework will include both new and old individual tree forest growth and yield models. The new framework will also include modules for estimating vegetation and soil carbon, and a module for life cycle analysis. The framework will allow for estimation of uncertainty through Monte Carlo simulations.

References

Berg LN, Jørgensen PF, Wilhelmsen G (2003) Bioenergiressurser i Norge. Norges vassdrags- og energidirektorat, Oslo

Bernard B, Bugge L (2006) Biomasse–nok til alle gode formål? Rapport KanEnergi 2006

Blingsmo KR (1984) Diametertilvekstfunksjoner for bjørk-, furu- og granbestand. Rapport fra Norsk institutt for skogforskning 7/84

Blingsmo KR, Veidahl A (1992) Funksjoner for bruttopris av gran- og furutrær på rot. Rapport fra Skogforsk 8/92

Braastad H (1966) Volumtabeller for bjørk. Meddelelser fra Det norske Skogforsøksvesen 21:23–78

Braastad H (1974) Diametertilvekstfunksjoner for gran. Meddelelser fra Norsk Institutt for Skogforskning, 31/1

Braastad H (1975) Produksjonstabeller og tilvekstmodeller for gran. Meddelelser fra Norsk Institutt for Skogforskning, 31/9

Braastad H (1977) Tilvekstmodellprogram for bjørk. Rapport fra Norsk Institutt for Skogforskning, 1/77

Braastad H (1980) Tilvekstmodellprogram for furu. Meddelelser fra Norsk Institutt for Skogforskning, 35/5
Braastad H (1982) Naturlig avgang i granbestand. Rapport fra Norsk institutt for skogforskning 12/82
Brantseg A (1967) Furu sønnafjells. Kubering av stående skog Funksjoner og tabeller. Meddelelser fra Det norske Skogforsøksvesen 22:695–739
Eid T, Hobbelstad K (2000) AVVIRK-2000: a large-scale forestry scenario model for long-term investment, income and harvest analyses. Scand J For Res 15:472. doi:10.1080/028275800750172736
Eriksen R, Tomter S, Ludahl A (2006) Statistikk over skogforhold og -ressurser i Møre og Romsdal. Landsskogtakseringen 2000–2004. Norwegian Forest and Landscape Institute, Ås
Gjølsjø S, Hobbelstad K (2009) Energipotensialet fra skogen i Norge. Norwegian Forest and Landscape Institute, Ås
Hobbelstad K (2007a) Ressurssituasjonen i Hedmark og Oppland. Norwegian Forest and Landscape Institute, Ås
Hobbelstad K (2007b) Skogressurser i Sør-Østerdal. Norwegian Forest and Landscape Institute, Ås
Hoen HF, Eid T (1990) En modell for analyse av behandlingsalternativer for en skog ved bestandssimulering og lineær programmering. Rapport fra Norsk institutt for skogforskning 9/90
Langerud B, Størdal S, Wiig H, Ørbeck M (2007) Bioenergi i Norge–potensialer, markeder og virkemidler. Østlandsforskning rapport 2007/17
Marklund LG (1987) Biomass functions for Norway spruce (*Picea abies* (L.) Karst.) in Sweden. Sveriges lantbruksuniversitet, Institutionen för skogstaxering, Umeå
Marklund LG (1988) Biomassafunktioner för tall, gran och björk i Sverige. Swedish University of Agricultural Sciences, Department of Forest Survey, Umeå
NOU (1998) Energi- og kraftbalansen mot 2020. Norges offentlige utredninger, Oslo
OED (1997) Det Interdepartementale arbeidsutvalget for bioenergi. Olje- og energidepartementet, Energi- og vassdragsavdelingen, Oslo
Petersson H, Ståhl G (2006) Functions for below-ground biomass of *Pinus sylvestris*, *Picea abies*, *Betula pendula* and *Betula pubescens* in Sweden. Scand J For Res 21:84–93. doi:10.1080/14004080500486864
Strand L (1967) Høydekurver for bjørk in Braastad, H. Produksjonstabeller for bjørk. Meddelelser fra Det norske Skogforsøksvesen 22:265–365
Tomter SM (2016a) Analyser av skogressursene i Hedmark. Basert på Landsskogtakseringens data. NIBIO Rapport, vol 2, Nr. 53, 2016
Tomter SM (2016b) Analyser av skogressursene i Oppland. Basert på Landsskogtakseringens data. NIBIO Rapport, vol 2, Nr. 52, 2016
Tomter SM, Hylen G, Nilsen J-E (2010) National Forest Inventories reports: Norway. In: Tomppo E, Gschwantner T, Lawrence M, RE MR (eds) National forest inventories – pathways for common reporting. Springer, Heidelberg, pp 411–424
Tveite B (1967) Sambandet mellom grunnflateveid middelhøyde og noen andre bestandshøyder i gran- og furuskog. Meddelelser fra Det norske Skogforsøksvesen 22:483–538
Tveite B (1976) Bonitetskurver for furu. Internal report
Tveite B (1977) Bonitetskurver for gran. Meddelelser fra Det norske Skogforsøksvesen 33/1
Vennesland B, Hobbelstad K, Bolkesjø T et al (2006) Skogressursene i Norge 2006. Muligheter og aktuelle strategier for økt avvirkning. Norwegian Forest and Landscape Institute, Ås
Vestjordet E (1967) Funksjoner og tabeller for kubering av stående gran. Meddelelser fra Det norske Skogforsøksvesen 22:539–574

Chapter 22
Portugal

Susana Barreiro and Margarida Tomé

22.1 Introduction

22.1.1 Descriptive Statistics

Portuguese forest area has been increasing since 1874. The main tree species used to be cork and holm oaks (*Quercus suber* and *Quercus rotundifolia*), but it was the expansion of maritime pine (*Pinus pinaster*) that most contributed to increasing forest area. A century later, over 2.2 milllion hectares had been afforested leading to a total forest area of 2838 million hectares (DGF 1975). Maritime pine represented 48% of the forest area followed by evergreen oaks (41%) and *Eucalyptus globulus* (6%). Over time, eucalyptus plantations gained ground over maritime pine. In 2010 a backward-looking study on land uses and forest species occupation between the years of 1995 and 2010 (Fig. 22.1) documented this change. The decrease in maritime pine forest area has been accentuated after the pine wood nematode (*Bursaphelenchus xylophilus*) became established in Portugal. The application of stringent measures consisting of harvest and destruction of woody material from trees affected or showing any symptoms of decline has led to a reduction of 263 thousand hectares of maritime pine forest. Conversely, the area of Eucalyptus increased at the expense of 70 thousand hectares of harvested pine forests, 12 thousand hectares of abandoned agricultural land and 13.5 thousand hectares of shrublands and pastures (ICNF 2013).

According to preliminary results of the sixth National Forest Inventory (NFI6), Portuguese forests cover around 3.155 million hectares making up about 35% of the land surface (ICNF 2013). Eucalyptus and maritime pine, both managed for wood production, represent nearly 50% of the forest area covering 812 thousand hectares

S. Barreiro (✉) • M. Tomé
Forest Research Centre (CEF), School of Agriculture, University of Lisbon, Lisbon, Portugal
e-mail: smb@isa.ulisboa.pt; magatome@isa.ulisboa.pt

S. Barreiro et al. (eds.), *Forest Inventory-based Projection Systems for Wood and Biomass Availability*, Managing Forest Ecosystems 29,
DOI 10.1007/978-3-319-56201-8_22

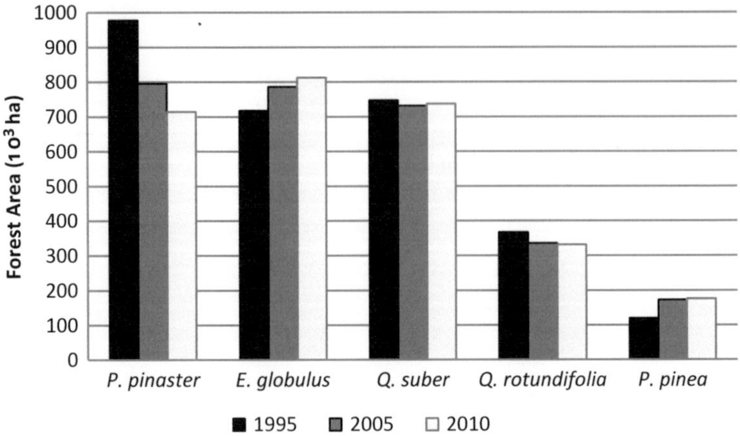

Fig. 22.1 Evolution of forest area by tree species over the period of 1995 and 2010; where P., E. and Q. stand for *Pinus*, *Eucalyptus* and *Quercus* genus, respectively

and 714 thousand hectares, respectively (ICNF 2013). About 40% of Portuguese forests are managed for non-wood forest products: 737 thousand hectares of cork oak forests managed for cork production often as an agroforestry system; 331 thousand hectares of holm oak managed as an agroforestry system; and 176 thousand hectares of stone pine (*Pinus pinea*) managed for fruit production for human consumption (ICNF 2013). The areas of cork and holm oaks are managed under severe legal harvesting restrictions and are not considered as Forest Available for Wood Supply (FAWS). In these areas, thinning is allowed to some extent, although thinned woody biomass from these forests, as well as from stone pine forests, is difficult to utilize and, when the selling price covers transportation costs, is frequently used by local communities. The species groups that make up the remaining 9% of the forest area are: other hardwoods (6%), other softwoods (2%), other oaks (2%) and chestnut (1%). In total, softwood species represent 31% of the Portuguese forest area. Maritime pine forests are mainly concentrated in the north of the country, while eucalyptus is generally found in north and central Portugal, mainly along the coast. Evergreen oaks dominate the central and southern areas. Portuguese forests are mainly managed as pure stands, although mixed stands can also be found with the most common mixtures being maritime pine and eucalyptus, cork and holm oaks and, more recently, cork oak and stone pine.

The Global Forest Resource Assessment 2010 considers Portugal among the 10 countries with greatest percentages of privately owned forests in Europe. In the mainland, 97.2% of the forests are privately owned, of which approximately 5% belong to industries.

Most forest properties are small and fragmented. Around 61% of private owners have less than 5 ha representing 26% of the forest area (Santos et al. 2013). The size of forest holdings increases from north to south. The small size of properties often makes it difficult to implement policy as well as preventive and protective measures against biotic and abiotic agents. One possible solution is organizing forest owners in associations to guarantee a minimum area for sustainable forest management (DGRF 2007). The number of forest owner associations has increased from 16 in 1977 to 177 in 2013 (ICNF 2014). Moreover, by the end of 2012, 330,000 hectares of forest were certified under FSC, and also by PEFC with some areas being certified by both systems (Santos et al. 2013).

Forest fires are the greatest threat to Portuguese forests. About 50% of fire ignitions are attributed to negligent or criminal behaviour. This factor, combined with hot and dry summers that facilitate fire spread, depopulation of the interior of the country which led to an increase in unattended shrublands and lack of forest management, the presence of fire prone species such as pine and eucalyptus, the small average holding size, and limited accessibility all contribute to a massive problem.

Since 1995, total standing stock increased from 153,600 m^3 (DGF 2001) to 172,600 m^3 (AFN 2010). In spite of the increase, a reduction in the maritime pine standing stock of 16.7 million m^3 was observed due to the pine forest area reduction, whereas a moderate increase was registered in eucalyptus standing stock (6 million m^3) mainly in pure stands. The average site index for eucalyptus is 18 m (base age 10) with an average gross mean annual increment of 12 m^3/ha per year calculated for the age of 12. As for maritime pine, the average site index (base age 50) is around 18 m, corresponding to a gross mean annual increment of 7 m^3/ha per year (Santos et al. 2013). The area of maritime pine and eucalyptus stands with a MAI higher than 22 represent 39% and 12.5% of each species forest area respectively.

22.1.2 NFI History

National Forest Inventories (NFIs) have been carried out more or less every 10 years since 1965. The Portuguese NFI involves both land cover and field data collection. Estimates of forest areas are based on photointerpretation of aerial photography and characterisation of stands with the most representative tree species in the country. However, the Portuguese NFI has not been based on permanent plots, and a different sampling design was used for each NFI until NFI5 (2005), when a 500 × 500 m systematic grid covering the whole country was established for photo-interpretation (Barreiro et al. 2010). The current NFI is based on the same grid system. Two subgrids derived from the base grid were established every 2 km and every 4 km and were used to select locations for forest and shrubland NFI field plots, respectively. The intensive sampling design was justified by the intention of implementing a system of continuous forest inventory that at the end was not implemented. NFI5 would represent the base year, for which all plots would be measured. In the course of NFI6, 8000 forest plots were remeasured. The plots were randomly selected in proportion to species abundance within the level two national subdivisions defined by the Nomenclature of Territorial Units for Statistics (NUTS II). However, an increase in sampling intensity was allowed for some species in the NUTS II subdivisions for which the number of plots led to unacceptable sampling error for those species. The scope of the current inventory (NFI6) has increased and now includes habitat identification and conservation status evaluation, soil characterisation and organic carbon evaluation, and collection of increment bores in two dominant and one average tree per plot. The Portuguese NFI provides official estimates at national level, NUTS II subdivisions, and Regions for Forest Planning (PROF) level. The last two levels do not include all statistics published at country level due to the small sample sizes for some species in some NUTS II and NUTS III subdivisions. NFI5 estimates can be accessed through the FloreStat software at the Portuguese Forest Services' website (http://www.icnf.pt/portal/florestas/ifn).

22.1.3 The Portuguese Forest Simulators

To comply with international commitments under the United Nations Framework Convention on Climate Change (UNFCCC), Portugal established a national Greenhouse Gas (GHG) inventory and regularly submits annual reports on GHG emissions and removals that require forecasting future forest development. Given the importance of the forest sector in Portugal, the Portuguese public administration has national reporting duties and therefore requires that wood availability studies be conducted.

In the beginning of 2000, stand-level forest simulators were made available separately for the three main tree species in Portugal, *Pinus pinaster*, *Eucalyptus globulus* and *Quercus suber* (FPFP 2001a, b, c). However, the programming for these simulators made updating difficult and made it impossible to include new models or updated model versions developed in the past years. It was therefore of utmost importance to support the development of European and national policies by relying on consistent national level forest information. To achieve this objective, new forest simulators were required that facilitated projecting forest development using the latest model versions while at the same time taking into account forest management, climate changes and hazards.

SIMYT, a SIMulator based on Yield Tables, was created to simulate forest resources in a region. This simulator uses as input NFI data aggregated in 1-year age classes' taking into account wood demand, hazards and Land Use Changes (LUC). This forest simulator is available for eucalyptus, maritime pine, cork and holm oaks (Coelho et al. 2012) and was used for 2009 Kyoto Reporting. Another regional forest simulator based on Markov chains was developed for eucalyptus and maritime pine forests (Rego et al. 2013). This forest simulator not only uses a high level of data aggregation (10-year and 4-year age classes for maritime pine and eucalyptus, respectively) but was developed using an NFI data series (from 1972 to 2005) to calibrate its runs, all combining to restrict flexibility in the simulations.

In Portugal, most of the industrial round-wood yield (just considering wood, furniture, pulp and paper products) originated from maritime pine and eucalyptus stands representing 77.9% of the gross economic value of forest products (Santos et al. 2013). Eucalyptus is mainly used for paper and paperboard production (45.7%), pulp (29.2%) and corrugated paperboard (10.1%), whereas maritime pine has been traditionally used for carpentry and construction (39%), sawnlogs (26%), fiberboard and particleboard (14%) (Santos et al. 2013). More recently, maritime pine wood has also been used to produce pellets that are mainly exported to Italy and the United Kingdom (Serra Ramos personal communication). Notwithstanding, over the last three decades, wood industries have been using woody biomass to produce thermal and electrical energy in Portugal. In 2005, the Portuguese Energy Strategy (PES) considered it necessary to increase the power infrastructure, and the construction of several thermoelectric plants fuelled by woody biomass was approved (Diário da República (DRE) 2005). Because the energy produced by these plants is competitively priced, there is a possibility for woody biomass to be used for bio-energy.

With the competing uses for wood, another tool was developed. SIMPLOT, a demand driven SIMulator based on NFI PLOTs, was created to overcome the disadvantages of using aggregated data. Initially developed for eucalyptus forests (Barreiro and Tomé 2011), SIMPLOT presently allows the simulation of maritime pine forests as well. SIMPLOT's projections are made separately for eucalyptus (using stand level data) and maritime pine (using stand and tree level data) and are mainly driven by wood and biomass demands. Different Forest Management Approaches (FMA), hazards (fire) and Land Use Changes (LUC) are also taken into account (Barreiro and Tomé 2012). Several forest management schedules per species can be described and assigned to each input plot. More recently, SIMPLOT was integrated into the StandsSIM forest simulator which has two alternative running options: (1) the demand driven option (StandsSIM.dd), corresponding to SIMPLOT; and (2) the management driven option (StandsSIM.md) described in Barreiro et al. (2016). Only StandsSIM.dd/SIMPLOT which was used for the last policy study carried out for the Portuguese forest sector will be described in the following sections.

22.2 Data

22.2.1 NFI Data

Field measurements are obtained for simple fixed-area circular plots of 500 m^2 for wood production forests and 2000 m^2 for cork and holm oak stands (non-wood production forests). For more details see Barreiro et al. (2010).

For each NFI plot, the diameters of all trees with diameters greater than the minimum diameter threshold are measured. Height is measured for all dominant and sample trees where dominant trees are the 100 largest trees per ha and sample trees are those whose diameters are closest to the mean diameter in each diameter class. All plot trees are classified in terms of their health condition (e.g. discoloration and defoliation), shape (e.g. broken top, diameter defect, bent and curved), status (e.g. dead, alive and stump), age and canopy position. Plot level information is also collected including accessibility, topography, hazards occurrence, natural regeneration, vertical structure by height classes and understory use (e.g. for pasture). Land use, stand composition and structure as well as evidence of recent silvicultural operations are described for the stand.

After field-data collection, and before data processing, a series of quality control procedures take place. The estimation of standing and growing stock volume and biomass is important for estimating woody biomass availability, but also for carbon evaluation, and includes several steps: (1) tree height estimation for each non-sampled tree using species-specific height-diameter models, (2) tree volume and biomass estimation for each tree on the plot using species-specific models, (3) summation of tree volume and biomass estimates for each plot scaled to a per hectare basis, and (4) aggregation of plot mean volume or biomass per hectare estimates by forest stratum and multiplication by the respective area to estimate total volume or biomass of standing stock and growing stock of each forest stratum. These variables combined with other stand level information are used as inputs for running StandsSIM.dd (e.g. Barreiro and Tomé 2011, 2012). Change estimates can also be obtained with StandsSIM.dd.

22.2.2 NFI Based Initial Forest Resources Input

Each NFI plot represents a stand with the characteristics of an NFI plot. Each stand represents an area corresponding to the total area of the species in the country divided by the number of NFI plots for that species. Most eucalyptus and maritime pine stands are managed as pure even-aged. Thus, mixed stands are simulated as pure even-aged stands by correcting the stand area based on the ratio between the species standing stock in the stand and the stands' total standing stock (Barreiro and Tomé 2011). Input requirements depend on the type of growth model used for each species. Eucalyptus stands are simulated using a stand level growth model which, apart from stand area, requires other numerical variables, namely stand age (in even-aged stands), dominant height, number of trees (stools and sprouts) and basal area. Volumes (volume under-bark with stump, volume of bark, volume of the stump) are not a compulsory input, but more accurate results are obtained if provided. In turn, maritime pine stands are simulated using a tree-level growth model and for this reason additionally require the plot area and number of trees, the measured diameter and height of the trees, as well as the status and dominance code for each tree. Additional stand level information is also required such as stand area, stand age and stand density. Non-numerical information such as stand type, composition and structure is also needed.

22.2.3 Simulation Parameters

Simulation runs depend on a series of parameters that can be modified by the user. Simulation parameters include the number of years to project, fire parameters (the minimum age for industrial use of wood after a fire and the proportion of salvage wood) and harvesting parameters (minimum age for harvesting and the harvesting probabilities by age and stand type). Additionally, StandsSIM.dd needs simulation parameters concerning short-rotation forestry (rotation length and the number of rotations) as well as parameters related to harvesting residues removal options (e.g. removal of tops, or branches, or bark or combinations of the three). Some of these parameters are obtained from published reports, such as the proportion of salvage wood. Parameters related to type of thinning, thinning intensity (the number of eucalyptus sprouts left per stool after the shoots selection, the Wilson Factor for maritime pine) and thinning periodicity need to be provided as well as assortments (number of assortments and their characteristics, namely the top diameter and log length).

22.2.4 Scenarios and Key Drivers

Scenarios can range from rather simple to quite complex depictions of the future. In StandsSIM.dd, the range of possible scenarios result from combining two or more key drivers for each species: (1) the demand of wood and biomass which has implications for the amount of felling per year, (2) the occurrence of hazards that takes into account the burnt area per forest type, (3) land use changes to and from other uses representing the afforested area per year and the deforested area per year, and (4) the application of Forest Management Approaches (FMA). The intensity of each driver is expressed by an area, volume or proportion.

The quantitative values that characterize the scenarios for each year of simulation throughout the planning horizon can result from published statistics and/or represent the expected trend of each key driver. It is important to state that the Portuguese NFI does not provide statistics on the amount of harvested woody biomass or its final use. In some cases the same statistics can be found published by different entities often presenting distinct numbers.

22.3 Methods

22.3.1 Running the Forest Simulator

StandsSIM.dd runs for a single species at a time. It "visits" each stand every year to update growth. The growth module includes different growth models that are used according to the species and stand structure (Barreiro and Tomé 2011). The drivers (Sect. 22.2.4) are applied after forest resources have been updated. To select which stands will be burnt, harvested or abandoned, the probability of each one of these events occurring in a single stand is compared to a pseudo-random number, and the event occurs if the occurrence probability is smaller than the random number. The annual intensity of each driver per year is defined in the scenario as an area, a volume or a proportion of an area. As the drivers' modules run, burned area, harvested volume, abandoned area and new planted areas are accumulated until the values defined in the scenario for each year of simulation are met (Barreiro and Tomé 2012). When running the tool for eucalyptus, in case both wood and biomass demands have been considered, the wood volume for pulp and the woody biomass for bio-energy are calculated. Similarly, two different afforested areas are also considered: newly planted areas managed for pulp production, and newly planted areas for bio-energy. Figure 22.2 shows the structure of the simulator.

22.3.2 Growth Module

StandsSIM.dd starts by running the growth module for all the stands at year j to update forest resources to year j + 1. This module comprises different sets of forest growth models according to the species, stand structure and forest management. Eucalyptus stands are managed using stand-level growth models: GYMMA for simulating uneven-aged stands (Barreiro et al. 2004), and GLOBULUS 3.0 model (Tomé et al. 2006) or GLOBEP model (Barreiro 2011) depending on whether the even-aged stands are managed for pulp production or as short-rotations for bio-energy production, respectively. Maritime pine stands are simulated using the tree-level models PINASTER for even-aged stands (Nunes et al. 2010, 2011a, b) and PBIRROL for uneven-aged stands (Alegria and Tomé 2013). Thinning is simulated as part of the growth module. Thinned pine wood is assigned to the Wood/Pulp pool, whereas the woody biomass resulting from the shoots selection operation in eucalyptus stands is assigned to the Woody Biomass pool.

22.3.3 Hazards Module – Fire

After forest resources have been updated, a pseudo-random number is generated for each stand and compared to the stands' probability of burning. In the current version of the simulator, the probability of burning is considered equal for all stands and is set based on the proportion of the area to be burned defined in the scenario and on the total forest area (Barreiro and Tomé 2011). As the module runs, burned area per species accumulates and a user defined amount of salvage wood by tree species is considered for the Wood/Pulp pool, adding to the wood resulting from thinning, whereas the remaining wood is assigned to the Woody Biomass pool. The module stops when the burned area defined in the scenario for each species in each year of simulation is met. If the area to burn is not met after all stands were "visited", stands which were not selected in the first round will be "revisited" before moving to the next simulation year. This operation can be repeated 1000 times until the area to be burned is met.

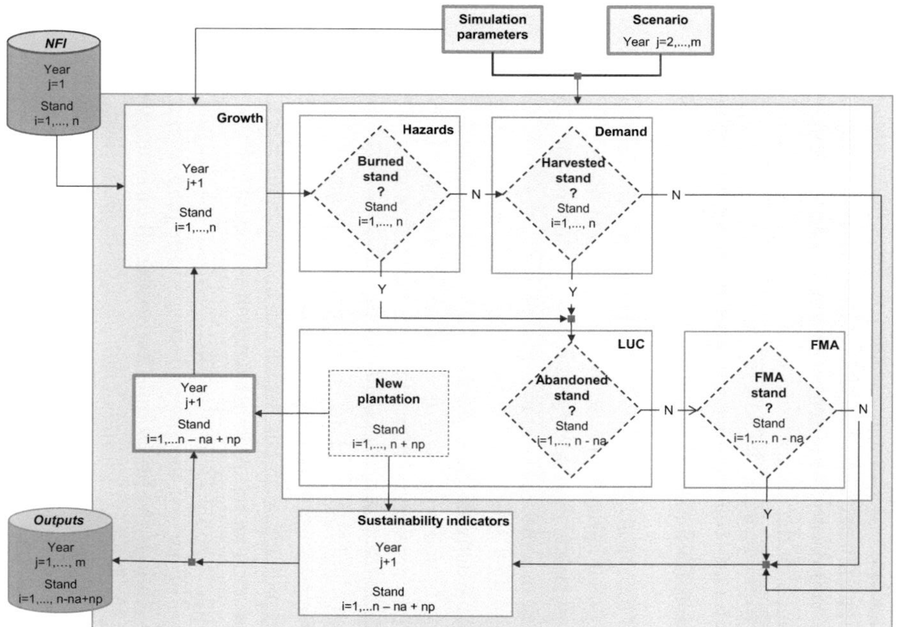

Where *m* represents the number of years in the planning horizon; *n* represents the number of initial stands; *na* the number of abandoned stands; and *np* the number of planted stands

Fig. 22.2 Schematic overview of standsSIM.dd (SIMPLOT) simulator

22.3.4 Harvest Modules

Stands which have not burnt can be harvested. The main harvest module (WP/W) operates on eucalyptus and maritime pine stands managed for Wood/Pulp Production by generating a pseudo-random number for each stand and comparing it to each stands' probability of being harvested. The stands' harvesting probabilities are defined by the user and depend on tree species and stand structure (even-aged stands allow assigning different probabilities according to stand age). The harvested wood volume of each species is added to the respective Wood/Pulp pool, while the harvest residues are assigned to each species Woody Biomass pool. In the case of maritime pine, the Wood/Pulp pool is sub-divided into different categories representing the assortments defined by the user. After a stand is harvested, harvested wood is compared with the wood demand defined in the scenario. Harvesting continues until wood demand is met. Harvest residues contribute to the Woody Biomass pool according to the harvest residues removal option defined by the user (see Sect. 22.2.3).

In the case of the eucalyptus, if biomass demand is considered as a scenario key driver, two additional harvesting modules are used: (1) the BP/E Harvest module, which operates on stands managed for Biomass Production for bio-Energy; and (2) the WP/E Harvest module that operates on stands initially managed for Wood/pulp Production, but which are used for bio-Energy. Both these modules contribute exclusively to the Woody Biomass pool.

Whenever the biomass demand is considered, the three Harvest modules are called to operate in a specific order: (1) BP/E Harvest module, (2) WP/W Harvest module, and (3) WP/E Harvest module. In this way we guarantee that the main source of woody biomass used to meet biomass demand comes from stands specifically managed for bio-energy and is complemented with the harvest residues from Wood/Pulp production stands, which can be left on the site to minimize nutrient extraction as soon as the biomass demand is met. After the first two harvest modules have run, if wood demand is met and biomass demand is not, the third harvest module will harvest the stands managed for pulp as the last option. The BP/E Harvest module is the first to run, harvesting all bio-energy stands for which the rotation length, defined as a simulation parameter, is reached. These stands are harvested regardless of whether the Woody Biomass pool already holds sufficient biomass to meet the demand. In case biomass demand is still not met by the time the module is done, the harvest residues resulting from stands harvested to meet the wood demand (WP/E Harvest module) will be used to increment the Woody Biomass pool. However if this is still not enough, the WP/E Harvest module is used. This module harvests wood production stands according to the same age/stand type dependent probability as in the Harvest WP/W module, but the difference is that all aboveground woody biomass is assigned to the woody biomass pool.

22.3.5 Land Use Changes Modules

Two LUC modules are responsible for planting new stands and abandoning existing stands according to the new plantation areas and the deforestation defined in the scenario for each year of simulation. All stands that were harvested, due to fire or to final felling, will be visited by the deforestation module. The decision on whether to abandon a stand is based on the generation of a pseudo-random number which is compared to a probability function that depends on the value of climatic variables that characterizes the region where the stand is located. Abandoned stands are no longer simulated. Conversely, new stands are assigned a climatic region so that the site index can be estimated and stand growth can be initialized. These new stands will only be disregarded if selected to be abandoned throughout the simulation period. When both wood and energy demands are considered, two afforestation scenarios, one for wood production plantations and another for bio-energy plantations, are required.

22.3.6 Forest Management Approach and the FMA Module

Similar to the deforestation module, the FMAs module operates only for stands that have been harvested following the percentage of changes expected between different Forest Management Approaches according to the scenario. This module is expected to be of extreme interest as soon as maritime pine and eucalyptus stands will run simultaneously for studying the impacts of possible tree species conversion in certain regions.

Several FMAs can be considered for eucalyptus and even- and uneven-aged maritime pine stands. Bio-energy production FMAs can only be considered for eucalyptus. All stands classified under a given FMA are managed in the same way throughout the simulation period, except for uneven-aged stands that are converted to even-aged stands after harvest. The user can define sets of silvicultural operations under each FMA by defining when in time and how often they take place. Additionally, the cost of each silvicultural operation is known, which allows calculation of a set of economic and some social indicators. The costs of operations are based on official statistics (CAOF 2014a, b). At present, the impact of carrying out some operations such as fertilization or weed control is directly reflected in production costs, but not on volume increment.

22.3.7 *Assumptions*

StandsSIM.dd assumes the same burning probability for all stands, regardless of stand type, considering bark and branches biomass reductions of 40% and 25%, respectively. Burnt stands are harvested, and a user defined percentage of salvage wood is considered. Because fires usually occur from March until October and final felling can take place throughout the whole year, fire and harvesting are applied when half of the annual increment has been achieved. Eucalyptus wood production stands can only be harvested for bio-energy if wood demand has been already met. As soon as the biomass demand is met, harvesting residues are assumed to be left on the site. The areas of gaps and clumps are assumed to be maintained constant throughout the simulation period. Only stands that have been harvested can be abandoned or replanted.

22.3.8 *Outputs*

For each scenario run, four (.csv) output files are exported: the totals file that contains annual values for some indicators along the simulation period, the socio-economic indicators file, the areas by age classes' file, and the plot level file. The main output is given in the totals output file containing the total standing stock, felled volume and woody biomass, forest area, burnt forest area, carbon stock and carbon sequestered among other variables before and after running each drivers module. The area output contains a transition matrix with the evolution of forest area between 1-year age classes over the simulation period (Barreiro and Tomé 2011). The plot output file allows tracing the history of each stand throughout the simulation period, whereas the socio-economic indicators (production costs and wages) are saved on a separate file.

22.4 Validation and Uncertainty

Projections obtained from running regional forest simulators are often used by policy makers to assist and support their decision making processes. To increase StandsSIM.dd's credibility and gain sufficient confidence for its outputs, the simulator was run using the NFI4 (1995–1997) data and projections were compared to NFI5 (2005–2007) data. This analysis showed reliable large scale projections for standing stock and reasonable results for forest areas distribution by age classes for all stand types except uneven-aged stands which were underestimated (Barreiro 2011). StandsSIM.dd can be run stochastically to allow for some variability, although sampling and model errors have not yet been included.

22.5 Applications and Future Challenges

StandsSIM.dd is described in two scientific papers (Barreiro and Tomé 2011, 2012) and has been presented at several national and international conferences. In 2014, with the increasing trend in wood imports and the rural development policy 2014–2020 program about to be launched, StandsSIM.dd was used by the Portuguese Agency for the Competitiveness of the Forest Sector (AIFF) for a wood availability prospective study intended to assist in defining policy proposals based on a cost-benefit analysis. Apart from eucalyptus and maritime pine wood production, the study also covered cork production, although a stand level simulator (SUBER, Tomé et al. 2015) was used for projecting cork growth.

StandsSIM.dd is under constant improvement and the next steps include: (1) validating the applicability of 3PG process-based model for the whole country, (2) calibrating 3PG for maritime pine in Portugal, (3) developing a new algorithm for converting even- to uneven-aged stands, (4) improving the tool so that all NFI data, including all species and strata, can be simulated simultaneously.

References

Alegria C, Tomé M (2013) A tree distance-dependent growth and yield model for naturally regenerated pure uneven-aged maritime pine stands in central inland of Portugal. Ann For Sci 70(3):261–276. doi:10.1007/s13595-012-0262-8

Autoridade Florestal Nacional (AFN) (2010) Inventário Florestal Nacional Portugal Continental IFN5, 2005–2006. Autoridade Florestal Nacional, Lisboa, 209 p

Barreiro S (2011) Development of forest simulation tools for assessing the impact of different management strategies and climatic changes on wood production and carbon sequestration for Eucalyptus in Portugal. Doctoral thesis, School of Agriculture, University of Lisbon, Portugal

Barreiro S, Tomé M (2011) SIMPLOT: simulating the impacts of fire severity on sustainability of eucalyptus forests in Portugal. Ecol Indic 11:36–45

Barreiro S, Tomé M (2012) Analysis of the impact of the use of eucalyptus biomass for energy on wood availability for eucalyptus forest in Portugal: a simulation study. Ecol Soc 17(2):14

Barreiro S, Tomé M, Tomé J (2004) Modeling growth of unknown age even-aged eucalyptus stands. In: Hasenauer H, Makela A (eds) Modeling Forest production. Scientific tools—data needs and sources. Validation and application. Proceedings of the international conference, Wien, pp 34–43

Barreiro S, Godinho PF, Azevedo A (2010) National Forest Inventories reports: Portugal. In: Tomppo E, Gschwantner T, Lawrence M, McRoberts RE (eds) National Forest Inventories – pathways for common reporting. Springer, Heidelberg, pp 437–464. ISBN:978-90-481-3232-4

Barreiro S, Rua J, Tomé M (2016) StandsSIM-MD: a management driven forest simulator. For Syst 25(2). doi:10.5424/fs/2016252-08916

Coelho MB, Paulo JA, Palma JNH, Tomé M (2012) Contribution of cork oak plantations installed after 1990 in Portugal to the Kyoto commitments and to the landowners economy. Forest Policy Econ 17:59–68. doi:10.1016/j.forpol.2011.10.005

Comissão de acompanhamento das Operações Florestais (CAOF) (2014a) Matriz de beneficiação 2014. Comissão de Acompanhamento das Operações Florestais, Lisboa

Comissão de acompanhamento das Operações Florestais (CAOF) (2014b) Matriz de (re) Arborização 2014. Comissão de Acompanhamento das Operações Florestais, Lisboa

DGRF (2007) Estratégia Nacional para as Florestas. Resolução do Conselho de Ministros nº114/2006 de 15 de Setembro. Imprensa Nacional-Casa da Moeda, Lisboa. p. 219
Diário da República (DRE) (2005) Diário da República Electrónico, Presidência do Conselho de Ministros, Resolução do Conselho de Ministros n° 169/2005, Diário da República-i Série-B, n° 204, 24 October 2005, Lisbon, Portugal. http://dre.pt/pdf1sdip/2005/10/204B00/61686176.pdf. Accessed 15 Nov 2016
Direcção Geral das Florestas (DGF) (1975) 1ª Revisão do Inventário Florestal Nacional 1975/80, Projecto da memória descritiva do Inventário, Lisboa, Portugal
Direcção Geral das Florestas (DGF) (2001) Inventário Florestal Nacional. Portugal Continental. 3ª Revisão. 1995–1998. Direcção Geral das Florestas, Lisboa, Portugal, 233 p
Federação Portuguesa de Produtores Florestais (FPFP) (2001a) Globulus v2.0 Modelo de produção para o Eucalipto. Manual do utilizador. Edição da Federação dos Produtores Florestais de Portugal, 35 pp (in Portuguese)
Federação Portuguesa de Produtores Florestais (FPFP) (2001b) Pbravo v2.0 Modelo de produção para o Pinheiro bravo. Manual do utilizador. Edição da Federação dos Produtores Florestais de Portugal, 47 pp (in Portuguese)
Federação Portuguesa de Produtores Florestais (FPFP) (2001c) SUBER v3.0 Modelo de produção para o Sobreiro. Manual do utilizador. Edição da Federação dos Produtores Florestais de Portugal, 43 pp (in Portuguese)
ICNF (2014) Estratégia Nacional para as Florestas - Atualização. Documento de trabalho para auscultação pública, Lisboa, p.94. http://www.icnf.pt/portal/icnf/docref/resource/doc/docref/enf-auscultacao. Accessed 11 May 2017
Instituto Conservação da Natureza e Florestas (ICNF) (2013) 6° Inventário Florestal Nacional, Áreas dos usos do solo e das espécies florestais de Portugal continental 1995, 2005 e 2010, Resultados preliminares, Ministério da Agricultura, do Mar, Ambiente e Ordenamento do Território, Portugal. http://www.icnf.pt/portal/florestas/ifn/resource/ficheiros/ifn/ifn6-res-prelimv1-1. Accessed 14 Sept 2015
Nunes L, Tomé J, Tomé M (2010) A compatible system for prediction of total and merchantable volumes allowing for different definitions of tree volume. Can J For Res 40(4):747–760. doi:10.1139/X10-030
Nunes L, Patrício M, Tomé J, Tomé M (2011a) Modelling dominant height growth of maritime pine in Portugal using GADA methodology with parameters depending on soil and climate variables. Ann For Sci 68:311–323. doi:10.1007/s13595-011-0036-8
Nunes L, Patrício M, Tomé J, Tomé M (2011b) Prediction of annual tree growth and survival for thinned and unthinned maritime pine stands in Portugal from data with different time measurement intervals. For Ecol Manag 262:1491–1499. doi:10.1016/j.foreco.2011.06.050
Rego F, Louro G, Constantino L (2013) The impact of changing wildfire regimes on wood availability from Portuguese forests. For Policy Econ 29:56–61. doi:10.1016/j.forpol.2012.11.010
Santos PM, Soares P, Mendes AMSC, Caldeira B, Brígido S, Praxedes J, Pina JP, Paulo JA, Tomé M, Barreiro S, Borges JGC, Palma JHN, Garcia-Gonzalo J, Sottomayor M (2013) Estudo Prospetivo Para o Setor Florestal – Relatório final. AIFF – Associação para a Competitividade da Indústria da Fileira Floresta, 295 pp
Tomé M, Oliveira T, Soares P (2006) O modelo GLOBULUS 3.0. Dados e equações. Publicações GIMREF RC2/2006. Universidade Técnica de Lisboa, Instituto Superior de Agronomia, Centro de Estudos Florestais, Lisboa
Tomé M, Paulo JA, Faias S (2015) SUBER model version 5. Structure, equations and computer interfaces. Publicações ForChange PT 5/2015. Centro de Estudos Florestais, Instituto Superior de Agronomia, Universidade de Lisboa, Portugal

Chapter 23
Romania

Marius Dumitru and Gheorghe Marin

23.1 Introduction

23.1.1 Descriptive Statistics

MECC is the central authority for forestry in Romania. The ministry elaborates policies, strategies and programs for the development of forests, forestry and all legislative forestry regulations.

According to official MECC data, the total forest area in Romania is 6.5 million ha (MM 2014), of which the state owns 3.4 million ha. Approximately 66% of forests are located in the Carpathian mountain region, 24% in the hilly area and 10% in plains (RNP 2009).

The preliminary NFI estimate of forest area is approximately 6.9 million ha (IFN 2011) based on the international forest definition by FAO and represents about 30% of the total country area. The difference between official data and NFI data comes from different definitions of forest. The official estimate is based on forests with management planning, whereas the NFI estimate includes all forest vegetation.

The main forest types are: beech; spruce; fir; mixtures of beech and conifer; mixtures of beech, oak and broadleaves; and oaks. In terms of species proportions, broadleaves represent 70% (beech 31%, oaks 18%, others 21%) and conifer 30% (spruce 23%, fir 5%, other 2%) (RNP 2009).

The total standing wood volume is estimated to exceed 1.350 million m^3, and the mean volume is 217 m^3/ha. In terms of volume, the main forest species are: beech (37%), conifer (39%) and oak (11%) (RNP 2009).

The total forest increment is estimated as 34 million m^3 per year in productive forests and protected forest areas. Until 1990 there was a trend to over-exploitation

M. Dumitru (✉) • G. Marin
Forest Research and Management Institute, Bucharest, Romania
e-mail: mariusd@roifn.ro; ghmarin@roifn.ro

S. Barreiro et al. (eds.), *Forest Inventory-based Projection Systems for Wood and Biomass Availability*, Managing Forest Ecosystems 29,
DOI 10.1007/978-3-319-56201-8_23

of forest. In addition, the lack of an extensive and uniformly distributed forest road network led to significant imbalances because forest harvesting which was concentrated in accessible areas. This condition led the national forest fund to suffer a significant disequilibrium which caused a decrease in allowable cuts as established by forest management plans after 1990.

In Romania, 48% of forests account for production and protection functions, whereas 52% of forests have special protection functions (RNP 2009). Of the special protection forests, soil and water protection forests represent 42% and 31%, respectively; followed by recreation forests with 11%, and scientific interest and conservation forests with 10%. The remaining 6% represents forests for settlement and cropland protection against climatic and industrial damages.

From 1948 until 1991, the entire national forest area was under state ownership. However, after 1990, application of laws for forest restitution led to the distribution of forests among different ownerships. The state forest (about 3.4 million ha) is managed by the National Forest Administration (ROMSILVA, NFA) which works under MECC. NFA aims at sustainable forest management in accordance with the forest law for forest state property with the intent of increasing the contributions of forest to environmental improvement and maintaining the national wood-based economy. NFA manages 23 national and natural parks for which state property is a significant percentage and ensures the conservation of biodiversity in the protected areas. Of the redistributed forest, approximately 1.1 million ha (17%) became municipality forests, 0.8 million ha (12%) were distributed to forest associations and 1.2 million ha (18%) were distributed to private owners. Percentages of forest area by management system are 89% for even-aged high-forest, 5% for uneven-aged forests, and 6% for coppice forests.

23.1.2 Forest Inventory History

Starting in 1948, a forest management planning system was implemented over the entire Romanian forest. Based on the information contained in the forest management plans, forest inventories for the forestry fund were conducted in 1959, 1965, 1973, 1980 and 1984 by the Forest Research and Management Institute (Marin et al. 2010). These forest inventories did not cover all the national forest resources (wooded land was not inventoried) and little information about other forest functions such as forest carbon sequestration, biodiversity, and forest ownership status was collected. Therefore, a new type of forest inventory was designed to include the most important information on forest and forest functions that previously had been ignored. After 1990, there were some attempts to implement a new standwise forest inventory, but they were not completed.

The National Forest Inventory (NFI) is currently the main tool for assessing forest resources at the national and regional levels and addresses the ever-increasing needs for forest information. The Romanian NFI was initiated in 2006, and in 2007 started the first pilot inventory. The first NFI cycle was conducted from 2008 to 2012, and the second NFI cycle has initiated in 2013 and will be finished in 2018.

23.2 Forest Inventory Data

23.2.1 The Standwise Forest Inventory

The first standwise forest inventory took place in 1959 at the end of the first campaign for management planning for all state forests. It was realized by aggregating data from management plans at country level. This was the usual way to implement forest inventories in all Eastern European countries. It was a cost-effective inventory, but provided little information about forest area, timber volume and increment.

The next four standwise forest inventories were carried out in 1965, 1973, 1980 and 1984 using the same basic methods, although the information scope increased from one inventory to the next another as variables such as age class structure, canopy cover, and production classes were added.

23.2.2 The National Forest Inventory

The NFI is designed as a continuous forest inventory with a 5-year cycle. The NFI covers the entire country and features a 4 × 4 km systematic sampling (Marin et al. 2010) grid which is intensified to a 2 × 2 km grid in the plains area where forest cover is less. Estimates are based on repeated measurements of permanent plots and measurements of temporary plots. It is a two-stage NFI that combines the analysis of aerial photos and field forest assessment. Orthophotomaps are used for assessing the presence or absence of vegetation for each sample plot. Sample plots are located in the field using GPS devices, orthophotomaps (1:5000) and topographic maps (1:25,000).

Each year the field teams collect data from approximately 2000 clusters (7000 permanent sample plots) with forest vegetation. In addition to these plots, approximately 700 clusters of temporary sample plots are also measured. Land cover categories and areas of forest vegetation are estimated via photointerpretation of points at the intersections of a 500 × 500 m grid on orthophotomaps.

The NFI plot cluster is a square with 250 m side length and with the sample plots located in each corner (Marin 2008). An NFI sample plot consists of three concentric circles with radius of 7.98 m, 12.62 m and 25 m, plus two concentric circles centered 10 m east and west of the sample plot center, with 1 m and 1.78 m radious. In the three concentric circles, trees are sampled according to their diameter at breast height (*dbh*). In the inner circle (200 m^2) all trees with *dbh* between 5.6 and

28.5 cm are measured. In this circle lying deadwood and ground vegetation are also assessed, whereas in the middle circle (500 m^2) only trees with a *dbh* greater than 28.5 cm are measured. The outer circle is used for collecting growth cores as well as information characterizing forest site, forest type, forest edges and soil characteristics. The two concentric smaller plots are used for evaluating young growth. Inside the 1 m radius circle all trees with 10 cm to 50 cm height are counted, whereas in the 1.78 m radius circle trees over 50 cm height and under 5.6 mm *dbh* are counted. Sample trees, including dead and fallen trees, are evaluated and measured. Multiple observations and measurements are recorded for each tree: *dbh*, height, tree status, sample tree identification, species, distance and azimuth to the sample plot center, crown length and crown shape, layers vegetation to which it belongs, social position, tree age, tree damage, living and death crown length etc.

Field data are collected by specialized teams and entered into field computers. Later, these data are sent to the NFI headquarters, verified and validated, and stored in an Oracle database in the central server for processing and analysis. In addition to tree and stand data, soil samples and growth cores are also collected. Soil data are collected from a single sample plot per forest cluster for the purpose of mapping forest soil types and sites. Growth cores are collected from trees located between the concentric circles with 12.62 m and 25 m radius.

If the concentric circle with radius of 12.62 m includes different ownership types, forest types and/or land uses, then the sample plot is divided into Sectors of Sample Plot (SSP). Because the area of the SSP should not be less than 100 m^2, a sample plot can have at most five sectors.

Of the approximately 300 variables that are measured or estimated, the most important are: tree species, *dbh*, height, age, soil, site, altitude, topography, origin, production class, and quality class. These variables are used to estimate the main NFI outcomes including forest area, growing stock increment and age class structures at regional and country level.

23.3 Data and Methods for Projecting Woody Biomass in Romania

The Romanian NFI has not yet made any predictions based on its own data.

Forest management plans, which are designed for each management unit of a forest district and are repeated every 10 years, can provide information that can be used for woody biomass projections. For elaboration of management plans, specific software is used for all the forest districts in the entire country. This software is used to calculate all stand level information (tree species composition, standing volume, increment, canopy cover etc.) and forest structure (forest area, timber volume and increment by tree species, age classes, productivity, wood quality, accessibility, etc.). The allowable cut for the next 10 years and the volume from thinning are also estimated through the management plans. All this information can be used for prediction of woody biomass.

23.4 Discussion and Conclusions

For the second NFI cycle, we intend to develop models and make predictions at different levels for forest area, growing stock, increment, age class structure, etc. The models will also be applied for prediction of various timber assortments (dimensional and industrial). The old system based on management plans will be replaced and new software developed so that NFI tree sample cores data can be used for projecting woody biomass.

References

IFN (2011) Romanian National Forest Inventory, About NFI. http://www.roifn.ro/site/about-nfi

Marin Gh (2008) Instructiuni pentru culegerea datelor de teren IFN – Forest Research and Management Institute Bucharest

Marin G, Bouriaud O, Dumitru M, Nitu D (2010) National forest inventory reports – Romania. In: Tomppo E, Gschwantner T, Lawrence M, McRoberts RE (eds) National forest inventories – Pathways for common reporting. Springer, Dordrecht, pp 473–480

MM (2014) Departamentul pentru ape, paduri si piscicultura, Paduri. http://apepaduri.gov.ro/paduri/

RNP (2009) Regia Nationala a Padurilor – ROMSILVA, Fondul forestier. http://86.34.169.35/categorie.php?id=4

Chapter 24
Spain

Sonia Condés, Iciar Alberdi, Fernando García-Robredo, and Roberto Vallejo

24.1 Introduction

The first Spanish National Forest Inventory (NFI1) was carried out in the period 1964–1977 and covered the entire forest area of the country. Each of the 50 Spanish provinces served as an assessment unit; mean surface area per unit is 1 million ha for a total area of 50.6 million ha. From the second NFI forward (NFI2, 1986–1996), sample plots were located at the intersections of a 1 × 1 km UTM grid, marked as permanent, and remeasured every 10 years. Nowadays, the fourth NFI (NFI4), which began in 2008, is being carried out.

According to the third NFI, the total forest area was 18.6 million ha of which about 34.5% was dominated by conifers with a total volume of about 531.5 million m^3 and about 46.4% was covered by broadleaf species with an approximate volume of 396.2 million m^3. The remaining 19.1% was covered by a mixture of conifers and broadleaf species with a volume of 927.8 million m^3. Regarding ownership, 29.0% of this forest land is public forest, 65.6% is considered private property and 5.4% is unclassified.

S. Condés (✉) • F. García-Robredo
Department of Natural Systems and Resources, School of Forestry, Technical University of Madrid, Madrid, Spain
e-mail: sonia.condes@upm.es; fernando.garcia.robredo@upm.es

I. Alberdi
Department of Silviculture and Forest Management, INIA, Forest Research Centre, Madrid, Spain
e-mail: alberdi.iciar@inia.es

R. Vallejo
Ministry of Agriculture, Food and Environment, Madrid, Spain
e-mail: rvallejo@mma.es

S. Barreiro et al. (eds.), *Forest Inventory-based Projection Systems for Wood and Biomass Availability*, Managing Forest Ecosystems 29,
DOI 10.1007/978-3-319-56201-8_24

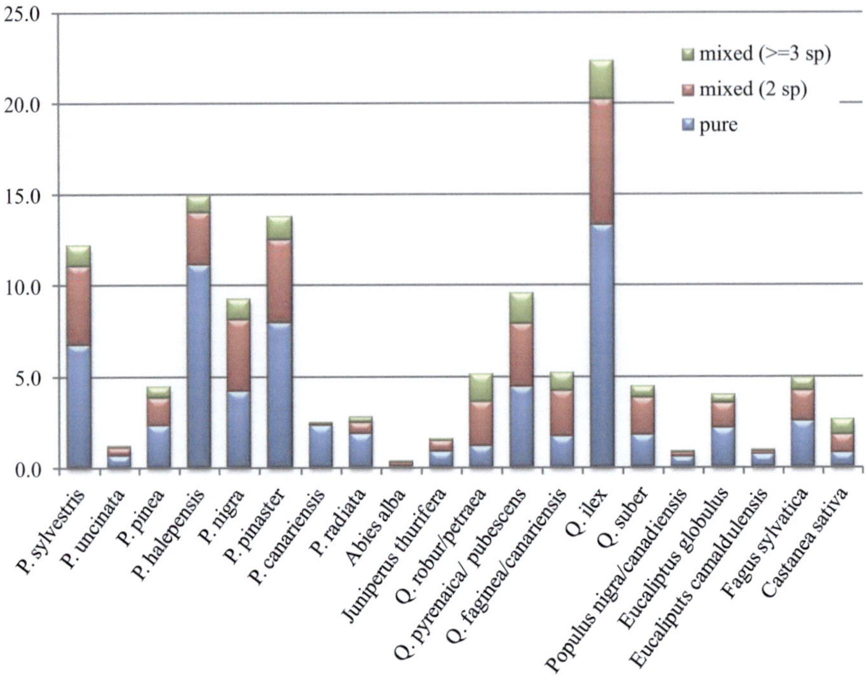

Fig. 24.1 Occurence of the main species in plots that are classified as pure or mixed forest, as percentage of the total number of plots. Note that mixed plots can occur more than once when hosting more than one main species

One of the main characteristics of Spanish forests is the large variability in species composition and structure. From a composition perspective, 70.5% of the NFI3 sample plots are regarded as pure stands with more than 90% of basal area in a single species and 29.5% as mixed stands, usually consisting of two (23.4%) or three species (5.2%). However, for biodiversity assessments the number of tree species per plot is much greater because for this purpose all trees were recorded, not only those with diameter larger than the inventory threshold (Magrama 2012).

In NFI3 more than 150 tree species were assessed of which about 20 were considered main species. These include eight *Pinus* species, five *Quercus* species, two *Eucalyptus* species and the genera *Fagus*, *Populus*, *Castanea*, *Abies* and *Juniperus*. The coverage of the main species and their occurences in pure and mixed stands is shown in Fig. 24.1. Species mixtures can be very variable, illustrated by the range of species occuring as the second most important species in *Pinus sylvestris* dominated stands (Fig. 24.2).

From a structural perspective, more than 50% of the sample plots were located in uneven-aged forest and about 15% of the plots were in mixed stands with two height layers. Productive species such as *Eucalyptus* spp., *Populus* spp. or *Pinus radiata* are frequently managed as even-aged, while Mediterranean species such as *Quercus ilex* or *Pinus halepensis* are more commonly managed as uneven-aged (Fig. 24.3).

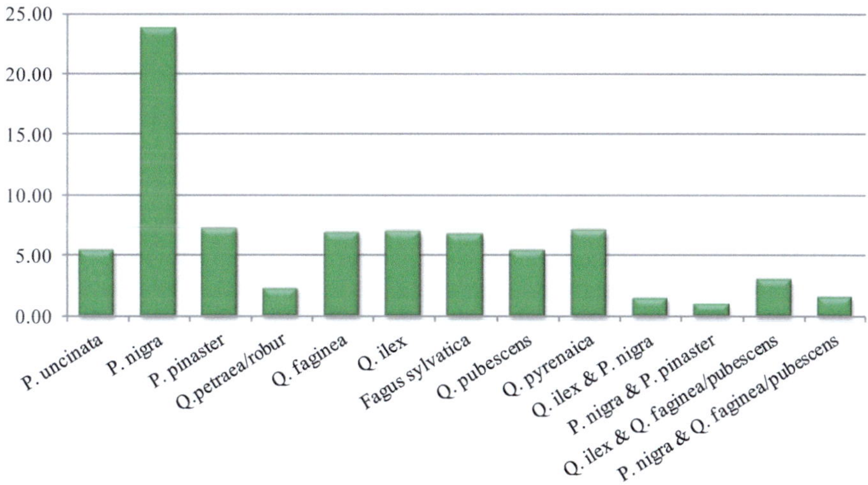

Fig. 24.2 Percentage of plots in *Pinus sylvestris* mixed stands for each main companion tree species

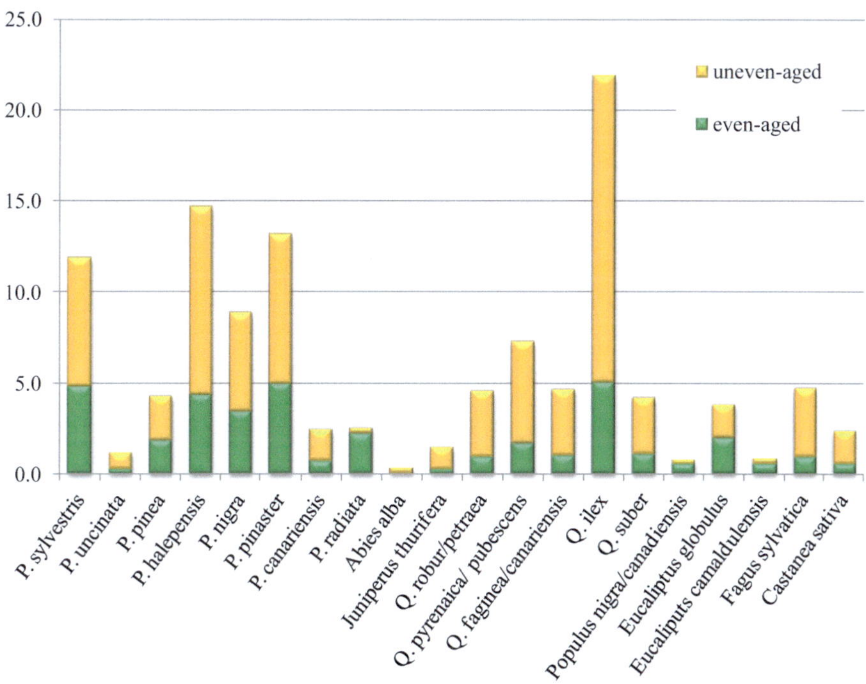

Fig. 24.3 Occurence of the main species in plots that are classified as even-aged or uneven-aged forest, as percentage of the total number of plots. Note that mixed plots can occur more than once when hosting more than one main species

24.2 Data Sources for Growth Modelling

24.2.1 National Forest Inventory

Since 1986, the NFI has been carried out in Spain using permanent sample plots consisting of four concentric circles, each with different a threshold for tree diameter at breast height (*dbh*). For each tree, *dbh*, height, taper and damages are assessed. In the smallest circle, trees with *dbh* between 2.5 and 7.4 cm are counted. In addition to the tree measurements, information such as forest type, number of height layers, mixture type, shrubs and some geographical variables such as aspect and slope are collected. Biodiversity is assessed through identification of the understorey vegetation, presence of dead trees and corresponding decay classes (Alberdi et al. 2014).

NFI data are an excellent source of information for developing growth models such as the individual tree-level models by Adame et al. (2008), Trasobares et al. (2004) and Condés and Sterba (2008) and the stand-level models for pure or mixed stands by Orois and Soalleiro (2002), Río and Sterba (2009) and Condés et al. (2013). Furthermore, the NFI data such as stem number, basal area and stand volume can provide the initial conditions for future projections as required by the growth models.

Unfortunately, the NFI database has four major weaknesses for forest modeling purposes (Álvarez-González et al. 2013): (1) excessively long inventory cycles for fast growing species, (2) a systematic inventory that does not permit comparisons of different silvicultural alternatives present in small forest areas, (3) estimates with large random errors such as ingrowth from the small radius plots, and (4) the unavailability of stand age which hinders development of dynamic stand growth models.

During NFI1, stem diameters at different tree heights were also recorded for sample trees. From these data, simple taper models in the form of fourth degree relative polynomials were constructed for the main species (ICONA 1979; ICONA 1980). These models facilitate estimation of merchantable volume as well as saw log or pulp log volumes at any stem diameter requested by industry.

24.2.2 Other Data Sources

In addition to NFI information, three other data sources are available:

- A database was constructed for all growth models for Spanish forests that have been developed by different research institutions. A review of the models developed before 2010, mostly for single-species even-aged stands, has been published (Bravo et al. 2011).

- Stem taper models for some commercial species have been published in scientific journals (Barrio Anta et al. 2007; Rojo et al. 2005) as have models for estimating the carbon sink capacity (Ruiz-Peinado et al. 2011; Montero et al. 2005).
- Meteorological data with complete series of monthly mean temperature and precipitation are available from the Spanish National Meteorological Agency which has meteorological stations spread throughout the Spanish territory. Because these stations are georeferenced, their data can be easily geo-referenced to NFI sample plot locations to facilitate inclusion of climatic data in new growth models (Condés and García-Robredo 2012).

24.3 Spanish Forest Growth Simulators

Developing large-scale, country-wide simulators in Spain is complicated due to the large variability in tree species, mixtures and stand structures. A large number of growth models (Bravo et al. 2011) and several growth simulators have been developed, but most focus on productive species in monospecific even-aged stands. Because no counry-wide growth models are available, no national forest resource projections exist. Here we give an overview of the best known Spanish simulators with emphasis on those with potential to be used to project resources at regional or national scales.

24.3.1 GesMO©

GesMO© (Diéguez-Aranda et al. 2009) was developed by members of the research group "Sustainable Forest Management Unit", from the *University of Santiago de Compostela* (Spain) and can be freely downloaded at www.usc.es/uxfs. It is based on stand-level growth models developed for single-species in even-aged stands. At the moment, the program implements dynamic growth models for the most productive species in the Galicia region in northwestern Spain which are *Pinus pinaster* (coastal and inland regions), *Pinus radiata*, *Pinus sylvestris*, *Betula alba* and *Quercus robur*. The program is able to incorporate new dynamic stand level growth models when they become available. However, it is restricted to single-species even-aged stands, which in Galicia represent about 65% of the forest lands.

GesMO© allows simulation of different silvicultural treatments and returns their economic assessment. GesMO© consists of six modules: a stand-level simulator for the main attributes of the forest (natural mortality can be included as well); a silvicultural scenario module allowing testing types and intensities of thinnings; a disaggregation module providing diameter distributions as well as volume, biomass and carbon content; a merchantable volume estimation module; an economic assessment module; and a module for generating tables, graphs and reports.

24.3.2 SIMANFOR

SIMANFOR (Bravo et al. 2012) was developed by a research group in the "Escuela Técnica Superior de Ingenierías Agrarias", University of Valladolid (Palencia, Spain), in close collaboration with modelers and end users. SIMANFOR is available for free to the forestry community at www.simanfor.org. The simulator includes two tree-level, distance independent models: one for *Pinus sylvestris* in Sistema Central and Sistema Ibérico, representing two of the most important natural distribution areas of Scots pine in Spain (Río and Sterba 2009), and the other for *Pinus pinaster* in the same area. The system can incorporate new models when available.

SIMANFOR is a forest growth simulator intended for different user types: administrators who manage the systems, modelers who can upload and develop models, and users who can upload, modify and download forest inventories and simulate different silvicultural scenarios to produce information at tree, size class or stand levels.

24.3.3 RODAL, MONTE, ESCEN

RODAL, MONTE and ESCEN are three commercial software products that have been developed by ForEco Technologies. More information on these products can be obtained through the company website (www.forecotech.com).

RODAL is an information system that supports decision-making at the stand level. It is suitable for both even-aged and uneven-aged stands, and, according to the authors, for both pure and mixed stands. The system is based on individual tree growth and survival models, predicts growth and yield, and calculates the economic profitability of the simulated management schedule. RODAL also includes an optimization algorithm to find the management schedule for maximizing wood production, net income or land expectation value.

MONTE is an information system for forest management planning that contains sub-systems for data management, stand development simulation, combinatorial optimization and visualization of forest landscape.

ESCEN is a forestry scenario model that uses NFI plots or similar data as input for regional forestry scenario analyses. The program permits long-term projections according to management alternatives.

24.4 Discussion

A deeper analysis of the NFI statistics and simulators shows that there is not an easy way to project biomass for the whole Spanish forest area. The problems encountered are described in this section.

24.4.1 Number of Species and Species Distribution

While there are about 20 main species in Spain, growth models are available only for a few conifers and some productive broadleaf species such as *Betula alba* or *Populus sp.* The growth of species that are widely distributed, such as *Quercus ilex*, have been poorly studied (Gea-Izquierdo et al. 2008). Furthermore, all growth models have been developed for a specific region only which is particularly important for a country like Spain where climatic conditions can be very different among regions. For instance, *Pinus pinaster* in Spain has two different varieties: *atlantica* and *mesogeensis*. Growth models are developed mainly for the first one in the Galicia region, or for the second one in central and northeastern Spain. Large areas of the southern and southeastern Spain remain unstudied, or the existing models are only developed for even-aged stands (Bravo-Oviedo et al. 2004).

24.4.2 Composition and Structure

Most of the growth simulators described above include models that can be used only for even-aged, single-species stands. However, a large proportion of Spanish forests is covered by uneven-aged stands, and even for even-aged stands, age has not always been recorded during NFI3 or the estimates are unreliable (Álvarez-González et al. 2013). This means that models based on stand age cannot be easily applied to NFI data.

Around 30% of the NFI sample plots are classified as mixed stands. In these stands, inter-specific relationships can result in competition or facilitation, which has to be taken into account in growth models (Condés et al. 2013; Río and Sterba 2009). Moreover, different companion species could have very different effects on target species growth (Río et al. 2013; Río et al. 2014). The necessity of using species-specific models for mixed stands emphasizes the requirement for development of new growth models.

24.4.3 Forest Growth Simulators

Of the forest growth simulators presented in this document, GesMo© is probably the simplest in terms of user interface and the required input data. However, it can only be used for mono-specific, even-aged stands in northwestern Spain. Because stand age data are required, this simulator is not entirely compatible with NFI data.

The simulator that is most adapted to NFI data is probably the one developed by ForEco Technologies. Because it is based on individual tree growth models and is age independent, it is more flexible and applicable to all stand types and species found in Spain. However, currently few models have been included in the simulator, and the high prices required by the company for updating or developing new models limits its use.

The Simanfor simulator has the advantage that simulations using NFI data are possible. However, further analysis is required to determine if new models can be easily incorporated or even developed by users.

A variety of other tools and simulators exist as well in Spain but do not meet the basic criteria for simulators to be applicable country-wide. For example, the process-based model GOTILWA (Gracia et al. 1999) has limited applicability because it not only requires soil properties and daily climate data as inputs, which are difficult to find for country-level simulations, but also because it is only applicable to pure stands. CUBIFOR is an EXCEL implemented tool for calculating volume by assortments, total biomass and carbon content, which would be valuable outputs in a country-wide projection system. However, it does not include a growth simulator and is currently only available for the main species in the "Castilla y Leon" region (Rodríguez et al. 2008, Ruiz-Peinado et al. 2011).

24.5 Conclusions and Future Challenges

Unfortunately, appropriate wood projections for Spain at the National or Autonomous Communities level cannot currently be provided. Rather, all that can be provided are volume increments for different forest types which are calculated through the comparison of NFI permanent sample plots. The required carbon sequestration projections could be obtained using the EFISCEN program, but the results may not be reliable because the program was developed for even-aged and managed forests and the deviations from these conditions make the application less suitable (www.efi.int). As has been described, more than half of Spanish stands are uneven aged forests and, in addition, little information is available about forest management. Approximately 70% of Spanish forest area is privately owned. Additionally, stand age is needed, but such information is not currently available.

However, due to the great interest in the subject, a research project is currently under way to develop country-wide growth models with financial support from the authorities. The initial idea is to develop growth models or annual volume change models for the different Spanish forest types based on NFI data. These data have the important advantage of having been acquired using rigorous and well-tested protocols from plots systematically distributed throughout the complete range of the forest types and therefore represent the most complete spatial distribution of stand and tree variables (Álvarez-González et al. 2013).

References

Adame P, Hynynen J, Cañellas I, del Río M (2008) Individual-tree diameter growth model for rebollo oak (*Quercus pyrenaica* Willd.) coppices. For Ecol Manag 255(3–4):1011–1022

Alberdi I, Cañellas I, Condes S (2014) A long-scale biodiversity monitoring methodology for Spanish national forest inventory. Application to Álava region. For Syst 23(1):93–110

Álvarez-González J, Cañellas I, Alberdi I et al (2013) National forest inventory and forest observational studies in Spain: applications to forest modeling. For Ecol Manag doi:http://dx.doi.org/10.1016/j.foreco.2013.09.007

Barrio Anta M, Diéguez-Aranda U, Castedo-Dorado F et al (2007) Merchantable volume system for pedunculate oak in northwestern Spain. Ann For Sci 64(5):511–520

Bravo F, Álvarez JG, Rio M et al (2011) Growth and yield models in Spain: historical overview, contemporary examples and perspectives. For Syst 20(2):315–328

Bravo F, Rodríguez F, Ordoñez A (2012) A web-based application to simulate alternatives for sustainable forest management: SIMANFOR. For Syst 21(1):4–8

Bravo-Oviedo A, Río MD, Montero G (2004) Site index curves and growth model for Mediterranean maritime pine (*Pinus pinaster* Ait.) in Spain. For Ecol Manag 201(2):187–197

Condés S, García-Robredo F (2012) An empirical mixed model to quantify climate influence on the growth of *Pinus halepensis* Mill stands in south-eastern Spain. For Ecol Manag 284:59–68

Condés S, Sterba H (2008) Comparing an individual tree growth model for *Pinus halepensis* Mill. in the Spanish region of Murcia with yield tables gained from the same area. Eur J For Res 127(3):253–261

Condés S, Del Rio M, Sterba H (2013) Mixing effect on volume growth of *Fagus sylvatica* and *Pinus sylvestris* is modulated by stand density. For Ecol Manag 292:86–95

Diéguez-Aranda U, Rojo Alboreca A, Castedo-Dorado F et al (2009) Herramientas selvícolas para la gestión forestal sostenible en Galicia. Xunta de Galicia, Santiago de Compostela

Gea-Izquierdo G, Cañellas I, Montero G (2008) Site index in agroforestry systems: age-dependent and age-independent dynamic diameter growth models for *Quercus ilex* in Iberian open oak woodlands. Can J For Res 38(1):101–113

Gracia CA, Tello E, Sabaté S, Bellot J (1999) GOTILWA: an integrated model of water dynamics and forest growth. In: Rodà F, Retana J, Gracia CA, Bellot J (eds) Ecological studies. ecology of Mediterranean evergreen Oak forests, vol 137. Springer, Berlin/Heidelberg, pp 163–179

ICONA (1979) Las coníferas en el primer inventario forestal nacional. Ministerio de Agricultura, Subdirección General de Protección de la Naturaleza, Sección de Inventario y Mapas

ICONA (1980) Las frondosas en el primer inventario forestal nacional. Ministerio de Agricultura, Subdirección General de Protección de la Naturaleza, Sección de Inventario y Mapas

Magrama (2012) Informe 2012 sobre el estado del Patrimonio Natural y de la Biodiversidad en España. Ministerio de Agricultura, Alimentación y Medio Ambiente, p 326

Montero G, Ruiz-Peinado R, Muñoz M (2005) Producción de biomasa y fijación de CO2 por los bosques españoles. Instituto Nacional de Investigación y Tecnología Agraria y Alimentaria

Orois SS, Soalleiro RR (2002) Modelling the growth and management of mixed uneven-aged maritime pine-broadleaved species forests in Galicia, North-western Spain. Scand J For Res 17(6):538–547

Río M, Sterba H (2009) Comparing volume growth in pure and mixed stands of *Pinus sylvestris* and Quercus pyrenaica. Ann For Sci 66(5):502–502

Río M, Condés S, Sterba H (2013) Productividad en masas mixtas vs. masas puras: influencia de la espesura en la interacción entre especies. In: Actas 6° Congreso Forestal Español CD-Rom. 6CFE01-121:13. Sociedad Española de Ciencias Forestales, Pontevedra

Río M, Condés S, Pretzsch H (2014) Analyzing size-symmetric vs. size-asymmetric and intra- vs. inter-specific competition in beech (*Fagus sylvatica* L.) mixed stands. For Ecol Manag 325:90–98

Rodríguez F, Broto M, Lizarralde I (2008) CubiFOR: herramienta para cubicar, clasificar productos y calcular biomasa y CO2 en masas forestales de Castilla y León. Montes: Revista de ámbito forestal 95:33–39

Rojo A, Perales X, Sanchez-Rodriguez F et al (2005) Stem taper functions for maritime pine (*Pinus pinaster* Ait.) in Galicia (northwestern Spain). Eur J For Res 124(3):177–186

Ruiz-Peinado R, del Rio M, Montero G (2011) New models for estimating the carbon sink capacity of Spanish softwood species. For Syst 20(1):176–188

Trasobares A, Pukkala T, Miina J (2004) Growth and yield model for uneven-aged mixtures of *Pinus sylvestris* L. and *Pinus nigra* Arn. in Catalonia, north-east Spain. Ann For Sci 61(1):9–24

Chapter 25
Sweden

Anders Lundström and Per-Erik Wikberg

25.1 Introduction

Sweden has a well-established tradition of making long-term analyses of the consequences of different courses of action in forestry. The results have been important for forming national forest policies and for taking strategic decisions within the forest industry. The first known estimates of the national potential harvesting levels were presented by Israel af Ström in the 1850s (Anonymous 1914). Af Ström showed that, at that time, the harvest by far exceeded the increment and predicted a catastrophic effect on the forests. He was right by the end of the century the forests were in a very poor state in most parts of the country. The poor state of the forest was partly a result of increasing harvest level driven by a corresponding increasing demand for sawn wood. Because the demand for forest products was likely to increase at the same time as the supply seemed to decrease, the need for knowledge among stake holders and decision makers about parameters such as standing volume, dimensions, species mixture, growth and geographical distribution became more and more obvious.

During 1923–1929, the first NFI was conducted with measurements obtained along straight 10-m wide lines. It was followed by a second NFI during 1938–1952 in which the lines were complemented with plots. During the third NFI in 1953–1962, the measurements were obtained for circular plots arranged along the sides of rectangles or squares called tracts, a design that is still used today. All tracts were temporary until the sixth NFI, 1983–1992, when permanent tracts were established.

A. Lundström (✉) • P.-E. Wikberg
Department of Forest Resource Management, Division of Forest Resource Data, Swedish University of Agricultural Sciences, Umeå, Sweden
e-mail: Anders.Lundstrom@slu.se; Per-Erik.Wikberg@slu.se

S. Barreiro et al. (eds.), *Forest Inventory-based Projection Systems for Wood and Biomass Availability*, Managing Forest Ecosystems 29,
DOI 10.1007/978-3-319-56201-8_25

At present, the whole sample consists of 2500 temporary and 4500 permanent tracts. Each tract consists of 4–12 plots. One-fifth of the whole sample is measured each year. Consequently, the annual sample consists of approximately 12,000 plots. The side-length of the tracts varies between 300 and 1800 m. The tracts are systematically distributed over all of Sweden, with a denser spacing in the south than in the north. The plot radius is 10 m for permanent plots and 7 m for temporary plots.

The inventory has been and still is concentrated on the tree population with other parameters added as important concerns about climate change and biodiversity have emerged. Thus, coarse woody debris has been measured since 1984, and soil samples have been collected and analyzed since 1983 to track attributes such as soil carbon and nitrogen.

The Swedish NFI has the task of describing the state and changes in Sweden's forests. The information collected is used for multiple purposes including as serving as a basis for forestry, energy and environmental policy in Sweden. Data from the Swedish NFI are part of Sweden's official statistics.

25.1.1 Descriptive Statistics

Sweden has 22.7 million hectares of productive forest land (growth >1 $m^{3/}$ha per year of stem volume above stump including bark and top and 0.7 million hectares of productive forest land within protected areas. Forest land according to international definitions covers 28.4 million hectares.

The largest ownership category is individual private owners who own 52% of the productive forest area. The rest is divided equally between companies and other owners, each with approximately 24%. The state comprises the majority of "other owners."

The most common forest type is Scots pine (*Pinus sylvestris*) forest which covers 39% of Sweden's productive forest area. Other important forest types are Norway spruce (*Picea abies*) forest (27%), mixed forest (22%) and deciduous forest (7%). The rest is *Pinus contorta* (2%) and bare land (3%).

Stands are classified with respect to maturity stages based on normal management measures. Thinning stage forest and final felling forest are the dominant maturity classes with 38% and 33% of productive forest area respectively. Young forest (including pre-commercial felling stage forest) and bare forest land account for 29% of the productive forest area.

Productivity is estimated with respect to the growth potential of a site. Productivity is greatest in the extreme south of Sweden (11.0 m^3/ha per year) and decreases in northern and then northwestern directions. The average site productivity for the entire country is 5.3 m^3/ha per year.

Since the 1920s, annual growth has increased by more than 100% from approximately 60 to approximately 120 million m^3. Mean annual growth for productive forest land in Sweden is 4.9 million m^3 per hectare.

Standing volume in Sweden's forests has increased consistently since the first NFI in the 1920s. In the middle of the 1920s the total standing volume was 1760 million m^3, and has increased to the current level of 3400 million m^3. This steady increase indicates a long-term trend of higher growth than loss (natural loss and felling) during the period. The increase was primarily in Norway spruce until the 1970s, after which the increase in standing volume has been in Scots pine and broadleaves. In recent years Norway spruce volume has stabilized with the increase due primarily to Scots pine and broadleaves.

Sweden's forests are dominated by Scots pine and Norway spruce, being typical boreal conifers. The proportion of broadleaves has increased from 15% of the volume at the end of the 1950s to 19% today. This means we are now back to the same proportion of broadleaves as was seen in the 1920s. Silver birch (*Betula pendula*) and hairy birch (*Betula pubescens*) are the dominant broadleaves. Other broadleaves include aspen (*Populus tremula*), rowan (*Sorbus aucuparia*), goat willow (*Salix caprea*) and grey alder (*Alnus incana*). Oak (*Quercus robur*), beech (*Fagus sylvatica*) and other southern broadleaves occur in the hemi boreal and nemoral zones.

The volume of dead and wind-thrown trees has been relatively constant, with a slight increase due to storms at the end of the 1960s. An increase can be seen, however, for the last 10–15 years with the proportion of dead and wind-thrown trees currently representing approximately 3% of the total standing volume.

The annual harvest has varied between 80 and 90 million m^3 during recent years.

25.1.2 NFI-Data in Future Projections

Data from the NFI have formed the base for several nationwide projections of future harvests levels and related features. In 1938, Jonson and Modin presented projections using data from the first NFI (1923–1929) (Anonymous 1939), showing a potential annual felling of 70 million m^3 in the 1970s. It agreed well with the actual average annual harvest of 67.3 million m^3 in 1970–1979. Since then, results from the NFI such as descriptive statistics and short term projections have been reported on several occasions.

Since the 1970s, countrywide studies of the development and potential of forest utilization in Sweden have been conducted every 5–10 years. Often the studies have been initiated by the Forest Agency, but in a multiple cases the initiative has been directly from the government. Estimates obtained for the twentieth century have indicated a constantly increasing harvest potential. This is partly due to improved silviculture, but also to annual increment exceeding the rate of harvest during most of the century.

The latest study was finished in 2008 (SKA-VB 08). The Swedish Forest Agency initiated these studies with estimation made in cooperation with Swedish University of Agricultural Sciences (SLU).

The objective of the SKA-VB 08 project was to give public authorities, organizations and industry a broad base for strategic decision-making. With this aim in view, the project team worked with scenarios that differed with respect to forest management, utilization and environmental considerations. None of the scenarios represent a "desired" development; rather they are merely possible alternatives, usually characterized as 'what-if' scenarios.

25.2 Data

The NFI collects tree, stand and site data for the main plot, sub-plots and an enlarged plot. Diameter is measured for all trees with a diameter at breast height (*dbh*) ≥10 cm on the main plot (7 m radius for temporary plots and 10 m radius for permanent plots), and for trees with *dbh* of 4–10 cm on a 3.5-m radius subplot. Saplings with heights from 30 cm and up to *dbh* of 4 cm are counted in height interval classes at two 1-m radius plots. Height, damage, age and other variables are measured on sample trees. The stand is described by height interval classes (species mixture, number of stems, etc.) for an enlarged 20-m radius plot. Dead wood is measured and classified by decay classes. Species occurrence and ground coverage and field vegetation are inventoried. Stumps on the plot are measured. If harvest operations have taken place at the plot during the last year, additional plots where only the stumps are measured are established between ordinary plots. Plots are divided along borders between stands or biotopes.

A large number of variables are calculated from the collected data including volume, growth, and biomass per tree component (stumps, branches, needles and stemwood), and is thereafter stored in databases together with climatic and geographical parameters.

SKA-VB 08 was conducted by entering NFI data into the computerized forest simulator HUGIN (Lundström and Söderberg 1996). Different assumptions about future courses of action (scenarios) were also entered into the simulator. The scenarios were based on information taken from multiple sources. The Swedish Forest Agency's annual questionnaire survey of forest owners was used to acquire information about regeneration methods and management programs. To acquire information about use of forest residues for energy, statistics on Sweden's total energy use and supply statistics were used. These statistics are compiled annually by the Swedish Energy Agency based on data from multiple surveys.

In HUGIN, approximately 32,000 NFI sample plots from 2002–2006 were assigned to the utilization and management schemes defined in the scenarios. The state of the forest and the greatest sustainable harvesting levels for 10-year periods from year 2010 to 2110 constituted the output from HUGIN.

Existing protected forest areas and areas planned for protection were included in the analyses. These areas were located using GIS information to identify forests likely to retain high nature values. Information for these areas was obtained from the Swedish Environmental Protection Agency.

25.3 Methods

The growth models in the HUGIN-simulator are constructed for all forest land in Sweden, for all types of stands and within a wide range of management alternatives. Different methods are used to simulate the development of young stands where mean height is less than 7 m ("stand establishment phase"), and for the remaining life of the stands ("established stand phase").

In the stand establishment phase, tree height is the dependent variable for the growth models (Elfving 1982). To obtain diameter distributions and volume growth, special models are used to estimate diameter from height, and volume from diameter. Special models to predict damage and mortality in young forests have been developed by Näslund (1986). These models may be revised depending on knowledge concerning damage agents; for example if moose numbers increase, the probability of moose damage can be increased.

During the simulations, stands are established after harvest by imputing a tree population from the HUGIN young forest database. Data for the database were collected from a countrywide survey of young stands during 1976–1979, the HUGIN young stand survey. The young stand survey was adapted to the HUGIN system, with greater detail than provided by the NFI for this developmental stage.

The method for forecasting growth changes when the average height of the trees on a plot reaches 7 m. The simulators for growth in established stands are based on NFI data. The dependent variable is basal area growth for individual trees (Söderberg 1986). The growth period for the simulators is 5 years and, after each period, volume is obtained using static form height models (Söderberg 1992). To obtain net growth for each period, special models for mortality (Söderberg 1986) and ingrowth (trees passing 4 cm in breast height) are used.

Models used to estimate tree biomass have been developed by Marklund (1988), Petersson and Ståhl (2006). The effect of thinning is estimated by means of thinning response models based on experimental data. If a plot has been fertilized, the effect of fertilization is estimated using models based upon results from experimental plots (Pettersson 1994a, b).

The effects of climate change have been incorporated in the simulation (Bergh et al. 2007).

25.3.1 Management Program

The effects of the following types of treatment can be investigated in the HUGIN system: (1) establishment of new stands by planting, sowing or by natural regeneration, (2) pre-commercial thinning (cleaning), (3) thinning, (4) fertilization, and (5) final felling (clear-cutting).

Because simulations are based on trees within sample plots and not on stands, special procedures are necessary to ensure more realistically simulated management.

Because the plots are too small to correctly describe management in practice, a specified proportion of plots is randomly selected for enforcement of thinning and final felling. The remaining plots are selected according to priority rules based on current standards for good stand-level management. The random treatment is intended to simulate mismanagement and the "stand-plot problem" mentioned above.

The simulator includes considerable flexibility with respect to how treatments are implemented. To make use of the simulator easier, default values for simulation parameters are included based on current good management standards.

The HUGIN-simulator generates results in standard tables for every 10-year period of which some are presented here. Harvested volume and area are presented in various classes, i.e. diameter, site index and age classes. The cut volume from final felling and thinning is presented separately in m^3 total volume (including bark and top) and solid volume (excluding bark and top), respectively and in kg dry weight.

The information is also saved in a database for further analysis, or for presentation of results from additional analyses presented in new tables. It is also possible to obtain more detailed results about growth and cutting methods.

25.3.2 *Scenarios*

Projections were made under different scenarios: Reference, Production, Protection, and Production + Protection. For all scenarios, the effects of climate change were included. The Reference scenario represents business as usual. For the Production scenario, measures for increased production are simulated, mainly by assuming improved management in general, more fertilization, and afforestation of agricultural land not in use. For the Protection scenario, the proportion of the forest set aside as reserves is increased. Restoration and adapted management of some protected forests have been simulated, and nature care has been considered to a greater degree in non-protected forests. The Production + Protection scenario is a combination of the above. The level of the simulated measures in the scenarios has been set according to guidance from policy documents issued by the Government and the Swedish Environmental Protection Agency.

25.4 Results

Examples of standard tables for the reference scenarios are presented here and include projected volumes for 2010, 2020, 2060 and 2100 for the entire country and by region.

In total, standing volume for all productive forest land is projected to increase from approximately 3100 to 4500 million m^3 until 2100 according to Reference scenario simulations. The proportion of the total volume on common forest land mainly subject to wood production (labelled 'production' in Table 25.1) was 84% at

Table 25.1 Standing volume by region and land-use class

	Standing volume (million m³ including bark and top)							
	2010		2020		2060		2110	
Region/land-use	Million m³	%	Million m³	%	Million m³	%	Million m³	%
Northern Norrland	644		678		828		1008	
Production	495	*77*	501	*74*	545	*66*	605	*60*
Environmental	90	*14*	108	*16*	181	*22*	260	*26*
Nature reserve	60	*9*	68	*10*	102	*12*	143	*14*
Southern Norrland	780		813		933		1080	
Production	677	*87*	686	*84*	720	*77*	793	*73*
Environmental	76	*10*	96	*12*	164	*18*	223	*21*
Nature reserve	27	*3*	31	*4*	48	*5*	64	*6*
Svealand	763		786		957		1136	
Production	648	*85*	642	*82*	712	*74*	819	*72*
Environmental	77	*10*	97	*12*	167	*17*	217	*19*
Nature reserve	39	*5*	47	*6*	78	*8*	101	*9*
Götaland	921		939		1129		1303	
Production	793	*86*	776	*83*	855	*76*	962	*74*
Environmental	97	*11*	126	*13*	214	*19*	266	*20*
Nature reserve	30	*3*	37	*4*	60	*5*	74	*6*
Whole country	3108		3216		3846		4527	
Production	2612	*84*	2606	*81*	2832	*74*	3179	*70*
Environmental	340	*11*	427	*13*	726	*19*	966	*21*
Nature reserve	156	*5*	184	*6*	288	*7*	382	*8*

the beginning of the time series and 70% at the end. Thus, the volume increase for unmanaged nature reserves or managed forests subject to environmental concern has been more rapid than in common forests (Table 25.1).

Tables 25.2, 25.3 and 25.4 contain future sustainable harvest levels presented in classes of cutting methods, tree species and assortments. The potential harvest level will increase from 90.5 m^3 to almost 135 m^3 in 2100. The increase will not be linear but will decrease somewhat in the first 10 years and then will increase. The harvested volume from thinnings will increase steadily while the cut volume from clear cuts will decrease approximately 10% and then increase (Table 25.2). The potential harvest of Scots pine will remain rather steady during the whole period while Norway spruce and birch harvest will increase substantially. A steady decrease in potential harvest of Scots pine can be seen in southern Sweden (Götaland) while the potential for Norway spruce will increase rapidly after the first 10-year period and onwards (Table 25.3). The reason for this change in tree species is due to the increasing use of Norway spruce in plantations during recent decades, which continues in the scenarios. The allocation between saw logs and pulpwood was not altered to any larger extent. A slight increase of pulp wood can be seen at the expense of saw logs.

The sustainable harvest will decrease about 7% in the Protection scenario in the short-term and decrease approximately 10% in the long-term. In the Production

Table 25.2 Annual sustainable harvest by region and harvest type for the Reference scenario and all land-use classes

Region/harvest type	Annual sustainable harvest[a] (1000 m³/year including bark and top)							
	2010–2019		2020–2029		2060–2069		2100–2109	
	1000 m³	%	1000 m³	%	1000 m³	%	1000 m³	%
Northern Norrland								
Clear-cut	10,751	*70*	9567	*65*	13,065	*70*	16,292	*68*
Thinning	4518	*30*	5231	*35*	5602	*30*	7771	*32*
Total	15,269		14,798		18,667		24,063	
Southern Norrland								
Clear-cut	14,672	*69*	13,900	*65*	17,869	*69*	21,468	*69*
Thinning	6511	*31*	7480	*35*	7876	*31*	9531	*31*
Total	21,183		21,379		25,744		30,999	
Svealand								
Clear-cut	16,489	*69*	14,717	*63*	19,287	*67*	24,432	*67*
Thinning	7555	*31*	8783	*37*	9664	*33*	12,009	*33*
Total	24,044		23,500		28,952		36,441	
Götaland								
Clear-cut	20,225	*68*	17,935	*64*	23,037	*65*	29,229	*68*
Thinning	9737	*32*	10,128	*36*	12,510	*35*	13,911	*32*
Total	29,962		28,064		35,547		43,140	
Whole country								
Clear-cut	62,137	*69*	56,120	*64*	73,257	*67*	91,422	*68*
Thinning	28,321	*31*	31,621	*36*	35,653	*33*	43,221	*32*
Total	90,458		87,741		108,910		134,643	

[a]Excluding cleaning

scenario the sustainable harvest is almost the same as in the Reference scenario during the first period, but will increase over time and will be approximately 15% greater at the end of the projection period.

25.4.1 *Bioenergy*

The potential for extracting forest residues (branches, twigs, needles, top) for energy production was simulated for three ecological and technical/economic restriction levels and is based on the Reference scenario. In Table 25.5, results for the first period (2010–2019) are presented.

- Level 1 means no restrictions, the estimates shown in Table 25.5 are the total available amount after performed final felling.
- Level 2 means ecological restrictions,
- Level 3 means ecological restrictions and technical/economical restrictions.

Table 25.3 Annual sustainable harvest by region and tree species for the Reference scenario and all land-use classes

	Annual sustainable harvest[a] (1000 m^3/year including bark and top)							
	2010–2019		2020–2029		2060–2069		2100–2109	
Region/tree species	Million m^3	%	Million m^3	%	Million m^3	%	Million m^3	%
Northern Norrland								
Pine	9.2	*61*	8.7	*59*	9.7	*52*	12.6	*53*
Spruce	4.3	*28*	4	*27*	4	*22*	5.9	*24*
Birch	1.3	*8*	1.4	*10*	3.9	*21*	4.8	*20*
Other	0.4	*3*	0.7	*5*	1.1	*6*	0.8	*3*
Southern Norrland								
Pine	8.8	*41*	9.2	*43*	8.7	*34*	11.8	*38*
Spruce	10	*47*	8.9	*42*	9.1	*35*	11.3	*36*
Birch	1.4	*7*	2	*9*	5.2	*20*	6	*19*
Other	1	*5*	1.3	*6*	2.8	*11*	1.8	*6*
Svealand								
Pine	10.2	*42*	9.5	*41*	8.2	*28*	8.7	*24*
Spruce	11.3	*47*	10.9	*46*	15.2	*53*	21.9	*60*
Birch	1.8	*7*	2.3	*10*	4.2	*14*	4.3	*12*
Other	0.8	*3*	0.8	*3*	1.4	*5*	1.4	*4*
Götaland								
Pine	8.5	*28*	6.6	*23*	5.7	*16*	4.7	*11*
Spruce	17.4	*58*	17	*60*	23.7	*67*	31.8	*74*
Birch	2.1	*7*	2.4	*9*	3.7	*11*	4.5	*10*
Other	2	*7*	2.1	*7*	2.4	*7*	2.1	*5*
Whole country								
Pine	36.7	*41*	34	*39*	32.3	*30*	37.9	*28*
Spruce	43.1	*48*	40.7	*46*	52	*48*	71	*53*
Birch	6.5	*7*	8.1	*9*	17	*16*	19.6	*15*
Other	4.2	*5*	4.9	*6*	7.7	*7*	6.2	*5*

[a]Excluding cleaning

At level 1 it was possible to harvest forest residues at 100% of the total clear cut area, at Level 2 it was possible to harvest forest residues from 88% of the clear cut area, and at Level 3 the area is reduced to 73% of the total clear cut area.

25.5 Discussion

25.5.1 Uncertainties

The model package in the HUGIN-simulator is almost entirely built on empirical data as previously described. As such, the models generate reliable results as long as the independent variables are within the range represented by the data used for

Table 25.4 Annual sustainable harvest by region and assortment for the Reference scenario and all land-use classes

	Annual sustainable harvest (1000 m^3 of solid wood under bark per year)							
	2010–2019		2020–2029		2060–2069		2100–2109	
Region/assortment	1000 m^3	%	1000 m^3	%	1000 m^3	%	1000 m^3	%
Northern Norrland								
Saw logs	7939	63	7369	60	9321	59	12,972	63
Pulpwood	4757	37	4915	40	6356	41	7650	37
Total	12,696		12,285		15,676		20,621	
Southern Norrland								
Saw logs	12,268	69	11,717	65	13,040	60	17,057	64
Pulpwood	5631	31	6286	35	8861	40	9776	36
Total	17,898		18,003		21,901		26,833	
Svealand								
Saw logs	14,041	69	13,163	67	15,688	64	21,192	68
Pulpwood	6165	31	6598	33	8998	36	10,111	32
Total	20,207		19,760		24,686		31,303	
Götaland								
Saw logs	17,674	70	16,122	68	19,681	65	25,162	68
Pulpwood	7569	30	7643	32	10,729	35	11,911	32
Total	25,243		23,765		30,410		37,073	
Whole country								
Saw logs	51,922	68	48,370	66	57,730	62	76,384	66
Pulpwood	24,122	32	25,443	34	34,944	38	39,446	34
Total	76,044		73,813		92,673		115,830	

building the models. Thus, simulating situations rarely occurring in reality using empirical models may be associated with considerable uncertainty. Examples of such simulations are catastrophic events, development of old growth forests and the effects of climate change. Validation of models used for forest growth is presented in Fahlvik et al. (2014).

25.5.2 *Climate*

The effect of climate change is included in the simulator and is probably the factor with the greatest uncertainty. Given that the climate effects are limited to temperature increases while other factors remain unchanged and that trees respond to temperature increase by increased growth, simulation uncertainty should not be particularly large, at least not during the first decades. However, the magnitude of climate change and how climate is affected in Sweden is quite uncertain. The negative effects of other climatic factors such as precipitation and wind and damage agents may compensate for the positive effect on growth resulting from an increase in temperature.

Table 25.5 Annual possible removals of harvest residues in clear-cut areas by regions and owner type for the different ecological and technical restrictions

Region/owner type	Annual possible removals of harvest residues in clear-cut areas (million ton dry wood (MtonDW)/year and TeraWatthours (TWh)/year)					
	Level 1		Level 2		Level 3	
	MtonDW	TWh	MtonDW	TWh	MTonDW	TWh
Northern Norrland						
Private	0.63	3.1	0.43	2.1	0.29	1.4
Other	0.7	3.4	0.49	2.4	0.33	1.6
Total	1.33	6.5	0.92	4.5	0.63	3
Southern Norrland						
Private	0.88	4.3	0.58	2.9	0.37	1.8
Other	0.93	4.6	0.66	3.2	0.4	2
Total	1.81	8.9	1.24	6.1	0.78	3.8
Svealand						
Private	1.02	5	0.71	3.5	0.45	2.2
Other	0.86	4.2	0.58	2.8	0.33	1.6
Total	1.88	9.2	1.29	6.3	0.78	3.8
Götaland						
Private	1.89	9.3	1.3	6.4	0.76	3.7
Other	0.52	2.5	0.35	1.7	0.21	1
Total	2.41	11.8	1.66	8.1	0.98	4.8
Whole country						
Private	4.42	21.6	3.03	14.8	1.88	9.2
Other	3.01	14.7	2.08	10.2	1.28	6.3
Total	7.42	36.3	5.11	25	3.16	15.5

25.5.3 Mortality and Development of Protected Forests

Mortality models in the HUGIN-simulator are based on average mortality according to the NFI and on self-thinning as investigated in scientific trials. This system seems insufficient for old growth forests. During the simulations, all plots in protected areas eventually reach volumes that seem unrealistically large. Undoubtedly, some plots will suffer from sudden mortality due to events such as storm, snow, fire, and insect outbreaks. Development of old growth forests generated by HUGIN should be compared against adequate data. Suitable trials have been established, and the NFI has included protected areas since 2003, but the existing time series is far too short. Possible improvements include incorporation of a maximum age for the different species and an increase in stochasticity when mortality is allocated among plot trees.

25.5.4 *Error Estimates*

The NFI is a sample survey and is subject to random errors whose effects increase as sample sizes decrease. Here, results are presented at a national level. Standard errors (SE) at national, regional and county level have been presented by Toet et al. (2007). For example, the SE for standing volume was 0.9% at the national level and 1.5–2.4% at the region level (one region constitutes about 25% of the total area).

25.5.5 *Heureka – New Projection System*

The HUGIN-simulator will be replaced by an application called RegWise developed by the research programme Heureka at Swedish University of Agricultural Sciences (SLU) (Wikström et al. 2011). In the next countrywide forest impact analysis, the new simulator RegWise will be used. The project started in 2013 and is planned to finish in 2015.

In RegWise, several components are added or improved compared to the HUGIN simulator. Examples are models for estimating soil carbon and nitrogen, volume, biomass, and carbon content of dead wood divided in decay classes, utilization of forest fuels (harvest residues) and its effect on carbon storage, income and growth, improved routines for continuous cover forestry, growth effects from retained trees and improved functions for ingrowth.

References

Anonymous (1914) Värmlands läns skogar. Betänkande avgivet av kommissionen för försökstaxering av virkeskapital, tillväxt m.m. av skogarna i Värmlands län, Stockholm (in Swedish)

Anonymous (1939) Utredning rörande skogsnäringens ekonomiska läge med förslag till åtgärder för höjande av näringens bärkraft. II. Åtgärder för främjande av en ändamålsenlig virkesproduktion. 1936 års skogsutrednings betänkande nr 2 avgivet den 26 november 1938, Stockholm (in Swedish)

Bergh J, Blennow K, Andersson M et al (2007) Effekter av ett förändrat klimat på skogen och implikationer för skogsbruket. Institutionsrapport nr 34 vid Institutionen för Sydsvensk Skogsvetenskap. ISBN:978-91-576-7231-5 (in Swedish)

Elfving (1982) Hugin's ungskogsinventering 1976–1979. SLU, Projekt Hugin, Rapport 27

Fahlvik N, Wikström P, Elfving B (2014) Evaluation of growth models used in the Swedish Forest Planning System Heureka, Silva Fenn 48 http://dx.doi.org/10.14214/sf.1013

Lundström A, Söderberg U (1996) Outline of the HUGIN system for longterm forecasts of timber yields and possible cut. Proceedings no. 5. In: Päivinen R, Roihuvuo L, Siitonen M (eds) Large-scale forestry scenario models: experiences and requirements. European Forest Institute, Joensuu, pp 63–77

Marklund LG (1988) Biomassafunktioner för tall, gran och björk i Sverige. Institutionen för skogstaxering, Sveriges lantbruksuniversitet, Umeå. Rapport 45, p 73. ISSN:0348-0496 (in Swedish)

Näslund B-Å (1986) Simulation of damage and mortality in young stands and associated stand development effects. Umeå, Swedish University of Agricultural Sciences, Department of Silviculture, report 18, p 147 (in Swedish with English summary)
Petersson H, Ståhl G (2006) Functions for below-ground biomass of *Pinus sylvestris*, *Picea abies*, *Betula pendula* and *Betula pubescens* in Sweden. Scand J For Res 2006:21 (s 84–93)
Pettersson F (1994a) Predictive functions for impact of nitrogen fertilization on growth over five years. Skogforsk, report no 3
Pettersson F (1994b) Predictive functions for calculation the total response in growth to nitrogen fertilization, duration and distribution over time. Skogforsk, report no 4
Söderberg U (1986) Funktioner för skogliga produktionsprognoser – Tillväxt och formhöjd för enskilda träd av inhemska trädslag i Sverige. Sveriges Lantbruksuniversitet, Avdelningen för skogsuppskattning och skogsindelning, Rapport 14, 1986, 251 s. ISBN:91-576-2634-0
Söderberg U (1992) Functions for forest management. Height, form height and bark thickness of induvidial trees. Sveriges lantbruksuniversitet, Institutionen för skogstaxering. Rapport 52, 1992. ISSN 0348-0496
Toet H, Fridman J, Holm S (2007) Precisionen i Riksskogstaxeringens skattningar 1998–2002. SLU, Institutionen för skoglig resurshushållning, Arbetsrapport 167/2007. ISSN:1401-1204
Wikström P, Edenius L, Elfving B et al (2011) The Heureka forestry decision support system: an overview. Math Comput For Nat-Resour Sci 3(2):87–94

Chapter 26
Switzerland

Christoph Fischer, Paolo Camin, Edgar Kaufmann, and Esther Thürig

26.1 Introduction

26.1.1 Results of Swiss NFI

In Switzerland, NFI data are available since 1983 when the first NFI was started. To date, data for three complete NFI's are available: 1983–1985, 1993–1995, and 2004–2006. These data comprise a long-term dataset that can be used to develop forest growth models. At the time of the fourth NFI (2009–2017), the continuous decadal inventory system changed to a continuous inventory for which one ninth of the sample plots are assessed each year. Forests are an important Swiss resource that is used not only for wood production but also for other functions such as protection against natural hazards, nature protection, and recreational purposes. Approximately 31% of Switzerland is covered with forests for a total forest area of 12,786 km^2 (Brändli 2010) of which 34% is located in the Alps. The greatest proportion of the forest is found in the Southern Alps with 51% of the forest total area. Between the second (1993–1995) and third (2004–2006) NFIs, forest area increased by 5% (60,000 ha) with the largest increase in the Alps and Southern Alps. Two thirds of the total forest area is in public ownership (69%) whereas one third (31%) is in private ownership (Brändli 2010).

The main forest functions in Switzerland are wood production and protection against natural hazards, accounting for 38% and 36% of the forest area, respectively.

C. Fischer (✉) • E. Kaufmann • E. Thürig
Swiss Federal Institute for Forest, Snow and Landscape Research WSL, Birmensdorf, Switzerland
e-mail: christoph.fischer@wsl.ch; edgar.kaufmann@wsl.ch; esther.thuerig@wsl.ch

P. Camin
Federal Office for the Environment FOEN, Ittigen, Switzerland
e-mail: paolo.camin@bafu.admin.ch

S. Barreiro et al. (eds.), *Forest Inventory-based Projection Systems for Wood and Biomass Availability*, Managing Forest Ecosystems 29,
DOI 10.1007/978-3-319-56201-8_26

Wood production areas are mostly located in the Plateau, the Swiss lowlands, whereas forest areas with protective functions (e.g. against landslides and avalanches) are located in the mountainous areas of the Alps and Southern Alps. Other functions such as nature conservation and landscape protection which are mostly located in the Alps make up 10% of the forest area. The NFI results (Brändli 2010) show that most of the annual yield of Swiss forests originates from forest with wood production functions (73%), whereas 17% comes from forests with protective functions, showing that all forest areas are subject to harvesting.

The conifer portion of Swiss forests is greater than the broadleaved portion and is greater than would be expected under natural conditions. Of the total forest area, 62% is dominated by conifers and 34% by broadleaved species (Brändli 2010). For the remaining 4%, no classification is possible because no trees satisfying the minimum diameter of threshold of 12 cm were found on the field plots (Brändli 2010).

In Switzerland, 19% of the forest area is uneven-aged forest (Brändli 2010). Nevertheless, the age structure of Swiss forests exhibits considerable regional variation with the proportion of uneven-aged forests varying between 9% in the Plateau and 27% in the mountainous Southern Alps. The large area of uneven-aged forests in the Alps has its origins in the traditional management regime.

The most important factor affecting Swiss forest productivity is altitude. Overall productivity is approximately 9 m^3/ha per year, but varies between 11 m^3/ha per year for altitudes (altitude above sea level) less than 600 m and 4 m^3/ha per year for altitudes greater than 1800 m, with further decreases until the timber line is reached for which tree productivity is negligible (Abegg et al. 2014).

26.1.2 Growing Stock, Growth, and Drain

Between the second and the third NFI, the proportion of wood subjected to drain (cut + mortality) relative to gross growth increased compared to the previous decade (NFI1 to NFI2). As reported by Brändli (2010), 94% of the gross growth was subjected to drain, 79% resulting from cut and 21% due to mortality. There are large differences for tree species and ownership. For conifers, 109% of gross growth is being harvested versus 63% for broadleaved species. The large rates of drain for conifers are strongly influenced by salvage logging due to the *Lothar* storm event (1999). Smaller drain rates are found for conifer forests which were not affected by *Lothar*. In public forests, 99% of gross growth is being used, whereas 87% of gross growth is being used in private forests.

In the third NFI, after *Lothar*, the average growing stock in Switzerland was 346 m^3/ha, of which 69% was allocated to conifers, and 31% to broadleaved species. Between the second and third Swiss NFIs, a 2% increase in growing stock was estimated with a fivefold increase registered in private forests compared to public forests. The average dead wood volume was 19 m^3/ha.

Regional differences in the proportion of cut and mortality were very important. In the pre-Alps and the Alps, mortality as proportion of drain was 27% and 35%, respectively. As a result of the *Lothar* storm, mortality as a proportion of drain increased from 14 to 21%. Between the second and the third NFIs, the proportion of wood drain relative to net growth was large with 93% of the net growth being utilized. In alpine regions, accessibility to wood resources is limited because of the lack of substantial infrastructure. Average road density in Switzerland is 27 m/ha, but in the Alps and Southern Alps densities are much less, 13 and 7 m/ha, respectively (Brändli 2010). As a result, alpine regions host the greatest growing stocks.

26.1.3 Scenario Models

The first version of *Massimo* was implemented during the second NFI in 1999 with detailed descriptions reported in Brassel and Lischke (2001). The *Massimo* tree-level simulation model is used to forecast the development of future forest resources (Kaufmann 2001a). The plausibility of the model is increased based on its use of empirical Swiss NFI data and its application to future management scenarios.

The main motivation for estimation of future forest development scenarios based on NFI data is the importance of forests as a multipurpose resource for which sustainably available wood and carbon sequestration play key roles (Kaufmann 2011). Applied management scenarios were defined by a group of experts from the forestry sector and grouped into "scenario lines". The expert group included representatives from the Federal Office for the Environment (FOEN), the Swiss timber industry, forestry scientists (e.g. from the Swiss Federal Institute for Forest, Snow and Landscape Research (WSL), and forestry practitioners (Hofer et al. 2011; Kaufmann 2011)).

Scenario lines of forest development based on NFI data cover three main topics: (1) wood resources, (2) carbon storage, and (3) close-to-nature forestry.

The three scenario lines developed serve as a basis for national forest policy discussions. With the 2011 decision to power-down the last nuclear power plant operating in Switzerland by 2034, the possible use of wood resources as a source of energy supply became more important. This decision makes woody biomass very attractive as an energy supply, especially for household heating purposes. For the second commitment period of the Kyoto Protocol (2013–2020), projections of future harvesting rates and forest development are required to estimate the projected forest management reference level as required for reporting national future carbon budgets.

26.2 Data Sources

Swiss NFI data are obtained from three sources: remote sensing, namely aerial photography, field measurements and interviews conducted with foresters for every field plot. Remote sensing data are used in the first-phase of a two-phase sampling approach to determine which field plots are to be measured in the field. In the third NFI, 20,638 sampling plots were selected by overlaying a 1.4 × 1.4 km sampling grid on aerial photography. Of the 20,638 possible sample plots, 6462 were classified as forest and measured in the field. Plots of multiple sizes are combined in each sample plot. First, a 50 × 50 m area serves as an interpretation area for the forest/non-forest decision during the first-phase photo interpretation. The centre of the sample plot is located at the centre of the square. In the field, the interpretation area is used for the classification of the forest type, forest structure, forest age, forest cover, and other variables (Keller 2005). In the centre of the sample plot, two concentric circular plots are located with areas of 200 m^2 and 500 m^2 on which all trees and shrubs are measured, depending on their dimensions. On the 200 m^2 nested plot, all trees and shrubs with diameters at breast-height (1.3 m, *dbh*) between 12.0 and 35.9 cm are measured, whereas all standing trees and shrubs with a *dbh* of at least 36 cm are measured on the 500 m^2 nested plot (Keller 2005). The variables collected on trees and shrubs include coordinates (distance and angle to the plot centre), species, *dbh*, tree height, tree damages, tree dominance, tree conditions (dead or alive). In addition to the two nested plots, two subplots with variable sizes depending on regeneration density and height class were assessed during the third NFI. The subplot centres are located 10 m from the sample plot centre in opposite directions (100 and 300 centesimal degree) (Keller 2005; Schwyzer and Lanz 2010). Methodological changes for the assessment of regeneration have been implemented in the fourth NFI, leading to one subplot with a fixed size (Schwyzer and Lanz 2010). Regeneration is assessed for species, height, quantity, browsing damage, and other variables (Keller 2012). For estimation of the volume of coarse woody debris (diameter ≥ 7 cm), line intersect sampling with three lines in each NFI plot is used. Line intersect sampling is applied as described in Gregoire and Valentine (2007). In addition to the field measurements, interviews are conducted within the Swiss NFI. Because there is no national management directive for forests in Switzerland, the interviews conducted within the NFI are of considerable importance, giving a detailed picture of Swiss management practices for the entire country.

26.3 Methods

26.3.1 Models

Biomass projections for Switzerland are based on the stochastic, empirical single-tree forest management scenario model *Massimo*; the current version is *Massimo 3* (Kaufmann 2011). The estimates are based on 10-year periods for accessible forest in Switzerland, not including shrub forest. *Massimo 3* is largely based on NFI data and runs on every NFI site in Switzerland. The model projects the development of tree growth by updating single-tree *dbh* information for all sites. Biomass is estimated by using allometric models with estimated *dbh* as the predictor variable (Perruchoud 1999; Kaufmann 2001b; Thürig and Herold 2013). Five different tree components can be estimated: (1) twigs, (2) branches, (3) bole wood, (4) stump, and (5) roots. By aggregating estimates for single trees, the stand variables growing stock, increment, and drain in m^3 of bole wood or biomass can be projected. The model output is normally a text file. The output of *Massimo 3* can further be used in combination with the soil carbon model Yasso07 (Liski et al. 2005; Tuomi et al. 2009, 2011) to estimate soil carbon changes over time (Didion et al. 2014). Carbon gains and losses for Swiss forests and the Swiss Forest Management Reference Level (FMRL) submitted to the UNFCCC-Secretariat for annual reporting are both estimated with the *Massimo 3* and Yasso07 models (FOEN 2015). *Massimo 3* is implemented in SAS 9.3 with input and output written to an oracle database (Oracle Database 11 g).

Kaufmann (2012, personal communication) describes *Massimo 3* as consisting of three major modules: (1) a single tree growth module, (2) a cut and mortality module, and (3) a regeneration module, in addition sub-modules for increment after thinning and in-growth are included.

The growth module integrates a single-tree basal area increment model, based on an allometric model using *dbh* as the explanatory variable. Further explanatory variables are basal area per hectare, basal area of trees with a larger diameter than the subject tree (serving as a competition index), fertility, altitude, stand age, a growth boost factor to account for the increase in growth after thinning operations (Fig. 26.1) (Thürig et al. 2005a). For uneven-aged forests, the diameters of the 100 thickest trees per hectare are used instead of stand age (Kaufmann 2011).

The wood harvesting module estimates the annual harvested and natural mortality amounts based on several assumptions. All assumptions and rules concerning losses are empirically derived from NFI data. Natural mortality was based on empirical data from the NFI1 and NFI2 and amounts to 15% of total losses (Kaufmann 2011). Rotation periods for even-aged forests as well as thinning criteria for all types of forests are defined. It is assumed that a rotation period largely depends on the site conditions: 90–110 year rotation periods on very good sites, 110–130 year periods on good sites, and 130–150 year periods on poor sites (Kaufmann 2012, personal

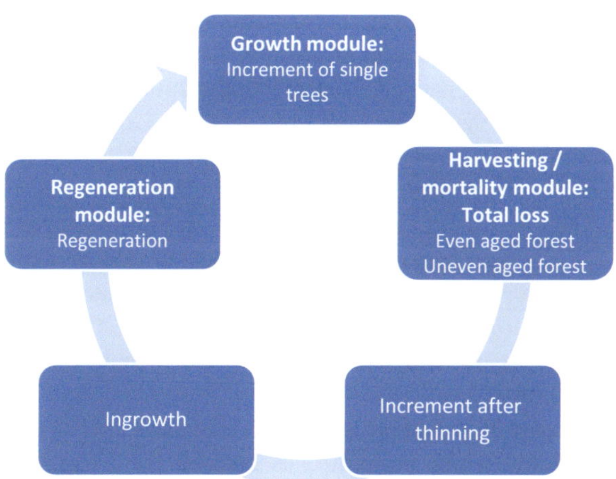

Fig. 26.1 Schematic overview of *Massimo 3*

communication). Because clear cutting practices are forbidden in Switzerland, mature stands are harvested over a time span of 20–30 years, thereby promoting natural regeneration; in Switzerland, 90% of the forests are regenerated naturally (Brändli 2010). Thinning is assumed to take place when a basal area increment of 10% has been reached compared to the basal area before the previous thinning. In *Massimo 3*, even-aged and uneven-aged forest thinning methods are distinguished. Basal area reductions of 30% and 25% are assumed for even-aged and uneven-aged forests, respectively (Kaufmann 2011). Trees to be thinned are selected according to their diameter. The selection probability for single trees is empirically derived by the difference between the diameter distribution of initial stands before thinning and the distribution of remaining stands after thinning (see Kaufmann 2001a). Hence, assumptions regarding harvesting are all based on NFI data. To define different forest management scenarios, the effective date of regeneration cuts, the diameter distribution after thinning, and the periodicity of thinning can be modified.

The regeneration module uses an imputation approach based on NFI data. For each plot requiring a regeneration prediction, a pool of plots that are similar with respect to site conditions is selected from the NFI database. One plot is randomly selected from this pool, and its regeneration and other stand data are assigned or imputed to the plot requiring the prediction. *Massimo 3* bases its predictions for the plot on this imputed data.

26.3.2 Management Scenarios

Many different management scenarios have been developed and published with *Massimo* (Kaufmann 2001a; Thürig et al. 2005b; Schmid et al. 2006; Werner et al. 2010; Thürig and Kaufmann 2010; Hofer et al. 2011; Kaufmann 2011; Thees et al. 2013). Here, we focus on the six scenarios used for national potential wood supply estimates (Hofer et al. 2011). Within the scenario line "wood resources", three scenarios are described here, a base scenario ($A^{1,2}$), a sustainable high growth scenario (B^1), and a high demand scenario (D^1). The wood resources-business as usual with constant wood supply scenario line ($A^{1,2}$) was designed to reflect the management practices (thinnings, etc.) assessed during the third NFI, taking into account the growing stock decreases resulting from "Lothar" (Brändli 2010). Within the scenario, growing stock was maintained at a constant level by assuming that the drain (cut and mortality together) equals the gross growth. Due to the extraordinarily high mortality during the period of the third NFI, the mortality rates (15%) are based on the second NFI as described in Brassel and Brändli (1999). The rotation length frequency and intensity of thinnings (Sect. 26.3.1) were applied as implemented in *Massimo 3* (Kaufmann 2011).

The "wood resources-sustainable high growth scenario line (B^1)" is based on growing stock of 300 m^3/ha which is defined as optimal for a maximizing increment according to the yield-tables developed by EAFV (1968). To achieve the recommended 300 m^3/ha, the existing growing stock was constantly reduced for a period of

20 years. Following Hofer et al. (2011), not only were the removals increased, but the sizes of harvested areas were also increased by 40%, compared to the base scenario.

The "wood resources-high demand scenario line (D[1])" assumed an increasing supply of wood for a period of 20 years. The increasing supply is achieved by shortening the rotation periods within even-aged forest by approximately 40% for the defined length of projection (Kaufmann 2011). As a result of shortening the rotation periods as applied in the base scenario, the areas of annually fellings increase, leading to a growing stock decrease of approximately 15%. The changes from the previous management approaches to new interventions and management intensities are implemented for the period 2006 to 2026. After the first management period with a duration of 20 years, all forests are managed following the base scenario from 2026 to the end of the twenty-first century. The growing stock then reaches equilibrium at a sustainable level as growth increases due to increased thinnings and fellings (Kaufmann 2011).

Under the "carbon storage-Kyoto scenario line (C[1])" the projection addresses carbon storage and carbon balance. This scenario line was used by the Swiss delegation attending the conference of the parties (COP) in Copenhagen, Cancún, and Durban. The aim of this scenario was to estimate the possible drain from the Swiss forests while meeting the defined yearly CO_2 offsets for Switzerland as defined by the Kyoto Protocol (for 2008–2012). Within this scenario, two important forest attributes were combined, wood production and carbon sink. Between 2006 and 2106, in addition to a steady wood supply, the carbon stock had to be increased by as much as 470 m^3/ha (1.5 m^3/ha per year).

For the "close-to-nature forestry scenario line" (Table 26.1) two scenarios are described here: (1) uneven-aged forests (E[2]), and (2) close to nature forestry (F[2]). In the past decades many forests were planted and then managed towards even-aged forests, which represent about 80% of the Swiss forest area (Brändli 2010). Nowadays, many efforts are undertaken towards development of uneven-aged forests. To assess the impacts of such management, the "uneven-aged scenario" was developed. In this scenario, young and middle even-aged stands (defined as stands with the end of the rotation period at the earliest in 30 years) are consequently converted to uneven-aged stands. The conversion is implemented by seeking a diameter distribution of typical uneven-aged stands after several thinning events. Selection probabilities for single tree removal are implemented accordingly. Kaufmann (2011) estimated this would affect approximately 30% of the even-aged forest area.

Table 26.1 Scenario lines and scenarios (A–F)

Scenario line					
Wood resources		Carbon storage		Close to nature forestry	
A[1,2]	Base	**C[1]**	Kyoto	B[2]	Forest reserves
B[1]	Sustainable high growth			**E[2]**	Uneven-aged forests
C[2]	Harvesting costs			**F[2]**	Close to nature forestry
D[1]	High demand				
D[2]	Shortening rotation periods				

Hofer et al. (2011)[1] and Kaufmann (2011)[2]

The "close to nature forestry scenario (F^2)", aimed to define the sustainable level of coniferous regeneration as a means of establishing a baseline for the sustainable yield of conifers (Kaufmann 2011). To achieve such a baseline, the same definitions as applied in the base scenario were used, thereby reducing coniferous regeneration.

26.3.3 Scenario Model Output Application

At the national level, the main use for *Massimo 3* outputs is the prediction of future sustainable wood supply. For this prediction, the "onion model" by Hofer et al. (2011) is applied in Switzerland. Each layer of the "onion model" reduces the actual available wood by subtracting ecological and socioeconomic losses, such as harvest losses, from the *Massimo 3* outputs. In the past, those reductions were largely based on expert estimates which, in the future, will be replaced by estimates based on NFI data. A second topic of interest is the estimation of future harvesting costs under different forest management regimes. Here, the *Massimo 3* outputs are important because harvesting costs strongly depend on the harvestable growing stock.

26.4 Discussion and Conclusions

26.4.1 General Remarks on Models

Model-based prediction always entails uncertainty, because a model by definition is an approximation and partly relies on assumptions. Thus, a model is only as good in terms of reproducing the actual situation as the input into the model. In the case of *Massimo 3*, we have two major data sources, the statistically correct empirical data obtained from field measurements and the scenario definitions which are largely based on expert knowledge. Because *Massimo 3* is based on NFI data covering the entire Swiss forest, we expect the model to be quite reliable for current Swiss conditions. However, changes in environmental conditions could have a negative effect on model prediction accuracy. In contrast to process-based models, empirical models are based on statistical relationships which can change with environmental alterations.

26.4.2 *Use of* Massimo 3 *for Political Implementation*

In 2009, the Federal Office for the Environment (FOEN) initiated a study with the objective of estimating the sustainable potential wood supply of Swiss forests. To estimate the potential under different harvesting regimes, four of the above described scenarios were analysed: (1) Base, (2) maximizing increment, (3) carbon storage, and (4) high demand on wood. Based on these scenarios and considering ecological and socioeconomic factors, the sustainable potential wood supply was estimated.

The results of the study provide a scientific basis for implementation of the Swiss forest policy and for the consolidation of the wood mobilization policy. Within the frame of the described political processes, the next step would be to apply the results at a regional level, differentiating and validating the scenarios. Further, regionalization would offer the possibility of comparing the available wood supply with the actual wood demand.

Under the new LULUCF (Land Use, Land Use Change and Forestry) accounting rules for the second commitment period (2013–2020) of the Kyoto Protocol, parties must calculate their respective forest management reference levels. Therefore, countries must estimate future (2013–2020) emissions and removals from forest for the six carbon pools: above and belowground biomass, dead wood, litter, soil, organic carbon, and harvested wood products. Switzerland estimated its future changes in living biomass using a well-defined harvesting scenario implemented with the stochastic-simulation model *Massimo 3* and Yasso07. Results are disseminated in form of national reports and scientific publications published by the Swiss Federal Institute for Forest, Snow and Landscape Research (WSL) and the Federal office for the environment (FOEN 2015).

26.4.3 Validation and Future Development of Massimo 3

The main advantage of empirical models based on large data sets is their statistically sound basis. Building a model using NFI data facilitates reliable model calibration because the NFI data cover the entire area, including its environmental and management conditions. However, because the environmental conditions are implicitly represented in the explanatory variables and the model parameter estimates, the models are prone to error if the environmental conditions change. Process models based on high resolution functional relationships might offer advantages.

As discussed in Thürig et al. (2005a), model evaluation is important and can be undertaken by different means. General model verification and accounting for random errors of *Massimo 3* was conducted as described in Kaufmann (2001b). The evaluation of the *Massimo 3* growth model was undertaken with a partial sensitivity analyses (Thürig et al. 2005a). An independent data set, the NFI data of Lichtenstein, which was not used for model calibration but is assessed following the NFI method of Switzerland, was used. Thürig et al. (2005a) tested the model to assess whether predicted diameter increments differ from the measured diameter increments for 1520 sample trees. Results of the Wilcoxon signed rank test indicate that measured and predicted diameter increments are not statistically significantly different. Nevertheless, Thürig et al. (2005a) observed that *Massimo 3* is very sensitive to stand age, an explanatory variable for the growth model. Small errors in the estimation of stand age lead to large deviations in the predicted increment. Currently, the growth function is being reconstructed using climate variables (Rohner et al. 2015). Plans for evaluating the new growth function include using NFI4 data and data from long-term forest growth and yield plots (Zingg 2009).

References

Abegg M, Brändli U-B, Cioldi F et al (2014) Fourth national forest inventory – result tables and maps on the Internet for the NFI 2009–2013 (NFI4b). http://www.lfi.ch/resultate/

Brändli U-B (ed) (2010) Schweizerisches Landesforstinventar. Ergebnisse der dritten Erhebung 2004–2006. Eidgenössische Forschungsanstalt für Wald, Schnee und Landschaft WSL/ Bundesamt für Umwelt, BAFU, Birmensdorf/Bern

Brassel P, Brändli U-B (eds) (1999) Schweizerisches Landesforstinventar. Ergebnisse der Zweitaufnahme 1993–1995. Eidgenössische Forschungsanstalt für Wald, Schnee und Landschaft WSL/Bundesamt für Umwelt, BAFU, Birmensdorf/Bern

Brassel P, Lischke H (eds) (2001) Swiss National Forest Inventory: methods and models of the second assessment. WSL Swiss Federal Research Institute, Birmensdorf

Didion M, Frey B, Rogiers N, Thürig E (2014) Validating tree litter decomposition in the Yasso07 carbon model. Ecol Model 291:58–68

EAFV (Anstalt für das forstliche Versuchswesen) (1968) Ertragstafeln für Fichte, Buche. Eidgenössische Anstalt für das forstliche Versuchswesen, Birmensdorf

FOEN (Federal Office for the Environment) (2015) Switzerland's Greenhouse Gas Inventory 1990–2013. National Inventory Report 2015 including reporting elements under the Kyoto Protocol Submission of 15 April 2015 under the United Nations Framework Convention on Climate Change and under the Kyoto Protocol. Federal Office for the Environment, Bern. Available online at www.bafu.admin.ch/climatereporting

Gregoire TG, Valentine HT (2007) Sampling strategies for natural resources and the environment, Applied environmental statistics. Chapman & Hall, Boca Raton

Hofer P, Altwegg J, Schoop A et al (2011) Holznutzungspotenziale im Schweizer Wald. Auswertung von Nutzungsszenarien und Waldwachstumsentwicklung, Umwelt-Wissen Nr. 1116:80 S. Bundesamt für Umwelt, Bern

Kaufmann E (2001a) Prognosis and management scenarios. In: Brassel P, Lischke H (eds) Swiss National Forest Inventory: methods and models of the second assessment. Swiss Federal Research Institute WSL, Birmensdorf, pp 197–206

Kaufmann E (2001b) Estimation of standing timber, growth and cut. In: Brassel P, Lischke H (eds) Swiss National Forest Inventory: methods and models of the second assessment. Swiss Federal Research Institute WSL, Birmensdorf, pp 162–196

Kaufmann E (2011) Nachhaltiges Holzproduktionspotenzial im Schweizer Wald. Schweizerische Zeitschrift für Forstwesen 162:300–311. doi:10.3188/szf.2011.0300

Keller M (ed) (2005) Schweizerisches Landesforstinventar – Anleitung für die Feldaufnahmen der Erhebung 2004–2007. Eidg. Forschungsanstalt WSL, Birmesdorf

Keller M (ed) (2012) Schweizerisches Landesforstinventar – Feldaufnahme Anleitung 2012

Liski J, Palosuo T, Peltoniemi M, Sievänen R (2005) Carbon and decomposition model Yasso for forest soils. Ecol Model 189:168–182. doi:10.1016/j.ecolmodel.2005.03.005

Perruchoud D (1999) 20th century carbon budget of forest soils in the Alps. Ecosystems 2:320–337. doi:10.1007/s100219900083

Rohner B, Thürig E et al (2015) Predicting tree growth as a function of site, stand, climate and nitrogen deposition representative for all of Switzerlands environmental variability (in prep.)

Schmid S, Thürig E, Kaufmann E et al (2006) Effect of forest management on future carbon pools and fluxes: a model comparison. For Ecol Manag 237:65–82. doi:10.1016/j.foreco.2006.09.028

Schwyzer A, Lanz A (2010) Verjüngungserhebung im schweizerischen Landesforstinventar. Jahrestagung der Sektion Forstliche Biometrie und Informatik im Deutschen Verband der Forstlichen Forschungsanstalten (22.) und der AG Ökologie und Umwelt in der Internationalen Biometrischen Gesellschaft – Deutsche Region. Klaus Römisch, TU Desden Fachrichtung Forstwissenschaften Tharandt. Arne Nothdurft, Forstliche Versuchs – und Forschungsanstalt Baden-Württemberg (FVA) Freiburg. Uwe Wunn, Forschungsanstalt für Waldökologie und Forstwissenschaft Reinland-Pfalz in Lippstadt, Göttingen

Thees O, Kaufmann E, Lemm R, Bürgi A (2013) Energieholzpotenziale im Schweizer Wald. Schweizerische Zeitschrift fur Forstwesen 164:351–364. doi:10.3188/szf.2013.0351

Thürig E, Herold A (2013) Recalculation of emission factors in Swiss forests for the Swiss GHGI Internal documentation of technical adjustments of data delivery and more recent data. 10 pp. http://www.bafu.admin.ch/climatereporting/00545/01913/index.html?lang=en

Thürig E, Kaufmann E (2010) Increasing carbon sinks through forest management: a model-based comparison for Switzerland with its Eastern Plateau and Eastern Alps. Eur J For Res 129:563–572. doi:10.1007/s10342-010-0354-7

Thürig E, Kaufmann E, Frisullo R, Bugmannc H (2005a) Evaluation of the growth function of an empirical forest scenario model. For Ecol Manag 204:53–68. doi:10.1016/j.foreco.2004.07.070

Thürig E, Palosuo T, Bucher J, Kaufmann E (2005b) The impact of windthrow on carbon sequestration in Switzerland: a model-based assessment. For Ecol Manag 210:337–350. doi:10.1016/j.foreco.2005.02.030

Tuomi M, Thum T, Järvinen H et al (2009) Leaf litter decomposition – estimates of global variability based on Yasso07 model. Ecol Model 220:3362–3371. doi:10.1016/j.ecolmodel.2009.05.016

Tuomi M, Laiho R, Repo A, Liski J (2011) Wood decomposition model for boreal forests. Ecol Model 222:709–718. doi:10.1016/j.ecolmodel.2010.10.025

Werner F, Taverna R, Hofer P et al (2010) National and global greenhouse gas dynamics of different forest management and wood use scenarios: a model-based assessment. Environ Sci Pol 13:72–85. doi:10.1016/j.envsci.2009.10.004

Zingg A (2009) Long-term forest growth and yield research: organizational and methodological problems and their consequences on the results for research and practice. Swiss Federal Research Insitute WSL, Birmensdorf

Chapter 27
United States of America (USA)

Ronald E. McRoberts

27.1 Introduction

27.1.1 Descriptive Statistics

In the United States of America (USA), forest land is defined with respect to four criteria: (1) minimum area of 0.4 ha, (2) minimum crown cover of 10%, (3) minimum width of 36.6 m, and (4) forest use. Total forest land currently exceeds 310 million ha and has generally been increasing since the 1920s, despite a near tripling of the human population. Currently, more than 30% of the country is characterized as forest land with 58% of it in various categories of private ownership. Unlike in many European countries, private forest land owners in the USA have considerable freedom to convert their land from forest to non-forest uses and vice-versa in response to varying commodity prices and other factors. This feature of private land ownership at least partially explains substantially varying local forest areas over time.

The Nation's many topographic and climatic zones define a large number and great variety of forest biomes ranging from boreal in the north to tropical in the southeast. In the eastern half of the country, the oak-history forest type group represents 34% of forest land and dominates in the central part of the region, pine groups represent 17% and dominate in the southeast, and the maple-beech-birch group with 13% and aspen-birch group with 4% dominate in the north. In the western part of the country, coniferous species dominate with Douglas fir representing 18%, pinyon-juniper representing 15%, fir-spruce representing 14%, and Ponderosa pine representing 11% of forest land. In Alaska, the fir-spruce forest type group is the largest and represents 34% of forest land.

R.E. McRoberts (✉)
Northern Research Station, U.S. Forest Service, Saint Paul, MN, USA
e-mail: rmcroberts@fs.fed.us

S. Barreiro et al. (eds.), *Forest Inventory-based Projection Systems for Wood and Biomass Availability*, Managing Forest Ecosystems 29,
DOI 10.1007/978-3-319-56201-8_27

The forest land use category Forest Available for Wood Supply (FAWS) is not commonly used in the USA, although the concept is prevalent. In particular, the definition of the American land use category Timberland includes the specification that the land must not be withdrawn from timber utilization. Thus, all timberland is in the FAWS land use category. More than 200 million ha of forest land are classified as timberland, meaning that they are capable of producing 0.57 m^3 of industrial wood products annually, are not legally reserved from timber harvest, and are similar to what is known elsewhere as productive forest land. Nearly 70% of timberland is privately owned (Oswalt et al. 2014). In contrast to timberland, the land use category Reserved Forest Land is defined to be land where management for production of wood products is explicitly prohibited. Reserved Forest Land constitutes 7–10% of forest land, includes mostly state and federal parks and wilderness areas, has doubled in area since the early 1950s, and is concentrated in western States.

27.1.2 Wood Resources and Their Uses

Wood resources are generally available for use on all private, non-reserved forest land and most public non-reserved forest land. Local exceptions include buffer zones established for aesthetic purposes and for filtering purposes such as near water sources. Net growing stock volume on timberland, one measure of wood resources, totals nearly 30 billion m^3 and has increased by more than 50% since 1953, mostly in the eastern USA (Oswalt et al. 2014). Over the past 50 years, growth has generally exceeded removals throughout the country. Although removal levels have stabilized in recent years, the source of removals has shifted decidedly from public land in the West to private land in the East. In 1996, coniferous removals in the South exceeded growth for the first time since 1952 when data were first reported. Between 2007 and 2012, the southern region of the country had 63% of American removals (Oswalt et al. 2014), hence the characterization of this region as the "woodbasket of the country." Nation-wide in 2011, timber harvested for industrial products and fuelwood totaled more than 360 million m^3 of which saw logs accounted for 39%; the combination of pulpwood and composite panel outputs accounted for 39%; fuelwood accounted for 14%; veneer production accounted for 5%; and poles, posts, and mulch accounted for 2%.

27.2 The National Forest Inventory

The Forest Inventory and Analysis (FIA) program of the U.S. Forest Service conducts the National Forest Inventory (NFI) of the USA for purposes of estimating the area of forest land; the volume, growth, and removal of resources; and the health of the forest (McRoberts et al. 2010). The FIA program and its predecessors have continuously assessed the Nation's forest resources since 1928. The current form of the inventory was initiated in the early 1990s and is consistent across the entire country with respect to major features such as plot configuration, sampling design, core variables, measurement protocols, and reporting requirements. The inventory is conducted in three phases. In Phase 1, remotely sensed data in the form of aerial photographs and satellite imagery are used for intial assessments of all plots and to stratify the area of interest in support of stratified estimation for purposes of increasing the precision of estimators. In Phase 2, field crews visit plot locations that include forest land, and observe or measure a broad array of site and mensurational variables. Phase 3 consists of more intense measurements of a subset of Phase 2 plots for purposes of assessing a suite of variables related to forest health, ground vegetation, and soils. Additional components of the FIA program include a survey of woodland owners, an emerging urban forest survey, and a mill survey using questionnaires designed to determine the locations, sizes, and types of mills; the volume of roundwood received by product, species and geographic origin; and the volume, type and disposition of wood residues generated during primary processing.

The FIA sampling design is based on a tesselation of the entire country into 2400 ha hexagons with a plot established at a randomly selected location in each hexagon. In the eastern USA, 14% of the plots are measured each year, and in the western USA where growth is slower, 10% of the plots are measured each year. An interpenetrating panel design is used whereby the plots measured in any particular year constitute a complete sample of each state. Each year, approximately 50,000 plots are assessed of which more than 17,000 are on forest land and are measured by field crews.

For all plots on forest land, FIA field crews obtain observations and measurements that describe individual trees, site quality, stocking, general land use, ownership, reserved status, and general stand characteristics such as forest type, stand age, and disturbance. For all trees with diameter at breast height (*dbh*, 1.37 m) of at least 12.7 cm, individual tree observations and measurements include species, *dbh*, height, and removals. Tree- and plot-level observations and measurements are used to predict additional variables including tree and plot volume and biomass, growth, and mortality. Following quality checks, plot data are made available via a publicly accessible Internet site.

FIA data are available to forest planners and managers to assist in managing the Nation's forest resources. However, the data are only sufficient for assessing the past and current state of the resource and for estimating trends. Today, forest planners and managers are increasingly expected to manage stands in ways that are ecologically and economically sound; that comply with a diverse array of objectives including criteria related to health and condition, scenic value, wildfire risk, invasive species, insects and pathogens, and biodiversity; and that produce sufficient additional revenue at harvest to justify intermediate treatments. Further, forest planners and managers have little first-hand knowledge or validated guidance regarding how to comply simultaneously with these often competing criteria. Thus, they need more than just data; they need techniques and tools that use the current state of the resource as a starting point for projecting the long-term consequences of alternative management strategies.

27.3 The Forest Vegetation Simulator

27.3.1 Overview

Multiple forest projection systems are used in the USA, depending on the region and the purpose. Of these, the Forest Vegetation Simulator (FVS) is the most widely used, particularly by federal and state government agencies and private landowners. FVS is an empirical, distance-independent, individual-tree forest growth modeling system constructed, maintained and updated by the U.S. Forest Service. The basic spatial analysis unit is a stand, but the system accommodates many thousands of stands simultaneously. Initial conditions for model predictions consist of a summary of current forest conditions, often in the form of a tree list that can be obtained from the FIA database. Minimal input requirements are individual tree species, diameter, height, and expansion factor. Predictions are possible for several hundred years with time steps of 5–10 years. Details of the FVS system are documented in Dixon (2002) and in an excellent review by Crookston and Dixon (2005).

More than 20 geographical FVS variants have been developed to accommodate regional and local conditions in the USA and parts of western Canada (Crookston and Dixon 2005). Extensions of the basic model functions in the forms of integrated modules or separate models that use FVS output files as input have also been developed. These extensions broaden the system to address specific applications such as the effects on forest growth and mortality of insects and pathogens, fire and fuels, and climate change. Additionally, the economic extension aids in the assessment of silvicultural alternatives, and the fire and fuels extension can be used for carbon accounting. The system includes a graphical user interface and post-processing programs that allow stand visualization and that customize output reports to meet user requests. Brief descriptions of the models and some of the extensions follow with additional details available in the Internet-accessible references provided.

27.3.2 Models and Extensions

27.3.2.1 Growth and Mortality Models

The most important components of the system are the large tree diameter increment and mortality models for the regional variants. These models drive much of the system. Overall, FVS has four primary model components: height growth, diameter growth, crown change, and mortality. The first three components consist of separate sub-models for small trees and for large trees. Within a projection cycle, the processing sequence is: large-tree diameter growth, large-tree height growth, small-tree height growth, small-tree diameter growth, mortality, and crown change. The small-tree model is driven by height growth, whereas the large-tree model is driven by diameter growth. This approach is intended to ensure a smooth transition in height growth as tree size increases. Logistic models are used to predict mortality based on a suite of site and tree conditions. Mortality predictions generally reflect typical background level with mortality due to insects, pathogens, and fire accommodated in the extensions. In addition, as stand basal area approaches the site potential, mortality rates increase (Dixon 2002).

27.3.2.2 Regeneration

Natural regeneration is predicted by summarizing species-specific small-tree attributes such as average frequency and height for seedlings and small saplings based on local FIA plot observations. Seedling recruitment is based on relationships with maximum stand density index. Small sapling regeneration is based on distribution patterns characteristic of current stand size class (seedling/sapling, pole, sawtimber) and density conditions. Apportioning the small sapling frequencies according to the observed distribution patterns provides an expected natural regeneration by stand size/density condition or vegetation state (Vandendriesche 2010).

27.3.2.3 Insects and Pathogens

Extensions to FVS have been developed to accommodate the effects of insects and pathogens (Crookston and Dixon 2005). These extensions are in form of modules that are integrated with the base FVS program. They modify the standard growth and mortality predictions to represent insect and pathogen losses and are implemented once per year within a projection cycle. Crookston and Dixon (2005, Table 1) document many of the separate modules that have been developed for individual insects and pathogens.

27.3.2.4 Fire, Fuels and Carbon

The Fire and Fuels Extension supports fuel management and post-fire treatment decisions by simulating fuel dynamics over time. The extension estimates changes in vegetation due to fuel treatments that include prescribed burns, multiple thinning schemes, piled fuel and burning, pile burning, and mastication. The extension evaluates the effectiveness of these treatments with respect to short- and long-term stand dynamics that are important for silviculture, wildlife habitat, and fuel hazards. The extension does not predict the probability of fire or the spread of fire among stands.

The fuel modeling and accounting approach used by the Fires and Fuels Extension can be used to account for stand carbon stocks and carbon in harvested products. With the exception of the litter and duff pools, carbon found in the living and dead biomass is converted to units of carbon by multiplying by 0.5. Stand C stocks are calculated and reported for multiple categories including total aboveground live biomass, standing dead biomass, forest down dead wood, forest floor litter and duff, herbs and shrubs, and belowground live and dead biomass (Rebain et al. 2015).

27.3.2.5 Climate Change

The climate change extension simulates stand-level impacts of climate change via three modifications of model predictors: linking species mortality and regeneration to climate variables that express climatic suitability, linking site index to climate and using it to modify growth rates, and changing growth rates to accommodate climate-induced genetic responses. Growth is affected when climate conditions at a given location change in relation to the optimal climate conditions under which the species is known to grow and thrive. Mortality is affected when the climate becomes inconsistent with the conditions where species are presently known to survive. Regeneration potential is limited by changes that cause an area's climate to become inconsistent with the known conditions where specific species survive; conversely, changes that cause climate characteristics to become more suitable for a species' survival facilitate successful regeneration of the species (Crookston et al. 2010).

27.3.2.6 Economics

The economic ramifications of individual stand prescriptions can be evaluated with any of several independent extensions that do not interact dynamically with FVS but rather run as separate programs using special FVS output files as input. These programs provide additional functionality for specifying rotation lengths, allow for scheduling activities based on economic parameters, and provide for enhanced reporting of revenues based on log dimensions (Dixon 2002 Sect. 8.10).

27.3.3 Uncertainty

Although FVS makes no provision for rigorous uncertainty assessment, random effects are incorporated into several system components. Diameter growth estimates for large-trees consist of the sum of the growth model estimates and random residuals. These residuals are correlated with residuals from the previous growth cycle to mimic the general finding that trees growing rapidly or slowly in the past tend to continue to grow rapidly or slowly, respectively, in the future. In addition, the component of the model that regenerates stands includes random variation in the heights of seedlings, stocking levels and species composition.

Uncertainty assessments are conducted as deemed necessary by external users and are of several forms. First, accuracy assessment consists of comparison of model predictions with ground observations. Russell et al. (2012) found that for the model variants for the northeastern portion of the country and for the region in the vicinity of the Great Lakes, mean deviations between observed and predicted 5-year diameter increments were in the range 0.025–0.625 cm, but that 10% error in diameter predictions produced 25% error in biomass estimates. Second, sensitivity analyses assess the effects of variability in inputs on the variability of outputs. For the Southern region, Vacchiano et al. (2008) found that diameter, site index, and competition-related variables were the most influential predictors of large-tree diameter growth. For the Southern region, Herring (2007) found that large-tree diameter growth predictions were sensitive to variability in only five predictors, whereas the variability in the 25 other predictors contributed less than 5% of total model uncertainty. Third, propagation of error techniques are used to assess the effects of the transmission of statistical uncertainty from sources such as sampling, parameter variances, and model prediction residuals through modelling predictions to the statistical uncertainty of system outputs. For complex models such as large-scale forest projection systems, assessment of the effects of uncertainty from all sources is a daunting task. Thus, assessments frequently focus on individual sources. For Washington State, Gregg and Hummel (2002) used bootstrapping techniques to propagate the uncertainty of in tree-list inputs through to the distribution of variables predicted by FVS. Investigations of the combined effects of uncertainty from all the major FVS sources should be considered.

27.3.4 Applications

The availability of FIA data in FVS-ready format has increased the number of FVS users to the point that it has become the tool of choice for many applications. In addition, FVS is the official tool for stand growth projection on National Forest lands owned and managed by the U.S. Forest Service. Forest managers have also used FVS extensively to summarize current stand conditions, predict future stand conditions under various management alternatives, and update inventory statistics. Crookston and Dixon (2005) document multiple specific applications. FVS was used to analyze the economics and feasibility of alternative timber harvest methods in South Dakota, to optimize management plans for a late successional reserve in Washington, and to develop an optimal forest rotation for reclaimed coal mines in the eastern USA. In response to potential loss of spotted owl habitat, logging in national forests in the Pacific Northwest of the USA was stopped by court order in 1991. FVS was used to identify plant communities in Oregon that would sustain the owl and to produce 300-year yield estimates to support owl habitat in California. Large-scale forest health assessments were conducted using FVS to assess the effects of dwarf mistletoe and root diseases on yield in Oregon and the effects of bark beetles on forest structure in Montana. The feasibility of applications has been further enhanced via linkages between FVS and other software systems, databases, and geographic information systems.

27.4 Conclusions

The FIA program collects information for a broad array of inventory variables using methods that are consistent across the entire country. The data are available to the public in formats that facilitate input to a variety of applications including forest projection systems. FVS is the most widely used forest projection system in the country and is based on integrated diameter growth models, mortality models, and stand regeneration methods. Extensions in the form of integrated modules or separate programs address specific issues such as insects, pathogens, fire, climate change, carbon accounting, and economic considerations. FVS has been successfully used for a wide range of applications including assessment of management alternatives, wildlife habitat analyses, and climate change effects.

References

Crookston NL, Dixon GE (2005) The forest vegetation simulator: a review of the structure, content, and applications. Comput Electron Agric 49:60–80 http://www.nrs.fs.fed.us/pubs/18474. Accessed 28 Nov 2016

Crookston NL, Rehfeldt GE, Dixon GE Weiskittel AR (2010) Addressing climate change in the forest vegetation simulator to assess impacts on landscape forest dynamics. For Ecol Manag 260:1198–1211 http://www.treesearch.fs.fed.us/pubs/35984. Accessed 28 Nov 2016

Dixon GE (2002) Essential FVS: a user's guide to the Forest Vegetation Simulator. Internal report. U.S. Department of Agriculture, Forest Service, Forest Management Service Center, Fort Collins, 226 p. (Revised: January 8, 2014). http://www.fs.fed.us/fmsc/ftp/fvs/docs/gtr/EssentialFVS.pdf. Accessed 28 Nov 2016

Gregg TF, Hummel S (2002) Assessing sampling uncertainty in FVS projections using a bootstrap resampling method. In: Crookston NL, Havis RN (eds) Second Forest Vegetation Simulator conference; 2002 February 12–14; Fort Collins, CO. Proceedings RMRS-P-25. U.S. Department of Agriculture, Forest Service, Rocky Mountain Research Station, Ogden, pp 64–167 http://www.fs.fed.us/rm/publications/titles/rmrs_proceedings.html. Accessed 28 Nov 2016

Herring ND (2007) Sensitivity analysis of the Forest Vegetation Simulator variant (FVS-Sn) for Southern Appalachian hardwoods. Master of Science thesis, Virginia Polytechnic Institute and State University

McRoberts RE, Hansen MH, Smith WB (2010) National Forest Inventories reports: United States of America (USA). In: Tomppo E, Gschwantner T, Lawrence M, McRoberts RE (eds) National forest inventories – pathways for common reporting. Springer, Dordrecht

Oswalt SN, Smith WB, Miles PD, Pugh SA (2014) Forest resources of the United States, 2012: a technical document supporting the forest service 2015 update of the RPA assessment. General technical report WO-91. U.S. Department of Agriculture, Forest Service, Washington office, Washington, DC. http://www.srs.fs.usda.gov/pubs/47322. Accessed 28 Nov 2016

Rebain SA, Reinhardt ED, Crookston NL et al (2015) The fire and fuels extension to the forest vegetation simulator: updated model documentation. Internal report. U.S. Department of Agriculture, Forest Service, Forest Management Service Center, Fort Collins. http://www.fs.fed.us/rm/pubs/rmrs_gtr116.pdf. Accessed 28 Nov 2016

Russell MB, Weiskittel AR, D'Amato AW (2012) Assessing the uncertainty of forest carbon estimates using the FVS family of diameter increment equations. In: Morin RS, Liknes GC (eds) Moving from status to trends: Forest Inventory and Analysis (FIA) symposium 2012; 2012 December 4–6; Baltimore, MD. Gen. Tech. Rep. NRS-P-105. U.S. Department of Agriculture,

Forest Service, Northern Research Station, Newtown Square. [CD-ROM], pp 378–382. http://www.treesearch.fs.fed.us/pubs/42785. Accessed 28 Nov 2016

Vacchiano G, Shaw JD, DeRose J, Long JN (2008) Inventory-based sensitivity analysis of the large tree diameter growth submodel of the Southern variant of the Forest Vegetation Simulator. In: Havis RN, Crookston NL (eds) Third Forest Vegetation Simulator Conference; 2007 February 13–15; Fort Collins, CO. Proceedings RMRS-P-54. U.S. Department of Agriculture, Forest Service, Rocky Mountain Research Station, Fort Collins, pp 49–159 http://www.treesearch.fs.fed.us/pubs/30977. Accessed 28 Nov 2016

Vandendriesche D (2010) An empirical approach for estimating natural regeneration for the forest vegetation simulator. In: Jain TB, Graham RT, Sandquist J (eds) Integrated management of carbon sequestration and biomass utilization opportunities in a changing climate: Proceedings of the 2009 National Silviculture Workshop; 2009 June 15–18; Boise, Idaho. Proceedings RMRS-P-61. U.S. Department of Agriculture, Forest Service, Rocky Mountain Research Station, Fort Collins, pp 307–327 http://www.treesearch.fs.fed.us/pubs/37338. Accessed 28 Nov 2016

Index

S. Barreiro et al. (eds.), *Forest Inventory-based Projection Systems for Wood and Biomass Availability*, Managing Forest Ecosystems 29,
DOI 10.1007/978-3-319-56201-8

K

L

M

N

O

P

MIX
Papier aus verantwortungsvollen Quellen
Paper from responsible sources
FSC® C105338
FSC
www.fsc.org

If you have any concerns about our products,
you can contact us on
ProductSafety@springernature.com

In case Publisher is established outside the EU,
the EU authorized representative is:
Springer Nature Customer Service Center GmbH
Europaplatz 3, 69115 Heidelberg, Germany

Printed by Libri Plureos GmbH
in Hamburg, Germany